WITHDRAWN

A CENTURY OF RUSSIAN BALLET

DOCUMENTS AND EYEWITNESS ACCOUNTS,
1810–1910

A CENTURY OF RUSSIAN BALLET

*Documents and Accounts,
1810–1910*

Selected and Translated by
ROLAND JOHN WILEY

CLARENDON PRESS · OXFORD
1990

Oxford University Press, Walton Street, Oxford OX2 6DP
Oxford New York Toronto
Delhi Bombay Calcutta Madras Karachi
Petaling Jaya Singapore Hong Kong Tokyo
Nairobi Dar es Salaam Cape Town
Melbourne Auckland
and associated companies in
Berlin Ibadan

Oxford is a trade mark of Oxford University Press

Published in the United States
by Oxford University Press, New York

© Roland John Wiley 1990

All rights reserved. No part of this publication may be reproduced,
stored in a retrieval system, or transmitted, in any form or by any means,
electronic, mechanical, photocopying, recording, or otherwise, without
the prior permission of Oxford University Press

British Library Cataloguing in Publication Data
A century of Russian ballet: documents and eyewitness
accounts, 1810–1910.
1. Soviet Union. Ballet, history
I. Wiley, Roland John
792.80947
ISBN 0–19–316416–7

Library of Congress Cataloging in Publication Data
A century of Russian ballet: documents and eyewitness accounts,
1810–1910 / selected and translated by Roland John Wiley.
Translated from the Russian.
"This book is . . . based on a dozen printed ballet libretti"—
Introd.
1. Ballet—Russian S.F.S.R.—History—19th century. 2. Ballet—
Russian S.F.S.R.—History—Sources. I. Wiley, Roland John.
GV1664.R87C46 1990   792.8'0947—dc20   89-71063
ISBN 0–19–316416–7

Typeset by Hope Services, (Abingdon) Ltd.
Printed in Great Britain by
Biddles Ltd., Guildford and King's Lynn

For
Natalia Ilyinichna Challis

PREFACE

This book is an account of the century of Russian ballet marked approximately by Charles Didelot's *Raoul de Créqui* of 1819 and Mikhail Fokine's *Le Pavillon d'Armide* of 1907, based on a dozen printed ballet libretti. As the libretti are expressions of style, they would by themselves produce an accurate outline of balletic conventions in nineteenth-century Russia: from Didelot's dramatic pantomime to Taglioni's romantic ballet and back again with Perrot, from Petipa's oriental extravaganzas to the sumptuous fairy tales of his later years and on to Fokine. Additional documents describe the technique of stagecraft, the social setting of ballet, its national components and identity, the rise of criticism, and the life of its artists. These provide a context for the libretti, and tell us something about urban, educated Russians of the time.

For the most part, the documents are about the imperial ballet in St Petersburg, but virtually all the works and artists involved also played in Moscow: Adam Glushkovsky spent his career there; the Taglionis did not actually perform in Moscow, but Filippo's ballets and imitators of Marie's dancing were quick to find their way to Russia's first capital. Fanny Elssler performed in Moscow the repertoire which made her famous in St Petersburg; the Bogdanov family was based at first in Moscow; and as balletmasters, Arthur St-Léon and Marius Petipa were officially responsible for the Moscow ballet, making several visits there in connection with productions of their works.

The selection has been guided by the following considerations: the interest and rarity of the documents, the authority of the contributors to write what they did, and the wish to avoid excessively restrictive sampling: only the longest sources have been excerpted. Emphasis has been placed on people and works for which there is not a sizeable bibliography in English.

In translating I have attempted to find an idiomatic English equivalent of the Russian, retaining the expansive gesture of nineteenth-century letters while avoiding word order, unnecessary emphatics, and redundancies which are awkward in English. I have, without comment, reparagraphed and provided first names and patronymics. The spelling of characters' names in libretti, which may vary from page to page, has also been made consistent. As this is not a critical edition, the documents with original footnotes have been left as they were, the footnotes, indicated by asterisks, placed in the translations at the point of citation. On occasion, clarifications within square brackets have been added. Footnotes giving source references, indicated by numerals, are my own. Unless indicated otherwise, dates cited in the nineteenth century are Old Style, which is twelve days behind the western calendar; they are thirteen days behind in the twentieth century.

In spelling proper names, it has been necessary to choose from a variety of possibilities. The rajah's daughter in *La Bayadère* and Désiré's tutor in *The Sleeping Beauty*, for example, are known as Hamsatti and Halifron respectively in west European languages, the former because it is Théophile Gautier's spelling, whence Petipa took it. In Cyrillic these spellings change to initial 'G', that is, Gamsatti and Galifron, yet even Frenchmen writing in their native language retain the 'G' in Russian sources—Petipa in his notes ('Galifron') and a reviewer for a French-language newspaper in St Petersburg ('Gamzatti'). There seems to be no preference. Other usages have been adopted arbitrarily: the French 'Alexandre' instead of 'Alexsandr', the Slavic 'Nikolai' and 'Pyotr' instead of 'Nicholas' and 'Peter'.

It is a pleasure to acknowledge the assistance to this project provided by the International Research and Exchanges Board, for arranging two exchanges with the Academy of Sciences of the USSR during which some of the documents were scrutinized for the first time. Mr Clement Crisp, Professor John M. Ward, and my wife Jitka have kindly taken time to read the text and to suggest improvements. Madame Elizaveta Surits of Moscow and Mr George Verdak of Indianapolis have also given time and provided assistance in various ways. And in the effort to find the right sense and the correct turn of phrase in English, to say nothing of providing an endless supply of Russian lore, I am indebted to Mrs Natalia Challis of the Slavic Department of the University of Michigan at Ann Arbor. English should have a word to connote such dedication and patience, intelligence and goodwill as come together in her. Alexandre Pushkin once referred to translators as 'the post-horses of enlightenment'. If anywhere in what follows the scales tip in favour of Pushkin's view, it is absolutely to her credit.

CONTENTS

List of Illustrations   xi

I   Charles Didelot in Russia   1

1. 'Recollections of the Great Choreographer Ch. L. Didelot and Some Deliberations Concerning the Art of Dance'
   ADAM PAVLOVICH GLUSHKOVSKY   5
2. Libretto of *Raoul de Créqui, or The Return from the Crusades*
   CHARLES DIDELOT   50
3. Libretto of *The Captive of the Caucasus, or The Bride's Shade*
   CHARLES DIDELOT   68

II   Marie and Filippo Taglioni in Russia   79

4. Libretto of *La Fille du Danube*
   FILIPPO TAGLIONI   90

III   The Life of a Russian Artist   105

5. 'Recollections of T. A. Stukolkin, Artist of the Imperial Theatres'   108
6. From the 'Recollections of the Artiste A. P. Natarova'   135

IV   Fanny Elssler and Jules Perrot in Russia   171

7. Libretto of *Esmeralda*
   JULES PERROT   178

V   Nadezhda Bogdanova   183

8. 'The Bogdanov Artistic Family'   188

VI   The 1860s   215

9. Libretto of *The Pharaoh's Daughter*
   VERNOY DE SAINT-GEORGES AND MARIUS PETIPA   220
10. 'The First Performance of the Ballet *The Pharaoh's Daughter*'
    SERGEI NIKOLAEVICH KHUDEKOV   234
11. Libretto of *The Little Humpbacked Horse, or The Tsar-Maiden*
    ARTHUR SAINT-LÉON   238
12. 'The Petersburg Ballet During the Production of *The Little Humpbacked Horse* (Recollections)'
    SERGEI NIKOLAEVICH KHUDEKOV   250

x   Contents

VII The 1870s — 277

13 From the *Memoirs of a Ballerina of the St Petersburg Bolshoy Theatre 1867–1884*
     EKATERINA OTTOVNA VAZEM — 280

14 Libretto of *La Bayadère*
     MARIUS PETIPA AND SERGEI NIKOLAEVICH KHUDEKOV — 291

VIII The 1880s — 305

15 From *In the Theatre World*
     KONSTANTIN APOLLONOVICH SKALKOVSKY — 309

16 Libretto of *The Vestal*
     SERGEI NIKOLAEVICH KHUDEKOV — 323

17 'Marius Ivanovich Petipa'
     MIKHAIL MIKHAILOVICH IVANOV — 350

IX The 1890s — 357

18 Libretto of *The Sleeping Beauty*
     IVAN ALEXANDROVICH VSEVOLOZHSKY — 360

19 Skalkovsky's Review of *The Sleeping Beauty* — 373

20 'A Musical Letter from Petersburg. Apropos *The Sleeping Beauty*, the ballet of M. Petipa, with music by P. Tchaikovsky'
     GERMAN AVGUSTOVICH LAROCHE — 377

21 'The Sleeping Beauty' from *My Reminiscences*
     ALEXANDRE NIKOLAEVICH BENOIS — 385

22 Libretto of *Raymonda*
     LYDIA ALEXANDROVNA PASHKOVA — 392

X The Twentieth Century — 403

23 Libretto of *The Magic Mirror*
     MARIUS PETIPA — 408

24 'Letter to the Editor'
     SERGEI PAVLOVICH DIAGHILEV — 417

25 Libretto of *Le Pavillon d'Armide*
     ALEXANDRE BENOIS — 422

Postscript — 429

Index — 431

LIST OF ILLUSTRATIONS

(Between pages 212 and 213)

1. Charles Didelot, from an engraving by Reimond (V. Vsevolodskii, *Theatre in Russia at the Time of the Patriotic War*)
2. Jules Perrot (*The Yearbook of the Imperial Theatres*)
3. Cesare Pugni, Composer of *The Pharaoh's Daughter* and *The Little Humpbacked Horse* (*Harvard Theatre Collection*)
4. Sergei Nikolaevich Khudekov (*World Illustration*)
5. Marie Taglioni, probably as Flora in Didelot's *Zéphire et Flore*. Lithograph by Chalon and Lane, London, 1831 (*Harvard Theatre Collection*)
6. Adam Pavlovich Glushkovsky in the ballet, *Raoul de Crézui*. Lithograph by Wilhelm (V. Vsevolodskii, *Theatre in Russia at the Time of the Patriotic War*)
7. Alexandre Mikhailovich Gedeonov, Director of Imperial Theatres, 1834–58 (*The Historical Messenger*)
8. Timofei Alexeyevich Stukolkin (*The Yearbook of the Imperial Theatres*)
9. Marius Ivanovich Petipa (*Barushkin State Theatre Museum*)
10. Mathilde Kshesinskaya as Esmeralda, in a revival of Perrot's *Esmeralda* in 1899 (*The Yearbook of the Imperial Theatres*)
11. *The Pharaoh's Daughter*, Prologue: The Interior of a Pyramid. Decoration by Kukanov for a revival of the ballet in Moscow in 1892 (*The Yearbook of the Imperial Theatres*)
12. Interior view of the Amphitheatre at Olga Island, Peterhof (*The Yearbook of the Imperial Theatres*)
13. Ekaterina Ottovna Vazem (*Bakhrushin State Theatre Museum*)
14. Virginia Zucchi (*Bakhrushin State Theatre Museum*)
15. Carlotta Brianza, creator of the role of Aurora in *The Sleeping Beauty* (*Bakhrushin State Theatre Museum*)
16. Pierina Legnani, creator of the role of Raymonda (*Bakhrushin State Theatre Museum*)
17. Decoration by Lambin for *Raymonda*, Act I, Scene 2 (*The Yearbook of the Imperial Theatres*)
18. Sketch of a decoration by Golovin for *The Magic Mirror*, Act IV, Scene 7 (*The Yearbook of the Imperial Theatres*)
19. Vatslav Nijinsky, creator of the role of the Favourite Slave in *Le Pavillon d'Armide* (*The Yearbook of the Imperial Theatres*)

I
Charles Didelot in Russia

INTRODUCTION

Charles Didelot (1767–1837) was the most important figure in Russian ballet during the first third of the nineteenth century. He arrived in St Petersburg in 1801, after making his name as a dancer and choreographer in Paris and London.[1] He excelled in creating the two types of ballet well established at the time: the anacreontic divertissement, in which Cupid intervenes in the flirtations of pastoral lovers, and the dramatic pantomime, an adventure story based on episodes from antiquity or on the plots of rescue operas popular in his day.

For a person of his fame, Didelot left behind little information about himself. He wrote no memoirs, his choreographies are lost, and such production documents relating to his ballets as survive are exceptional bibliographic rarities. This scarcity of data makes all the more welcome the 'Recollections Of the Great Choreographer Ch. L. Didelot, and Some Deliberations Concerning the Art of Dance' by Adam Pavlovich Glushkovsky—the most informative first-hand account of Didelot's work.[2]

Didelot's student and protégé Glushkovsky was born in 1793 or 1794, studied in St Petersburg from 1799, finished school in 1811, and was sent to Moscow the next year. In 1817 an injury ended his career as a classical dancer, and that year he produced the first of thirteen ballets by Didelot for the Moscow theatre.[3] He died in about 1870.

Glushkovsky's memoir is less important for statistics—there are inaccuracies in his lists of ballets and dates—than for what he would remember best as a dancer and balletmaster: the action and artistic impression of Didelot's works. The 'Recollections' are a panorama of

---

[1] The standard work on Didelot in English is Mary Grace Swift's *A Loftier Flight; The Life and Accomplishments of Charles-Louis Didelot, Balletmaster* (Middletown and London, 1974); in Russian (among numerous shorter essays) it is Yurii Iosifovich Slonimskii's *Didlo; vekhi tvorcheskoi biografii* [Didelot; Landmarks of a Creative Biography] (Moscow and Leningrad, 1958); the most important 19th-century biography of Didelot is that of N. P. Mundt, 'Charles Louis Didelot, Former Balletmaster of the Imperial St Petersburg Theatres', in the literary supplement to *Russkii invalid* [Russian Invalid], 1837, No. 50, repr. with minor changes in *Repertuar russkogo teatra na 1840* [Repertoire of the Russian Theatre for 1840], vol. i, bk. 3, pp. 1–8; other valuable surveys of Didelot's work appear in A[bram Akimovich] Gozenpud, *Muzykal'nyi teatr v Rossii ot istokov do Glinki* [Musical Theatre in Russia from its Origins to Glinka] (Leningrad, 1959), pp. 450–550, and V[era Mikhailovna] Krasovskaya, *Zapadnoevropeiskii baletnyi teatr; ocherkii istorii; preromantizm* [Western European Ballet Theatre; Historical Essays; Pre-Romanticism] (Leningrad, 1983), pp. 286–[339].

[2] Glushkovsky's work was published in *Panteon i repertuar russkoi stseny* [Pantheon and Repertoire of the Russian Stage] in 1851 as follows: vol. ii, bk. 4 (Apr.), sect. 3, pp. 1–24; vol. iv, bk. 8 (Aug.), sect. 2, pp. [15]–28; vol. vi, bk. 12 (Dec.), sect. 3, pp. 1–20. It was reprinted, with the omission of two poems and the scenario of *Euthyme et Eucharis*, in A. P. Glushkovskii, *Vospominaniya baletmeistera* [A Balletmaster's Recollections], ed. P. A. Gusev, with introductory article by Yu. I. Slonimskii (Moscow and Leningrad, 1940), pp. 163–[198]. According to the editors of this volume (p. 234), a fourth instalment of Glushkovsky's memoirs once existed, and a fifth was contemplated.

[3] This information about Glushkovsky has been taken from *Vospominaniya baletmeistera*, pp. 37–9, [214].

Didelot's ballets and of the social dance of the early nineteenth century. And Glushkovsky's remarks about régisseurship, the composition of specialist ballet music, the relationship of dance and spectacle, and stage effects are as valid at the end of the century as at the beginning.

As a stylist in prose, Glushkovsky is sincere, forthcoming, digressive, and sometimes a little awkward. Words were surely not his most expressive medium. His homespun turns of phrase, refreshing lack of pretension, and proud devotion to his former master and to his art bring us nevertheless to overlook his faults.

1

'Recollections of the Great Choreographer Ch. L. Didelot, and Some Deliberations Concerning the Art of Dance'

(Article by A. P. Glushkovsky*)

From the Author

I

The thought of compiling my memoirs about the talent of the choreographer Didelot and several of his contemporaries came to me from the late Prince Alexandre Alexandrovich Shakhovskoy. In the last years of his life the Prince intended to write a comprehensive history of the Russian theatre. On one occasion he said to me: 'Dear Adam Pavlovich, you are an experienced theatre campaigner. Help me remember something of the Petersburg ballet. The history of legs has altogether quit my head. Besides, you are the expert when it comes to dancing.' These memoirs were begun and finished at that time, but our celebrated dramatist did not manage to use them. I offer them now in the same form as they were written, that is, as material for the history of the choreographic art in Russia. I write not as a poet who strives for glory, but as an artist for whom it is pleasant in old age to glance back at a career well spent and to recall times past. I ask that my memoirs be received with this in mind.

Charles-Louis Didelot studied dance in Paris with the celebrated balletmaster Dauberval. He was a graceful dancer, and performed each *pas* with great purity, but did not possess great elevation in entrechats and jumps. For himself he created a special genre of dance: graceful poses, gliding suavity, cleanness and speed in *pas à terre*, a pleasant position of the arms, and lively pirouettes.

For a long time he was a balletmaster in England, and in 1801 was engaged for the Petersburg theatre in the capacity of balletmaster, teacher

---

* A. P. Glushkovsky is retired from the Imperial Theatres. He is a veteran of the Russian ballet company. He was a dancer in Petersburg and a student of Didelot, and then for a very long time worked in Moscow as a balletmaster. Moscow is indebted to him for the production there of all Didelot's best ballets—Ed.

of the theatre school and first dancer. He came with his first wife, Rose Didelot, who was also one of Europe's best dancers.

The first of the grand ballets he created was *Apollo and Daphne*, which delighted the Petersburg public with its charming dances and picturesque groupings. These were the chief virtues of all Didelot's ballets. Following *Apollo and Daphne* he produced *Zéphire et Flore* and *Amour et Psyché*, in which there were enchanting aerial flights and groupings, which he introduced on the stage for the first time, and used afterwards in the theatre at Paris. Audiences were so taken by these ballets that they forgot they were in the theatre, imagining themselves transported to some other, fantastic world.

Didelot never lightened his scenic labours by producing other men's ballets. Having been a balletmaster in the leading capitals, he considered aping the works of other choreographers beneath his station; he used to say that it was excusable only for young students to produce ballets of their teachers; but they too had to produce their own ballets. 'A given ballet might not be very good at first, but it must be one's own, so it cannot be said that the balletmaster is empty in the *bel-étage*' (he would say, pointing expressively at his brow).

The dances of Didelot's ballets were amazingly varied. Serious dances were performed to the music of *adagio* and march; Didelot always composed smooth, fluent dances for the principal character, with various *attitudes* but rarely mixed with entrechats or fast pirouettes. The demi-caractère dancer performed to the music of *andante grazioso* and *allegro*; for him Didelot composed graceful dances with a completely different position of the body and arms, made use of quick and delicate steps and pirouettes of a different kind from those of the serious *pas*; and for the comic dancer he made a *pas* to *allegro* music—for the most part with different kinds of jumps, namely: *tours en l'air* with a different movement of the body and arms. Thus the public saw in each dancer a particular type of dance, not as in some ballets where you encounter different genres of dance which are all similar to one other: one dancer exits, having performed several entrechats and pirouettes, and after him a second and a third perform exactly the same thing.

Didelot called dancers who performed many entrechats and pirouettes 'steeplechasers'. Of course one cannot do without jumps in dances, but moderation is necessary in everything. The chief virtue of a dance resides in the graceful position of the body, arms, and in the expression of the face, because the face of a dancer, which transmits all shades of passion, takes the place of the actor's words, and from its expressions an audience more easily understands the dramatic content of a ballet.

At this very time entrechats came into fashion in Paris, and since

Didelot had left the Petersburg theatre and newly arrived dancers took his place on the stage, entrechats and pirouettes triumphed absolutely and finished off sensible, reasoned dance on the spot. At the time many ignoramuses called Didelot's dance *old-style*. But soon entrechats and pirouettes fell out of favour from over-use, and Didelot's method again began to be praised. It has now been fully vindicated by Marie Taglioni: she replaced endless pirouettes and entrechats with gracefulness of bodily movements and diversity in her *pas*, and no other dancer in Europe has matched her success in ballet.

In general Didelot avoided all affectations which, not satisfying genuine taste, can nevertheless catch the attention of an inexperienced public, and on which artists in our time so often depend. To draw superfluous applause from persons uninitiated into the secrets of the art, some women dancers indulge in immoderate twisting of the body, and while dancing lift their legs above their heads. There is no gracefulness in this, and it rather approaches affectation. Consider for a moment the pictures and statues of the best artists: where among them will you see figures in such corrupted poses? And a dancer, in taking her pose, must imitate the best picture or statue because these, in turn, imitate nature in all anatomical strictness.

Didelot, returning a second time to London, filled the post of choreographer there. In 1815, when the allied monarchs were in Paris, Didelot conceived a desire to visit that city and had the good fortune to produce his incomparable ballet, *Zéphire et Flore*, in its grand opera theatre. At the first performance of this ballet the allied monarchs favoured the theatre with their presence. Despite a thousand intrigues and obstacles to its production *Zéphire et Flore* had an enormous success, and Didelot was called to the box of His Majesty King Louis XVIII, from whom he received the most flattering praise for his talent. Not one Parisian artist was favoured with such an honour, and not one Parisian ballet sustained a hundred performances with such success as *Zéphire et Flore*.

Judging from the success that *Zéphire et Flore* enjoyed in Paris, nobody can seriously consider refuting Didelot's talent. France is the cradle and the chief judge of the art of dance. Ballet there has always been at the highest level of perfection: the French are so used to dancing that they all dance, in sorrow and in joy; for them theatre is like the air they breathe, without which they cannot survive. Taking into consideration the passion and understanding of the French in this sphere, nobody would suggest that they could be wrong about Didelot's true merits as a choreographer.

On his return to Petersburg Didelot enriched the ballet repertoire with varied new creations. He produced the celebrated ballet, *The Hungarian Hut*. One sees an element of high tragedy in this ballet: in many scenes tears welled up unbidden in the audience's eyes, which is difficult to achieve in wordless scenes, that is, in ballet; but after sad impressions the spectator shifts to other, lighter ones, and finally laughs to the point of exhaustion in the amusing places. The Hungarian and characteristic dances were overflowing with liveliness and beauty. Looking at another ballet of his—*Raoul de Créqui*—one is swept unwilled back to the memorable epoch of the crusades, when men took up arms to save Jerusalem.

Didelot was an amazing master at giving his ballets the character of the epoch, its local colour, and most important, unaffected, sensible meaning. His dances were in strict accord with the characters and with the principles of art and elegance.

I think every spectator in the theatre can sooner judge the music, the dramatic art, the painting, and the machines of a ballet with appropriate precision than the dances, because among the public one is more likely to encounter excellent dilettante painters, violinists, singers, engineers, and literary people than balletmasters and dancers. The spectator can judge a dancer's mime because mime is close to human nature; he can observe a dancer's lightness, agility, and dexterity in the execution of a *pas*. But a master of this art analyses dances as a skilled surgeon does the human body, in the composition of its parts. Any exaggeration immediately catches his eye, as does immoderate bending of the body, untidiness in the execution of a *pas*, incompatibility of the dances in their make-up with the abilities of the artist who is performing them, an ungraceful stance within a grouping, incorrectness of gestures, inappropriateness of pirouettes—in a word, everything which, despite its contradiction to nature and to the laws of art, produces an effect and delights the inexpert. Likewise, only a balletmaster can absolutely judge the libretti of ballets. Often libretti have been undertaken by writers, people with talent who did not know the art of dance; their work was excellent in subject-matter but not suitable for ballet because much of the action could not be realized in pantomime, as for example accounts of battles, plots to abduct some beautiful girl, and so forth.

In its grammar ballet will not tolerate future and past times, but demands the present; otherwise, spectators will have difficulty understanding its content. Old Didelot was commissioned to make a ballet out of the celebrated opera *The Caliph of Baghdad*. A master of his art, he dealt with the production in the following way: in the opera the caliph appears on stage in a bedouin's costume; one name uttered by him explains to the audience that he is caliph. In a ballet this [method of identification] was not feasible, for which reason Didelot introduced a new episode: at the beginning of the ballet he presented a room where the caliph is changing clothes from his rich costume into bedouin's dress; during this time his ministers bring him papers, which he signs. Having changed, the caliph shows the ministers a ring intended to serve instead of a password, a substitute for the word 'Il-Bondokani', used in the opera as a means of recognizing the caliph. This scene gives the spectator a key to the comprehension of the subsequent events of the ballet.

In the opera there is a narrative explaining how the caliph saved Zetulba and her mother from robbers. In the ballet this could not be realized in pantomime at all, so Didelot introduced another episode. He presented a suburb of Baghdad, where at night, from a minaret, a muezzin calls Moslems to prayer. As Zetulba and her mother are making their way to the mosque, robbers run out from the ruins and try to abduct them. The caliph with his vizier Jafar, both in disguise, fall upon the robbers and save Zetulba and her mother. For ballet this is both effective and comprehensible to the public without a [written] programme. In the opera, when the disguised caliph is in Zetulba's hut, his retinue brings her gifts: a box with zecchinos [Venetian gold coins], a brilliant turban, a rich table with desserts, and transparent lanterns with her monogram. With all this the impression was still not magnificent enough. Didelot imparted a more Asiatic splendour to this scene: he introduced slaves into it, who come in with ladders, climb them and entwine all the walls with gold brocade; the entire hut changes aspect; at this moment eunuchs and slave girls appear and, dancing, bring in a fancy mirror, a dress, diamonds, fruit in baskets, flowers in vases; then in front of the mirror the slave girls dress Zetulba in magnificent attire and adorn her with diamonds and flowers. Slaves appear in Chinese costume, with various musical instruments, and amuse Zetulba with their dances and music, form a tent for her out of umbrellas and shawls, and waft peacock feathers above her to cool her; others kindle censers with fragrant herbs. There is much ado, and all the bustle is intermingled with dances, which makes a great effect.

Almaida, Zetulba's mother, is astonished by such luxury and imagines that her guest is surely some rich pirate; she runs to the *cadi* [a magistrate] and brings him with the guard. The *cadi* rushes angrily at the supposed pirate, the caliph shows him the ring known to the audience, whereupon both *cadi* and guard tremble and fall down before the caliph. He immediately orders the *cadi* to stand up, who then retires with the guard. Almaida and Zetulba's feelings change from astonishment to horror. During dinner the caliph gives the women a sleeping draught: they quickly fall asleep. The caliph orders his slaves to remove the rich adornments and to restore the hut to its former condition: maids place the sleeping Zetulba and Almaida on to some straw; the entire retinue withdraws. The caliph and Jafar hide in the hut in order to see what pleases Zetulba more—

wealth or the caliph's heart. Upon awakening, Zetulba and Almaida do not believe their eyes; they think it was all a dream. Almaida bewails the dream, but Zetulba places her hand on her heart and recalls her beloved with tears, looks for him everywhere, and not finding him, faints in her mother's arms. The caliph, seeing Zetulba's pure unselfish love, commands Jafar to take her and her mother to his palace on a palanquin; magnificent palanquins are brought in and the two women are carried into the caliph's reception hall, where he comes down from his throne, takes Zetulba into his embrace, and elevates her to the throne with him.

Behold how artfully the experienced choreographer makes use of a story for ballet which seems wholly unsuitable for it! In Didelot's work, meanwhile, the spectator's interest is sustained from the first scene to the last, and often in unexpected places the audience expressed spontaneous delight.

The ballet *Alceste*, based on myth, is also distinguished by the riches of its story. It is an embodiment of the ideal of conjugal love. To save her husband, Alceste offers her own life with truly heroic self-sacrifice. In the fourth act there is a beautiful [divided] scene which represents both hell and the Elysian Fields, where Pluto is seen sitting on his throne and the three judges—Minos, Rhadamanthys, and Sarpedon—scrutinizing the book of fates for the verdicts against transgressors who find themselves in hell.

On a rock the three Parcae—Clotho, Lachesis, and Atropos—are spinning the thread of life. At the entrance the guard of these horrible places, the three-headed Cerberus, is rushing about. The river Styx is visible in the distance, along which Charon in a boat is transporting guilty souls into hell. Around Charon shades anxious to cross the river hover in the air, but the gloomy boatman fends them away with his oar. In various places furies are forging chains, sabres, and arrows on an anvil; others press poisoned grasses into a cup and feed the serpents crawling up to them; yet others in Tartarus are stirring up red-hot coals with pokers. On the opposite side the Elysian Fields are visible, with shades in white dress in various groupings, taking pleasure in amusements near cascades, in bowers of roses and fruit trees.

At the end of this act Hercules appears, and with Mercury's help frees Alceste from hell and with her rises up into the clouds. Mercury flies ahead of them, pointing their way with his staff. Just then, at Pluto's order, all the furies on earth fly off in pursuit; but Hercules battles the furies in the air and drives them off with his club. How much thought, variety and gracefulness there is in these scenes! Besides action, the visual representation itself must powerfully move the viewer's soul! And in precisely this respect Didelot distinguished himself in his ballets: he chose stories not for zephyr-like dancers alone, who move about excellently on their legs but who in pantomime cannot transmit elevated feelings and thoughts. Such dancers, however, did have a place in his ballets: when dancing was at issue he composed charming *pas* for them, so that the spectator was doubly satisfied, by the high tragic acting and by the gracefulness of the dance, and not just by jumps, by people spinning like a top and breaking the [line of the] body with a danseur's support, as has become a habit in the newest art.

The plot of the ballet *Cora and Alonso* Didelot took from Marmontel's *History of the Incas*. In this ballet Didelot depicts in masterly fashion the customs of the Peruvians, the festivals of the Incas, their worship of the sun, the representation of their rituals, the consecration of young maidens to the sun, the style of their dances, their warlike amusements, battles, the throwing of lances, and racing.

It can be said that each of Didelot's ballets had its character, historically true and poetically complete. In each ballet he delighted the spectator as much by the charming choice of story as by accommodating, to every extent possible, each nation in its national dances and customs, and likewise by his carefully thought-out pantomime, which everyone could understand easily without recourse to a programme.

The second act of the ballet *Cora and Alonso* is especially picturesque, recalling Bryullov's painting, *The Last Day of Pompei*. It is unfortunate that the ballet was composed long before the celebrated picture, or else Didelot would surely have used it to achieve an even greater effect. The first and second acts represent a picturesque locale: on the right is the temple of the sun with a beautiful staircase leading from the portico. The terrace and garden of the temple continue along the entire stage. The mystic column of the sun stands separately in the middle of the stage. In the distance a lake is surrounded by cliffs from which cascades fall. On the left is a volcano; in front of the temple, a consecrated garden.

In Act I the Spaniard Alonso is invited by Attaliba, the Peruvian king, to a festival at which young maidens are consecrated to the sun. The beautiful Cora, who is being consecrated against her will at her stern father's command, trembling pronounces the vow and relinquishes the world. During this ritual Cora and Alonso encounter each other for the first time; a fiery passion is ignited in their hearts.

In Act II Alonso is wandering around the temple, wanting to see Cora again, and stops in delight on the terrace. Cora is praying to the sun that it expel the image of sweet Alonso from her heart, when suddenly she sees him reaching out to embrace her. 'Cora,' says Alonso, 'I have come to fight, perhaps I will die; tell me, do you love me?' 'What!' exclaims Cora, 'you speak to me of love before the very sun which gives us light? Ah!

Alonso, if you love me, leave here, flee and fear the vengeance of the angered heavens; flee, leave me alone as a sacrifice to my unhappiness.' A peal of thunder is heard, the harbinger of a threatening storm. Cora shudders and runs through the doors of the temple. A frightening subterranean tremor rocks the earth as the volcano begins to erupt; it pours forth sparks and soon a stream of fire and lava makes it white hot. The temple shakes, the earth in many places forms clefts from which issue flames and smoke, the column of the sun crumbles to pieces. Alonso runs to the temple to save Cora, but the doors are locked. He goes in search of another entrance. Meanwhile the doors of the temple are flung open: the high priest, maidens, and servants of the temple rush out in horror and are dispersed about the stage; the people in terror run in from all sides; mothers with infants seek cover, youths lead aged fathers; wives, having fainted, are surrounded by husbands and crying children; in the air birds fly by in flocks, seeking refuge; even wild animals, abandoning their bloodthirst, mingle with the crowd of people.

Soon part of the temple falls away to reveal Cora on a fragment of the column; Alonso rushes to her, overcoming all obstacles, takes her into his arms, and disappears with his precious burden. Cora's father and mother appear and call their daughter, but in vain. Stricken with grief, they fall on to the ruins as the storm and the earthquake gradually quieten and everything becomes calm again. The temple maidens return, and accompanied by the high priest, bewail their unhappy situation and the destruction of their temple. The groans of the dying are heard beneath the ruins; the high priest commands that they be helped. They are extricated from the ruins and brought on to the stage; the others surround them, offering help. Cora's father and mother search in vain among them for their daughter; they despair. The priest admonishes them to place their hope in the will of Providence. King Attaliba appears in the midst of this sad spectacle; the people lament their misfortune, kneel before him and beg for his mercy. The king reassures all and promises his protection; surrounded by his retinue and people, he orders the maidens of the sun into his mansions until the temple is restored.

The denouement of this ballet presents a thought both ingenious and poetic.

Cora has been found. She is chosen as the sacrifice of reconciliation to the sun. The pyre on which she is to be burned is ready. They lead her to it. Cora runs to embrace her mother and tries for the last time to mollify the high priest. But superstition makes him adamant. The mother calms her daughter and counsels her to submit to fate. The priests separate them. They are dragging the girl to her execution when suddenly cries of joy announce the arrival of Attaliba and his victory over Huaskar, his brother, who was taken prisoner with the Antises. Alonso, Rocca, Zorai, Palmor, and the entire army surround the monarch. Suddenly Alonso sees the frightful ceremony of sacrifice and Cora being dragged to the bonfire. He flies to her aid and wrests her, despite the priests' resistance, from the arms of the barbarians. But the high priest is unshakeable. He places his hand on Cora and demands her death. At that moment the face of the sun pales and an eclipse begins; a pervasive horror stuns the people. The eclipse inspires Alonso with an idea of how to save Cora and reconcile the brother-kings. 'Kings, priests, people!' he cries, 'Look at the darkened brow of your god, who gazes down in horror at the abominable sacrifice with which you wish to placate him, and on the ruinous discord of blood brothers, which is tormenting his land with a terrible war. Listen to my words, instilled in me by my God, the only one, true and eternal. He commands you to abolish the terrible law against maidens of the sun, and to cease this ruinous war immediately, whereupon His almighty voice will restore the radiance to your god!' Dumbstruck, all listen in silence and emotion. The Inca, recalling the rescue of his son, abolishes the horrible law, returns Cora to Alonso and her family, and reaches out to embrace his brother. An all-encompassing, indescribable delight; seeing that Alonso's words are coming true, and that the sun is returning to its former brightness, the people, amazed, think that Alonso is the ambassador of a divine being, and fall on their knees before him. But the Castilian rejects their signs of deference, and raising his arms to heaven, says: 'There, nations, sits the almighty God of gods; there, far above the sun and stars, is His eternal throne—kneel before Him and revere Him in all humility!' A sacred delight overcomes the people: solemnly they fall prostrate to the earth and acknowledge the true God, raising their prayers to heaven. The last scene is solemn and touching. It leaves the spectator with a most pleasant impression.

Didelot did not let pass the opportunity to make use of Pushkin's *The Captive of the Caucasus*, and in truth, never has one poet translated another into new forms as fully, closely, as eloquently as Didelot did, adapting the beautiful verses of our national poet into the poetical wordless prose of pantomime. The locale, the customs, the wildness and the warlike nature of the people were all accurately transmitted in this ballet. There were many truly poetic groupings. Thus you saw a soaring eagle with a child it had kidnapped. Everyone was terrified . . . The eagle landed with the child on a high rock and placed it in its nest. Like a snake the despairing mother crawled along the top of the rock and crept up to take back her child. The frightened eagle let go its prey; the mother was overjoyed. All rushed to her on the rock, and, having woven a litter out of twigs, carried her on to the stage. At the same time a *burka* [felt cloak] was seen stretched out on the ground with a Circassian on it, who with a fierce look sharpened his weapon against a rock; next to the rock, in the shade of the trees, a woman sat rocking her child. What cradle to devise for a Circassian girl in her lowly hut? Didelot imparted to the grouping a wild and warlike character appropriate to the people being represented: a thin sabre was driven into a tree; from its hilt a horse's harness was hung, enclosing a broad saddle in which the baby slept, protected not by a coverlet but by a jackal's hide. Games, wrestling, archery—everything was copied by him truly and naturally from life, but everything was covered by the coloration of grace and poetry. In Didelot's hands the ballet was made into a magnificent illustration of the poem.

The fantastic-heroic ballet *Roland and Morgana, or the Destruction of the Enchanted Island*, was taken from Ariosto's poem *Orlando furioso*. From it Didelot borrowed the episode in which Angelica charges her beloved Roland to free one of her friends, confined by the sorceress Morgana on the enchanted island; the first act of the ballet, beginning with the enticement of Astolfo, was the balletmaster's own invention. Alongside other temptations he placed the temptation of glory, which spurred Roland on to the most noble goal of his journey, namely the freeing of his friends from Morgana's captivity. In a magic mirror she shows him the unhappiness they are suffering. Didelot took the transformation of the young knight [from the poem], but replaced the adventures by which Roland is tested in the poem with completely different ones, more appropriate to the stage. He thought to transfer the action to the bottom of the sea. To Morgana he gave the character of Ogilia. Finally, all the magic scenes in this ballet are the fruit of his inexhaustible and graceful imagination; in a word, on to the subject of the long-celebrated fable Didelot composed another, in the highest degree absorbing and poetic, having taken from the famous Italian poet only the freeing of Roland's friends, the love and transformation of Astolfo and the destruction of Morgana's enchanted island.

This ballet is filled with theatrical effects of various kinds, such that it could not help but please every spectator without exception: the educated by the rightness of the subject, the graceful dances, the picturesque groupings in the water, in the air and on the ground and the artful mechanics; and the less educated by the battle of Roland with the giant, with furies, with a dragon, the dwelling of Erebus or the night, the magic sepulchres, the will-o-the-wisps and the knights riding about on horseback. As the story of this ballet was not chosen by the choreographer but worked out by him at the behest of the management, in the foreword to its libretto he expressed himself as follows: 'I sense all the disadvantages in my ballet because the costumes, decorations, [and] accessories are without novelty in the eyes of the public; for I have already composed so many ballets in this genre that my indebtedness to them is everywhere to be seen; but if, in the end, I came to be accused of being everywhere too much like myself, then I would answer that it is better to be like oneself than to steal from others!'

The Chinese ballet, *Ken-si and Tao, or Beauty and the Beast*, was taken from the opera *Zémire et Azor*. Perhaps it will seem strange to the spectator that Didelot changed the title of so famous a libretto when one sees on stage the same story as in the opera; Didelot transferred the action to China better to vary the effects of the decorations and the costumes, and to avoid presenting the same sights. In this ballet the spectator sees not only excellent decorations, machines and costumes, but also charming Chinese character dances with flounces (*volants*), parasols, little bells, and various coloured lanterns. Although there were not so many lanterns with lights in this ballet as in other, newer Chinese ballets—as for example in *Kia-King*—to compensate for that, common sense is everywhere in evidence, and there was more fire in its pantomime and dances.

Theseus and Ariadne, Phaedra, Laura and Henry, Acis and Galatea, The Shade of Lybas, Carl and Lisbetta, Hercules and Omphale, The Descent of Hercules into Hell, and other ballets excelled one another in the merit of their stories, in their charming dances, each with its own character, its special picturesque groupings, and its characteristic element in relation to plastique and poetic coloration.

Later on many 'experts and judges' concluded that Didelot had become obsolete in the theatrical profession, but they are cruelly mistaken, imagining that only a young balletmaster can compose excellent ballets. As if the gift of creation depends on youth! In a young person, of course, more imagination is presupposed than in an old one, but is everything in artistic creation really confined to imagination alone? If a person has no powers of conception, no poetical tendency of taste, no creative discipline, then despite the most flourishing imagination he cannot produce anything respectable. Let us take as an example the famous balletmasters: Noverre, Gardel, Picq, Milon, Aumer; they composed beautiful ballets deep into their old age, they conceived for them the best dances and produced an effect. In their old age Vestris and Coulon, the first dancing teachers in Paris, trained the best students: Duport, Albert, Paul, Battiste, and others. The best dancers flocked to Vestris, asking his advice about the art of dance despite the fact that he was getting on to eighty.

Here is another example: the choreographer Taglioni is now about sixty, but he is composing ballets which, although weaker than Didelot's ballets, nevertheless please audiences everywhere. To the last minute of his life Didelot was composing ballet scenarios, of which each one was better than its predecessor; unfortunately, they were not used on stage. Didelot ought to have made a voyage to London or Paris for diversion, in order to illumine those places anew with his talent and then come back with fresh glory. Every public in the world is alike, their motto being inconstancy. No artist ought to stay in one theatre for very long because his public is like a child: give a child the most charming toy, he will play with it a little, then come to be bored with it; then give him a new one, albeit one much worse than the first, and the child seizes it with delight, and throws the first away.

Didelot had one other important virtue, rarely encountered now: he did not try to sustain his ballets by machines alone, by magnificent costumes and new decorations as compensation for a deficiency of talent, as other balletmasters do. In place of all luxuriance of staging, of all false magnificence, he substituted richness of imagination. In his stories one could always do without velvet, brocade, and gold leaf: the depiction of life, interest, and gracefulness served in their stead. 'I do not want the glory of a work to fall upon anyone else but me, and to have a spectator say, "I was at the theatre yesterday, and saw the most charming decorations, wonderful machines, magnificent costumes," and have not a word to say about the ballet.'

I list here several such grand ballets, which did not involve large expenditures: *The Hungarian Hut, Raoul de Créqui*, and *The Captive of the Caucasus*. They were all given for benefit performances of Mr Auguste, the first dancer of the Petersburg theatre, because the theatre management found no advantage in making significant outlays for benefit performances.

Unfortunately, Didelot's ballets of genius will be completely lost with time, because few artists are still alive who could reproduce and mount them anew in the same form as they were composed by Didelot himself. Didelot's ballets were very complicated in their scenes, groupings, tableaux, and dances.

Perhaps the 'experts and judges' of the present will say that Didelot's ballets are old-fashioned. But do we not use things now—furniture, carriages, and costumes of the eighteenth century—which probably would have been laughed at some thirty years ago? Believe me, what is truly excellent can go out of fashion but not grow old; fashion changes, and what is excellent comes back anew. And they will want to see Didelot's ballets again. Raphael, Racine, Haydn, Shakespeare, Mozart—however old, serve as models of their genre to this day. Didelot, in his sphere, was a model and will live again when good taste prevails.

One old dancer was saying: 'The past is reproached, but it had its good points. I recall long ago that when two friends met they would embrace each other with heartfelt emotion, in the Russian manner, then kiss each other. Nowadays the foreign custom of shaking hands has become fashionable. As long as something can be squeezed from your hand, handshaking will continue, but if your hand is empty, they will do a *pas de zéphyr* away from you and exit the stage.'

I am not saying that the present time in its relation to art is worse than before; but every era has its experts and judges, some knowledgeable, some not. I myself have heard contemporary judgements which I would not have encountered in the past. In one circle there is much talk about Rubini, regretting that he visited Russia so late, when his talent was already in decline. 'Are *you* the one to reject Rubini's talent,' said one pretender to a profound knowledge of music, 'when all Europe recognizes him as its first singer?!' 'I do not deny his talent: in his singing there is much art,' said one of the persons in the debate, 'but it is unfortunate that he has lost his voice: he has a fistula.' 'There you see! And he sings *that* well with a sore throat? How beautifully he must sing when he is well!' There is a burst of laughter. 'You see how your odd opinions make everybody laugh,' continues the musical expert. 'On the contrary; it seems to me that they are laughing at you; "fistula" means a head tone, it is a musical term, and not a sickness as you take it to be.' And both remained quite satisifed: one by the fact that he reproached Rubini for his falsetto, which constitutes a particular achievement of every singer, the other because he demonstrated that Rubini sings well even with a frightful illness. There you have our current experts and judges!

It may be said that Didelot established mimed drama on the Russian stage, because he first developed mime and raised it to perfection.

Didelot would say: 'In order to be an excellent balletmaster, a man must devote the greater part of his time to the reading of historical books, must extract scenarios for future creations from them, and must apply all possible effort to the success of his students that they, by their talents, could enhance these creations. The balletmaster must also know the mores and customs of various peoples, study their national inclinations and costumes, and have a poetic gift in order to set forth his thoughts pleasantly in scenarios. He must know painting and mechanics so that he can produce all kinds of picturesque groupings, and more effectively explain his needs to the decorator and machinist; and music is a most necessary talent for the balletmaster, both for the composition of ballets and for assisting the composer who is writing music for them.'

Didelot possessed this knowledge in the highest degree.

There are balletmasters who, although they did not receive a brilliant education when they were young, become excellent second-class balletmasters through long practice, reading books, advice from excellent choreographers, and innate intelligence.

The excellent balletmaster can never be a burden on a theatre although he might receive a high salary: even on his departure the theatre uses the fruits of his labours for a long time, that is, his excellent students and the repertoire of his ballets. This is not true of a dancer, however excellent, who takes everything with him when he leaves the theatre. It is good to have an excellent dancer in the theatre, but he ought not stay on more than two or three years, because in the course of time he grows boring, or, to put it simply, he wearies the eyes. In one year a dancer can show all his well studied *pas* and *tours de force*, and after that the public will be seeing the same thing all the time. Besides, dances demand youth; strength, lightness, and agility in the body are required to show off one's talent: these wane with the years. An old dancer is not the same as a good balletmaster, who does not age and in his activities can be useful in many ways: if he no longer *creates*, then he *teaches*.

Everybody knows that the art of dance entails great labours; each dancer, man or woman, must practise two hours every day at the very least, otherwise the limbs begin to stiffen and they will fall out of the routine. They must dance without fail—*pour ne pas perdre l'haleine*, as the French say. Only through daily exercise does the dancer acquire the skill to perform long *pas* in such a way that the public does not notice any fatigue or exertion. A want of breath is immediately noticeable in the lazy dancer—from lack of practice. Even during a performance, before the beginning of a ballet, the dancer must practise some ten minutes to stretch the limbs, *pour se mettre en train*, as the French say. If the dancer begins to perform a demanding *pas* without any warm-up, he can suffer leg cramps, and even risks back sprain or breaking some leg tendon.

It is very disadvantageous for the theatre if the balletmaster occupies the positions of first dancer and ballet teacher at the same time. How many examples have there been of such artists always trying to advance themselves, taking the best roles and the effective dances in ballets, so that not one of their students is given an opportunity, with the self-serving justification that a student could not replace the teacher in the event the latter had a disagreement with the theatre. Such an exclusive right places the balletmaster-dancer in a completely independent position. He dances when and in what he pleases, and in this way inhibits the repertoire itself.

If the balletmaster-dancer, out of some whim, claims to be ill, no ballet can be given; but if he were only balletmaster the ballet would always go on, led by an excellent régisseur, and the theatre would not incur any loss. In this situation the balletmaster tries to make excellent dancers out of students because he has nothing to fear from competition; on the contrary, every success of his student works to the advantage of his ballets and lightens his own labours.

It is essential for each balletmaster to have as régisseurs people with talent, drawn from the ranks of second-class dancers. The régisseur must know every ballet absolutely: in the event of illness or some caprice of the balletmaster he must stand in for him. The régisseur must also be able to compose a corps de ballet or character dance for an opera or a drama. In a word, the well-organized ballet company has a régisseur with talent, a second leader after the balletmaster: what the choreographer creates and produces the régisseur preserves, supports, and brings to life.

The fine ballet régisseur is important to the theatre in another respect: with his knowledge of the art and pre-eminent taste, he can be sent abroad to choose a balletmaster or dancer for the company. The régisseur, always better than an agent *in absentia*, evaluates prospects' talents and the degree of benefit they can bring to his theatre. He can more suitably [than an agent] win over artists to conditions advantageous to the management because the serenity and comfort of his own position is inextricably linked with the quality of the company's personnel, whereas agents, people foreign to the theatre and having no artistic connection with it, for their own profit can sometimes act in bad faith and not on behalf of the management.

The excellent régisseur serves as a curb to the careless balletmaster: he is closer to him than all others, can always follow his actions, and seeing something incorrect, can stop him or point out the problem. For the meticulous balletmaster he is a most useful and necessary assistant.

A knowledge of mime is necessary for every actor and dancer. Although an actor has words for explaining things, he will always be cold in a role without excellent mime. The dancer on stage is a mute person: mime is his language. To obtain the skill of speaking with one's face, eyes, gestures, bodily movements, and poses, he must study good paintings, in which the facial expressions that convey various human feelings can be observed, and make a study of poses. He must practise in front of a mirror to train his face truly to express every nuance of the soul.

I have followed the art of dance for more than forty years, have seen many of the famous ballet artists who came to Russia, but in none of them have I seen a talent like the one possessed by Evgenia Ivanovna Kolosova I, dancer of the Petersburg theatre. Every movement of her face, each gesture was so natural and comprehensible that it completely replaced the language of speech in the perception of the audience. At that time the rules of mime were strictly observed, and the artist was not tolerated who performed in mime as if flicking away a fly. Mlle Kolosova was incomparable in all serious ballets; amid a superior company she shone like a diamond. Perhaps it will be said that in her time ballets were in the 'old method', and now everything has been improved: powerful emotions are projected into spoken drama, and the *naïveté* of smiles and the fluttering of legs are left to ballet. But the art of expressing human feelings, it seems to me, can never vary: joy, tears, anger, jealousy, sorrow, love remain the same in all ages and through all possible changes of taste. If in ballet the characters are still people, then there must still be passion in their actions, and if there is passion there must be mime which expresses it. Neglect of mime is the principal cause of the downfall of contemporary ballet.

Besides Mlle Kolosova, the famous Duport adorned Didelot's ballets from 1808 onwards. When he appeared on the Petersburg stage he charmed everyone with his wonderful talent. He enjoyed glory all over Europe at the time. Duport was a genius in his way who had no rival in Paris, Vienna, Berlin, or Naples.

Duport was not tall but well built, and had a pleasant, playful, agile physiognomy for the stage. He was a demi-caractère dancer. Duport possessed everything a dancer needs: unusual gracefulness, lightness, speed, and purity in dances; he perfected pirouettes of astonishing variety; he always performed them on pointe (*pirouette-filée*), and having finished always stopped in a pleasant pose, which constitutes the principal merit of the pirouette. Other dancers, men and women, perform endless pirouettes, their *tour de force*, but perform them without grace and incorrectly, now rising while turning on to their toes, now falling back on the heel. Such pirouettes can serve only buffo dancers in comic ballets; but people who do not understand the art are astonished by them and applaud them with gusto. Duport was like a well-built machine, the movement of which is definite and always true. Despite the fact that he executed the most difficult *pas*, every dance lay well on him because he was the soul of every one. Difficulty notwithstanding, he was always equally fresh both at the beginning and the end of a ballet. It was impossible to notice a trace of fatigue in him, as is often seen in men and women dancers who in the middle of a ballet can hardly breathe, which is very unpleasant for the spectators, as they see before them not an artist, but a simple human being burdened by the most difficult labour.

Duport's principal triumph was in the role of Zephyr. I have seen many danseurs in this role, but for the most part these Zephyrs were cumbersome, lacking Duport's talent and lightness. It was as if Duport were especially created for Didelot, but only a choreographer like Didelot could utilize this artist's multifaceted talent. If Duport inspired an idea in Didelot, he then served as its best interpreter.

In 1814, 1815, and 1816, in the Petersburg and Moscow theatres, Russian national dances reigned supreme.

Ballets in the French manner were little performed in those days, but instead, grand Russian divertissements such as *Love for the Fatherland*, *Festival in the Camp of the Allied Troops*, and *The Cossack in London*, composed by Messrs Auguste and Valberkh. The following ballets were produced in Moscow by Mr Ablets and myself: *Semik* [the week of celebrating at Pentecost], *The Fair at Makarievsk*, *The Sparrow Hills*, *Celebration in the Camp*, *Filatka and Fedora, or The Fair at the Village of Novinsk*, and *The First of May, or The Fair in the Sokolniki*.

At that time men and women danced mostly in the folk Russian and the Cossack manner, the best singers in every divertissement sang arias, duets, and quartets made up of Russian songs, the music for which was normally composed by the Kapellmeisters Cavos, Kashin, and Davydov. The most remarkable of these spectacles was the divertissement given on the occasion of the visit to Moscow of His Majesty, King Friedrich Wilhelm III. For a long time they pondered what to perform, and decided on a national divertissement.

His Majesty's favourable judgement fully vindicated the choice of *Semik, or The Fair at the Maria Woods*. It was given at a matinée performance in Ostankino, at the dacha of Count Dmitry Nikolaevich Sheremetev in suburban Moscow. The stage was constructed in a large hall, and in place of a decoration the hall was lined with birches which represented the actual woods.

On this occasion the best artists danced in the divertissement, trying with such zeal to outdo each other in Russian dances that they took lessons from Russian peasants; many went purposely to gypsies and paid them a lot of money just to imitate their manner in folk dance. And in fact, with what other dance does the Russian compare, if a young girl is seen in it, well rounded, rosy, with a shapely waist and excellently carried torso and arms, with an expressive face, who moves smoothly, like a swan gliding along a lake? Such languor, feeling, movement, and nobility there is in this dance! The very steps of this dance are modest and distinguished by a kind of maidenly demureness. There is nothing bacchic in it, nothing sharply sensual or like the French *recherché manner*. It is a Russian minuet, only graceful, free of starched mannerisms and extensions. His Majesty the King of Prussia conferred his favour on this dance.

The best singers of the time took part in *Semik*. Davydov composed an excellently made Russian soldiers' chorus expressly for them in honour of the Russian army. In addition, the famous Russian singer Lebedev was used in the divertissement, appearing in a *khorovod* before a circle of other singers playing simple Russian folk instruments: spoons, rattles, fifes, and the like. This chorus fulfilled its role so masterfully that even the oldsters' bones began to stir. Add to this the superb corps de ballet, who performed evenly [and] precisely—and the excellently conceived, diverse national costumes, and you will have some understanding of what impression this national ballet must have produced on the highborn visitors to the festival. Its success began to be spoken of everywhere, and soon I was commanded to Petersburg to convey the story and details of this divertissement to the balletmaster Auguste for its production in the Hermitage Theatre. One must give Mr Auguste his due: in his production he enriched this little ballet a great deal; he invented many details and several scenes which added greater interest, but mainly he adorned it with the Mlle Kolosova's incomparable Russian Dance, which he performed with her. Both artists were then first-class dancers and mimes to whom Didelot gave the best dramatic roles in his ballets, but in national dance they stood at that high level which no other Russian artist has reached to this day. If at present we have graceful, perfected Russian dances—bold, lively, energetic Cossack and gypsy dances—we are indebted for them to Evgenia Ivanovna Kolosova and Mr Auguste.

Whoever saw Didelot's ballets in Moscow cannot have a complete understanding of them because at the time these ballets were mounted there the local theatre still lacked the means to produce them in the grand manner, that is, with magnificent costumes, decorations, and machines. Nor did Moscow have such excellent artists as the St Petersburg theatre: the first dancers of Paris—Duport, Antonin, and Battiste—the tragic mimes Mr Auguste and Mlle Kolosova, Mr André, the grotesque, and most of all, Moscow did not have the creator of these ballets himself. To produce someone else's ballet on the stage is hardly a trifling matter, especially a ballet by Didelot. Ballets with dances and groupings are extremely complicated in their details. It is insufficient to see them once, twice, or even four times. Without the help of the choreographer himself it is difficult to remember all the details of a work, and the libretto does not much help; it contains only a brief explanation of the story. In such circumstances one balletmaster can rarely mount the ballets of another in the form that they issued from the hands of their creator. This is not true of opera, which contains a complete rendering of its story and parts for singing: the fate of an excellent opera depends primarily on its performers. Such is not the case in ballet: the entire responsibility falls on the balletmaster, who must show the dancer each gesture, every *pas*. To do this he must have all the details of the ballet in his head. But in most cases, balletmasters mounting others' works, for want of artists or room on stage, must modify them, that is, must operate on them, shortening them arbitrarily. Didelot's ballets so overflowed with effects, that if another balletmaster preserved only part of Didelot's ballet in the way it was composed, it could still please the public.

[II]

Some balletmasters, commissioning a composer to write music for a ballet, grant him complete discretion to compose what he wants. The excellent composer can, of course, write music wonderful in all respects to a given libretto if one considers the music separate from the ballet. But when it must conform to the requirements of ballet and of dances, this music may turn out to be quite unsuitable. For the music to be well matched with the acts of a ballet, the composer must have the same knowledge of the art of ballet production as the balletmaster himself. As this occurs so rarely, it is best if the balletmaster takes part in the composition of the music. Of course, it is not to the composer's advantage to compose in collaboration with the balletmaster, because he would then become completely dependent on the balletmaster, and limits would be set on his imagination. The proverb says: 'Dance to another's tune!' but that can work the other way around: 'Play the fife to a different dance!'

To write music for opera is completely different: the composer is completely uninhibited, he is master of the ideas of his music. The librettist gives him a manuscript, and he makes music at will according to the words and the story. In ballet one gesture of pantomime can take up an entire phrase, and if the Kapellmeister in his inexperience writes 32 bars for this pantomime instead of the necessary 4, he will spoil the whole scene, having deprived it of speed and expressiveness. In opera yet another long duet or aria can perhaps be endured, but in ballet superfluous arm-waving is too unpleasant for the spectators.

Didelot had his way of linking music with the argument of a ballet: one can say that he was always the composer's accomplice in the writing of the music. All large scenes of interest Didelot thought out by himself at home, determining the music of each number: its length, tone, tempo, orchestration, in what places pantomime and dances should be intensified through orchestration, and where the music ought to get along without trombones, timpani, and Turkish drum. Motif and melody were left to the composer's discretion. The music of a ballet is often beautifully written, the motifs are pleasant, the instrumentation artistic, but in the entire score there is almost no relief from loud instruments. From this one could conclude that in the music, unlike the dances and pantomime, no characters are represented; some ethereal being dances or some artless peasant girl, and you hear the roar of trombones and kettledrums more appropriate to a military celebration. Lovers of noise and fuss will perhaps say that loud instruments produce a special effect in music, but surely it is possible for an ethereal and pastoral dance to use clarinets, oboes, solo violins, harps, and other soft instruments, which express a pleasant melody and tenderly caress the ear. I agree: why not at the end of and in the middle of a *pas* use somewhat louder instrumentation, as the French say: *pour animer un peu la musique*. But how is it possible after that, when all its powers are expended on light dances, to write music for a battle or a storm? I think that ballet spectators, sitting in the front rows, will be most satisfied with my observation; loud music probably does not provide them particular pleasure, but produces a headache with its incessant noise. I am amazed: have they not thought to put cannons into the orchestra to give ballet music a still greater effect? In fact this happened once, only not in the theatre but in a garden where cannons accompanied the orchestra. The well known Russian composer Stepan Ivanovich Davydov, student of the celebrated Sarti, told me that in 1780 or 1790 Sarti composed music for a triumphant occasion in Petersburg; it was played with the accompaniment of cannons, in a garden though I don't remember which one, and Davydov conducted the cannons. This could be effective, triumphant, and the thought was beautiful: the roar of cannons recalled the glory of brave soldiers, not the dance of peasants to its martial sounds.*

Sometimes composers are innocent of the noise in their scores—they are asked to write that way. One must give ballet music its due: it has progressed within the time of my memory. I was wrong only to blame composers for poor ballet music—balletmasters who cannot manage very well are also guilty.

Didelot, having made a detailed plan of the music in agreement with his programme, came to Messrs Cavos or Antonolini, the theatre composers. The Kapellmeister sat at his fortepiano and Didelot, without music, went through the pantomime of several scenes, and explained to him that for this pantomime so many bars at such a tempo were necessary, with such orchestration. Sometimes Didelot, tired from great labours at rehearsal, sang musical motifs to the Kapellmeister in a hoarse voice, motifs which Cavos then filled out; afterwards he put all his music into order in agreement with the libretto and the wish of the balletmaster. Sometimes an argument arose between them because the composer was completely hidebound; but Didelot offered him clear proofs that, composed otherwise, the music could not be suitable for ballet, and Cavos had to agree with him. Often Didelot, to make pantomime more comprehensible in his ballets, chose motifs of music from favourite operas frequently performed in the theatre; both music and words were still fresh in the public's memory. From the opera *The Caliph of Baghdad* he made a ballet, and the

---

* It worked thus: trumpets and kettledrums played an attack on the enemy; after the music came the cannonfire, which represented a battle, then an orchestra played again, and in the pauses from time to time the volleys of the cannons, and at the end a chorus of singers was employed.—Ed.

music borrowed from it was a benefit to the pantomime, and clarified it for the spectators.

Didelot, making dances for the opera *Telemachus on the Island of Calypso*, greatly helped the librettist and the composer with his flights of poetic imagination. In this opera there is a passage where Telemachus betrays Calypso and falls in love with one of her nymphs; Mentor, to draw him away from his new passion, wants to bear him off the island. Despite Telemachus' betrayal, Calypso regrets his imminent departure, and creates a storm to prevent it. Without Didelot's assistance this episode would have been realized by Calypso waving her magic wand in the air and the storm beginning, a normal theatrical storm of course, in which thunder, lightning, and rain would do the job very well. But Didelot gave the librettist a truly poetical idea; he conceived the following realization for the passage: Calypso calls Aeolus, the god of winds, and asks him to create a storm on the sea to hold back Telemachus. Aeolus pushes aside a huge stone from the cave in which the spirits of the winds are held. They rush out, with long hair, large wings, puffed-out cheeks and protruding lips with which they made a blowing movement. They burst the chains which hold them captive, and rise into the air in various groupings, flying, waving their wings and causing the wind to blow; others run down to the sea in a frenzy, flap their wings and create waves; a third group shakes the trees and compels them to sway back and forth in the wind; some run in fury on to the clifftops and with movements of their wings and arms bring on threatening clouds, with flashing lightning and powerful thunder; finally, the last of them lower themselves down into the depths of the earth, produce a frightening roar, whereupon small fires appear from beneath the earth. All the elements are in frightful movement. At this time a ship appears on the sea in the distance; in it Telemachus and Mentor are making their way; waves toss it from side to side; now it disappears in an abyss, now it appears again on the crest of the waters; lightning strikes its mast, and the ship begins to burn. Telemachus and Mentor are saved on a plank broken off from the ship. At this moment Calypso, seeing the blazing ship, triumphs: Telemachus must return to the island and to her. Aeolus and the spirits of the winds surround her in groupings and express their joy at seeing the destruction of the ship. The entire action of the storm was arranged with dances; each grouping had its character; frenzy and fury were expressed in fast movements, co-ordinated with bursts of wind. The public was delighted with this scene. Didelot, constructing the storm scene, also gave the composer the idea of representing the hiss of the winds by means of man-made whistles. The subterranean volcanic roar was also apropos, an effect created by means of bellows which were constructed beneath the stage.

The first ballet Didelot produced on his return to Petersburg was *Acis and Galatea*. It was mounted for the name day of His Imperial Majesty, Tsar Alexandre Pavlovich, on 30 August 1816. I give here a partial account of this ballet.

The stage represents the sea, which, in the form of a bay, washes on to part of Mount Aetna. The bay is separated by hills, one of which seems to have been torn away from Aetna by violent earthquakes, though part of it is still joined to the mountain by a broken arch. The hill is joined to the other islands by a bridge. Lindens, beeches, and acacias on the island form vaults of green, beneath which nymphs customarily gather to bathe. At the audience's left is a broad forest, in which an old oak stands out. On the right, trees and gazebos of rose bushes are scattered—a charming place for the meeting of Acis and Galatea. The open sea is lost over the horizon and completes the picture. Polyphemus loves Galatea, who, however, pays no heed to his passion because she loves the shepherd Acis. Venus' son looks after them. Polyphemus knows this and intends to take vengeance on them.

On the day consecrated to the god of love, shepherds and shepherdesses gather to celebrate, Acis among them. Galatea represents in expressive dances the feelings which stir within her. Venus' son unites them. But the arrival of Polyphemus disturbs this happy moment. A loud cry reveals his presence and fury to all. Frightened nymphs disperse; others surround Acis at Cupid's command, and he disappears. Galatea wants to run with Cupid to the sea, but the cyclops prevents her: unable to reach Acis, he wants to vent his rage on Cupid. 'Villain! you made me miserable, now my vengeance shall befall you.' Galatea and the nymphs, seeing the plight of Cupid, on whom their own well-being depends, rush to his aid despite their fright. Polyphemus has caught Cupid, lifts him into the air and threatens him with his mace; but Galatea, with an entreaty and tears in her eyes, restrains the cyclops' hand. Trembling, Cupid asks Polyphemus to hear him out, and, showing him an arrow which he draws from his quiver, promises to use it to the giant's benefit, to cause him to be loved. The simple-minded cyclops, pacified by this promise, puts Cupid down. In joy Galatea and the nymphs comfort the cherub and rock him in their arms. Without letting Cupid from his sight, the cyclops demands that he keep his promise; soon the proud son of Aetna shows his impatience.

Sly Cupid signals to Galatea to feign being in love, and lightly assures the cyclops that the arrow has been let go. Polyphemus, thinking he sees love's agitation in Galatea's eyes, falls at her knees; but Cupid, always wicked and crafty to those he does not favour, wants to be entertained at the cyclops' expense, to take revenge for the fright Polyphemus has caused him, and to show that force is nothing in the face of love. The fierce cyclops, transformed into a gentle creature, is ensnared by Cupid with a garland; the god puts a staff into his hand and a wreath on his head. What a shepherd this is! In this mirthful attire Cupid presents him to Galatea, who is astonished at the happy change and declares that now he pleases her.

Thrilled by his bliss, Polyphemus falls at Galatea's feet. 'Well, I haven't tricked you!' Cupid says to him, and as the cyclops thanks him and strews endearments about, Cupid and Galatea call Acis. He appears. Polyphemus hymns his imaginary victory, and the shepherd steals off among the nymphs, who attempt to hide him. Galatea, better to deceive the giant, dances before him; Cupid, afraid of his jealous gaze, takes his bandage and places it on Polyphemus' eye, while Galatea rushes into her beloved's embrace. Soon the dance is taken up by all. Acis and Galatea, hiding from the cyclops' eye, express their love more with each encounter.

Suddenly Polyphemus senses that his rival is there; in a fury he rips off the bandage, and how great his astonishment when he finds only Amatheia (one of the oreads). He has erred and begs forgiveness; but Cupid, wishing to be done with this savage and tiresome lover as soon as possible, calls to the bacchantes to pour wine into cups and add the essence of poppy. Galatea offers Polyphemus the goblet and he drains it. Soon his eyelid grows heavy and he cannot fight off sleep; he hardly sees Galatea through thick haze, and finally falls asleep beneath the ancient oak tree, where Acis and Galatea have made their monogram out of roses, and form delicate, graceful groupings. General rejoicing is shown in dances. But Cupid commands all to withdraw. Each grouping disappears, taking leave of the loving couple, who also disappear in the waves together with Cupid.

Polyphemus awakens. Returning, Cupid takes back his bandage and the proud son of Aetna, freed from the charm, shakes with rage that he was deceived and that he finds himself wearing such adornment. He rips off the garland, and noticing the monogram on the tree flies into a horrible rage. His fury is inexpressible. Everything must feel his vengeance, and the first tree which bears the brunt of his shame he breaks and uproots. Finally, he climbs on to the mountain and from there sees Galatea and Acis in each other's embrace. His fury approaches frenzy; he cannot possess Galatea and so their fate is decided: the lovers both must perish. He unearths a huge stone and throws it; it slides down and threatens to crush Acis and Galatea, but suddenly it stops, collapses, and from it emerges Cupid, who flies to the lovers. The astonished Polyphemus sees that he has thrown to his rival not death, as he had intended, but the god of love.

Nereids rise up from the bottom of the sea with Neptune and Amphitrite.

Proteus and old Nereus, Galatea's father, appear beside the god of the sea. Tritons and all the inhabitants of the waters fill the shores of the islands. Several tritons shoot brilliant fountains from their curved seashells at the zephyrs floating above them, who shower them with flowers. The mountains are covered with fauns, satyrs, bacchantes, shepherds, and oreads, who have come at Cupid's command to celebrate the marriage of Acis and Galatea. Mercury descends from the clouds and presents the shepherd with a wreath and an oar—the mark of divinity. Venus, daughter of the waves, appears and congratulates the new demigod of the sea. To celebrate her presence the tritons bring forth fountains and make a hall of moving waters for the goddess. Graces and cupids accompany her and hold light screens in the air, which form a decoration for this moving architecture.

Polyphemus has hardly come down the mountain when Aetna suddenly disappears. He returns, fury in his eye, rage in his heart; he rushes to destroy Acis and Galatea and even Cupid, running to the sea to overtake them. But Neptune with his trident stops his son: 'Acis numbers among the gods; Jupiter commands you to desist, my son; withdraw and do not disturb this joyous celebration by your presence.' The spiteful cyclops threatens everyone, but thunder roars and announces Jupiter's will: the lightning which strikes at Polyphemus' feet tells him that only at his father's request will the mercy of the gods spare his life. Perplexity shows on the cyclops' face, but quivering with fear he quickly withdraws and runs to hide his despair and shame in the depths of his cave.

Apollo descends from a radiant chariot; Thetis leads him over to share the joy of the new husband and wife and to grace the celebration with his presence. General rejoicing. Meanwhile night is falling: Diana illumines the horizon with her silver disc. Drunk with happiness and merriment, witnesses of the marriage fall asleep in embraces of love and friendship, while Venus' son puts out his torch and with the nymphs flies off in his mother's wake to the island of Cythera.

The conception of this ballet is beautiful; each grouping has its idea, so that the action of the ballet is closely tied to the story. Nastasia Nikolaevna Novitskaya, who took the role of Acis in this ballet, was beautiful despite the fact that she was playing a male character. Of Avdotia Ilyinichna Istomina it is impossible to speak: she was divinely excellent in every ballet. After *Acis and Galatea*, Pushkin hymned her talent in *Eugene Onegin*.

> The theatre is filled already; the boxes glitter;
> The parterre and the stalls, all are seething;
> In the 'paradise' they clap impatiently,
> And flying up, the curtain rustles.

> Splendid, half-ethereal,
> To the enchanting bow obedient,
> By a crowd of nymphs surrounded,
> Stands Istomina; she,
> Touching the floor with one foot
> Lets the other slowly turn,
> And suddenly a leap, and suddenly she flies,
> Flies like down from the lips of Aeolus,
> Now turning inward at the waist, now turning out,
> And beats one small foot rapidly against the other.

Speaking of composing music for ballet, I returned automatically to Didelot and his ballets; at this point I shall relate several events from his life, which can serve as a supplement to the description of his character. He is perhaps the best example for young artists of a person who devoted his entire life to art and who maintained a high level of morality.

I lived in his home and could observe his domestic life.

Didelot got up early and occupied himself with the study of history books, from which he took stories for his ballets; being excellent at drawing, he sketched groupings for his ballets, then went to the theatre school where he gave lessons in dance, thence to the theatre where he either directed rehearsals of a ballet or actually mounted one. After dinner he returned either to the theatre school, or to the theatre when his ballets were being given. He spent every day of the week this way except Sunday, which he devoted completely to rest. In the morning he went to church; his close friends and associates paid a call for dinner; his circle of acquaintances was made up for the most part of foreigners, scholarly people, and artists. I recall that among his acquaintances were engineering officers, who at that time were sent from France into Russian service and subsequently filled important posts. Didelot loved to converse with such people, and for most of dinner they engaged in scholarly arguments, so that dinner with coffee and liquor often lasted no less than four hours; in my youth I always wanted to take to my heels after eating, for a walk on Kamennyi Ostrov or in some other place, but I had to sit and listen. On his second trip to Russia Didelot bought a beautiful dacha on the Karpovka riverlet, where he placed skittles and billiards in his garden under an excellently constructed awning, so that one could play beneath it during summers in any weather. For those who loved cards he would arrange a game of Boston. Didelot himself never played cards, and did not even understand them.

Later on, in the service of the Moscow theatre, I often went to St Petersburg in 1822 and 1823, and while there I visited Didelot almost daily, working with him to learn his ballets in order to mount them on the

Moscow stage. One day at Didelot's dacha I met the Governor-General of St Petersburg, Count Mikhail Andreyevich Miloradovich, who favoured Didelot that day with a visit. I heard his gracious conversation with Didelot; Alexandre Alexandrovich Shakhovskoy, then an official of the St Petersburg theatres who helped determine the repertoire, came with the count.*

The count described various artists he had seen in foreign theatres, recalled when he was with the Russian army in Paris and saw the magnificent ballet *Zéphire et Flore*, discussed the improvement of the Petersburg theatre, and, seeing billiards in the garden, said: 'Ah, Mr Didelot! I see you are a lover of billiards; let us play.' Didelot answered: 'Excuse me, Count, I have never mastered this art. I could not even hit a ball.' 'Do you know that this game is very good for your health ... but I forgot that you get enough exercise without billiards.'

I think that many Petersburg artists of his time will recall that Didelot had an irascible temper. If something went amiss with his ballets—if the conductor played a cadence poorly or artists performed their roles badly or if the dances or the machines did not work well—he simply lost his temper and forgot everyone and everything around him. But to make up for it he was very good in other situations. I myself was witness: sometimes foreigners here whom he knew well came into serious need, making it impossible for them to return to their homeland, and he gave them money for the trip. Sometimes, hearing from the pastor about a poor family burdened with quite a few children, he hastened to help them. Here is an example of Didelot's unselfish love and goodness which touched me personally. When I was in the Petersburg theatre school, Didelot, having noticed my aptitude for dance, and having grown to like me, wanted to ensure my future; in 1806 he requested permission from the Director of the Imperial Theatres, Alexandre Lvovich Naryshkin, to educate me privately, in his home. This request was granted, whereupon 500 roubles was given to him for my upkeep. When Pyotr Ivanovich Albrecht, the chief bursar of the theatre at that time, informed him of this, Didelot flew into a rage: 'Do you really think I asked to educate *le jeune Adam* for money?' (That is what he called me at that time.) Mr Albrecht said that the money was for clothes and other necessities. 'If I take him to live with me, he will need for nothing.' Albrecht, who knew him quite well, said: 'If you do not want to use the sum set aside for him now, complete your good deed in this way: put aside the money, and when your student must leave your home for whatever reason, this money could be very helpful to a young man who is not rich.' This proposal alone could compel Didelot to

---

* At this time the Imperial Theatres were under the jurisdiction of military governors.

accept the 500 roubles for my upkeep every year. Later, when he left for England, he called me and gave me the money he had accumulated. Not often in our lives do we encounter cases like this.

I have spoken at length of Didelot's talent and his ballets. I must also do justice to his contemporary balletmasters, Messrs Valberkh and Auguste, who also composed ballets for the Imperial Petersburg Theatre. Their ballets, although they yielded much to Didelot's nevertheless had their merits and pleased the public both in Petersburg and Moscow. I shall give as an example the ballet, *Raoul Bluebeard, or The Mysterious Chamber*. The story of this ballet, produced by Messrs Valberkh and Auguste, is taken from the famous opera of the same title, and the music is also taken from the opera, arranged and recomposed by Cavos.

One can always watch this ballet with great pleasure. It has everything: interest, a good story, fast action, effective scenes and beautiful groupings and dances.

If in response to this it is said that the story of the ballet is too serious, we shall answer: when once you grant that ballet is a dramatic presentation, why limit it to the sphere of mythology and deprive it of human drama? The stage portrays life: on it tears as well as laughter must reign. Besides, dances are extremely appropriate in this ballet; for example, in Act I at the wedding of Carlos and Annette, one can dance as much as one likes. During the tournament one can also produce magnificent dances for the courtiers, ladies, and pages of Raoul's court. Even Isaure can dance with a cavalier who has come to Raoul's for the celebration, or with the ladies.

Valberkh and Auguste must be given their due; they produced the ballet beautifully. Some will say there is little new in it. But is the only good thing something new? I saw Fanny Elssler several times in *La Fille mal gardée* in the Moscow theatre in 1851, and the public received this ballet with delight despite the fact that it was composed by Dauberval, a teacher of Didelot, an old man! To this they will say that Fanny Elssler shored up the production with her talent, that she was charming in it. True! But in order to be charming the ballet and the role must correspond to the talent of the artist. The artist's choice of this ballet for her début before a new public clearly shows that sometimes the old days have their merits, though the ballet didn't have machines or magnificent costumes or effective decorations or loud instruments. In it you see only the pure talent of the balletmaster.

Evgenia Ivanovna Kolosova was superb in the role of Isaure, especially at the end of Act IV. When Raoul demands the key, in her fright she thinks that she has it at her waist: with trembling hands, in terrible anxiety she looks for it on her person. Not finding it, she is at a loss to answer. Finally she tells him that it is in another room. Raoul wrathfully commands her to get it. She calms him with a smile mixed with fear. Angrily Raoul orders her a second time to find the key. She goes with lowered head, hardly able to walk: her knees give out, and she nearly falls from weakness; from time to time sighs issue from her breast. With fast steps, impatiently, Raoul walks up to her pointing at the hilt of his sword. In terror she stops in a touching pose: her every limb is trembling, her expression one of despair mixed with fright. Kolosova performed all this with astonishing naturalness; every passion expressed by her in pantomime came directly from the soul; to each she gave its full coloration. If someone should decide to play this role, my detailed description of it could be useful, because ideas can change but not the feelings of the soul: in mime there is no new method. When Isaure appears on stage—pale, with half-dishevelled hair, holding the broken key to the chamber—she stops, exhausted, leans on an armchair, then walks up to Raoul, her head turned away. Trembling, she gives him the key. Seeing that the key is broken Raoul is furious. Isaure falls to her knees before him in a swoon, begging forgiveness; he repulses her, she clings to his legs. Raoul wants to step back, but drags her with him because her arms are frozen to his knees. With an effort he frees himself and orders her to prepare for death while he, in haste and anger, goes into the chamber. This scene by Evgenia Ivanovna Kolosova with Raoul was the height of balletic perfection. Describing it I am transported in thought back more than forty years; it is as if I still hear the roar of the ecstatic applause, and still feel the inner excitement which I felt then, while watching her acting! I have heard from people who saw Madame Chevalier and Madame Bertin-Boïeldieu, first-class French actresses who played the role of Isaure in the opera *Raoul Bluebeard*, that Evgenia Ivanovna Kolosova much excelled them in the ballet despite the fact that an actress has the greater advantage because she commands the word, while the dancer must express herself only in mute language. But Kolosova, by virtue of her inspired feeling, by the mime of her face and correct poses, managed without words.

I consider dancers of either sex who lack the gift of pantomime no more than good, ennobled acrobats, because the excellent pirouette, the light, clean entrechat, the picturesque pose on very tiptoe—these are all pleasant tricks for the eye, but not what good pantomime is, which comes closer to human nature and provides nourishment for the mind and heart. In pantomime one sees human passions, whereas lightness in a turn and aplomb on pointe are nothing but dexterity. On the St Petersburg stage I saw the first-class artists Mr Battiste and his wife, who possessed a remarkable dance technique and who could mount ballets well but who were cold as ice, for which reason they were received by the public largely with indifference. Nothing is better for ballet than a good dancer, man or woman, who possesses excellent pantomime in addition to a fine technique, as did Didelot's student Maria Danilova some time ago; in dances she was Duport's partner. I think that many of Petersburg's inhabitants even now recall these two astonishing talents, the ornament of the ballet *Amour et Psyché*. Of the enchanting Psyché at that time many said that her love for Amour brought her to her grave; it is unfortunate that she so quickly disappeared from the stage in the very flower of her youth, and with such a great talent.

On the occasion of Maria Danilova's death the following strophes were written by M. Milonov:

> Where are you, young companion of Terpsichore?
> Yesterday you stood triumphantly before me;
> Yesterday your charms met with thirsting looks,
> Enchanted by miraculous speed;
> A zephyr's lightness, charm and movement,
> Flower of tender springtime, ever hidden,
> Yesterday you saw my clapping and delight,
> Today I joined the throng to greet your early grave!—
>
> When midst the crowd of sylphs and diverse oreads,
> They marvelled at your peerless art,
> Did e'er they dream that you, alive with gaiety,
> Would fettered be at morn by cold of sudden death?
> Admiring you in sweet languor,
> Did e'er they dream your flow'ring beauties
> Will have lost their praise and captivation,
> So early buried in eternal cloisters?
> O maiden! In oblivion perhaps this very hour,
> (Hidden is the moment of demise) and I, like you,
> Having spoke these solemn lines in sorrow at your dust,
> Will hear the voice of death, which also calls to me!

These verses were found in the *Messenger of Europe*, a journal published by Mr Kachenovsky, in the year 1812.

[III]

In 1836, at the beginning of December, at 9.00 in the evening, a messenger from the Hotel Leblanc came to me and said that a foreign lady had arrived from Petersburg, Mlle Cafarelli, who urgently needed to see me and for which reason requested that I make my way to her immediately. I was astonished, hearing a name unfamiliar to me, but the strangeness of the invitation much aroused my curiosity; what need would some foreign woman have of me? I thought for a moment and immediately went to the Hotel Leblanc with the messenger.

I enter; the messenger opens the door and ... imagine my astonishment!

In front of me stood not Mlle Cafarelli but Didelot, who bowed very low, bursting with laughter at my surprise. We rushed to embrace each other, for we had not met for seven years, though we had not stopped corresponding. My first item of business was to ask him the reason for his trip to Moscow. He said that he had come for his diversion and to see Moscow, so rich in historical events, but that he would not stay for long, as he was in a hurry to spend the summer in Kiev, there to restore his physical and mental health. The doctors had advised him to choose a good climate and he chose Kiev in order not to be far from his son, who had important reasons for staying in Petersburg.

As it was already quite late, we parted quickly on this occasion. The next day we went together to see the Moscow cathedrals and the Palace of the Facets; we climbed the bell tower of Ivan the Great, from which Didelot admired the view of Moscow, and examined the bell cast in the era of Anna Ioannovna and damaged by fire in 1733, and the cannons opposite the Senate, which were taken from the enemy in 1812.

During our review of the various rarities of Moscow, I thought that I would be playing the role of guide for Didelot, but it turned out otherwise: then as before I remained his student! He knew Russian history well and apropos the antiquities we were looking at he recalled the most distant epochs of our nation's life. He spoke in detail of the birth of Moscow, of the times of Ivan the Terrible, the False Dmitris and the interregnum. The great monuments of the Kremlin—towers, palaces, chambers, gigantic cannons—all called forth in him a multitude of the most witty and intelligent archaeological observations, which clearly revealed not a casual observer but a profound savant of historical facts.

The next day Didelot lunched with me and met my wife and children for the first time. He reproached me for having been to Petersburg several times without once bringing them, to show them one of the best cities in Europe. Noticing his portrait hanging on the drawing-room wall, he was deeply moved, embraced me and said: 'I see that you have a good heart, and that you have understood me!' But he was curious as to where I acquired his portrait. I responded that the painter Baranov, when he was making a portrait of him for the theatre management, kept a copy for himself, which I made a great effort to obtain after the painter's death.

On this very evening we went to the theatre together: they gave the ballet *Fenella*, produced by Mme Hullin. Didelot was pleased with the ballet; it was in his style: it had content, dramatic interest, and a message. During the whole time of his stay in Moscow we saw each other every day and were almost inseparable. On the day of his departure for Kiev he told me that he was thinking of living in Kiev only one summer, and after that, if he did not go back to France, he intended to live in Moscow permanently because he loved Russia very much; for his years and condition, however, the Petersburg climate was not healthy. It was obvious from everything that Didelot was bored without some occupation—to him theatre was the very breath of life. There was talk at the time that he might again be offered the post of balletmaster in Petersburg. It seems to me that Didelot on his return to Moscow intended to offer his services to the Director of the Moscow theatre and to mount several of his old and new ballets on the Moscow stage, in which he would probably have succeeded, to the great benefit of the stage, of the public and of the theatre school.

But alas! Man proposes, God disposes. From Kiev Didelot often wrote to me, going into archaeological discussions about the antiquities there. But he missed the son from whom he was separated: this only son was the apple of his eye and, it seems, occupied all his thoughts. In the first days of November I received a letter from him that he had already sent his library and several other things to Moscow by dray carriage, and that on 15 November he thought he would come himself. In a few days I received another letter, from his valet, with the news that Didelot had died on 7 November 1837 from an abcess in the throat, which suffocated him, and that he had been ill no more than a week. Didelot's sudden death struck me deeply: it had not been two weeks since I received a letter from him, and now he was no more. How many thoughts and plans are born in a man's head, and one breath of wind is enough to topple this entire house of cards. And so, our famous choreographer was gone, who among his kind not only numbered with the artists of the very first rank, but who has had no equal for the last half century. Although in the first part of my memoirs I pointed out Didelot's principal ballets, I did not enumerate them all; here I am setting forth, on the basis of his letters to me, a detailed list of all his works. In 1801 Didelot on his arrival in Petersburg composed for his début the ballet *Apollo and Daphne*. The ballets *Zéphire et Flore* (in one act), *Psyché et l'Amour*, *Laura and Henry or the Routing of the Moors* he composed for the Hermitage Theatre: the first in 1805, the second in 1809 on the occasion of the visit of Their Majesties the King and Queen of Prussia, and the third in 1810. They were subsequently produced also at the Imperial Bolshoy Theatre. Besides these, the following small ballets in one act were produced at the Hermitage Theatre: *Roland and Morgana*, and *The Shepherd and the Hamadryad*. I do not recall the years in which they were composed, and in his letter to me Didelot did not mention these small works.

In 1811, on 5 March, Didelot left Petersburg for London and produced several new ballets there: *The Courageous Alonso and the Beautiful Imogene*, *Alina, Queen of Golconda*, *Sappho*, *The Fortunate Shipwreck, or the Scottish Witches*, *L'Amour vengé*, *Richard the Lion-Hearted*, *The*

Origin of Painting, Zéphire et Flore, Kensi and Tao, a ballet in its content and production completely different from the one he produced in Petersburg, having [for Petersburg] taken the story from the tale of 'Beauty and the Beast'. Not one of these ballets was produced here, including *Zéphire et Flore*. On his return to Russia in 1816, he composed the following ballets.

In 1816: *Acis and Galatea,* performed for the first time on 30 August, the name-day of Emperor Alexandre I.

In 1817: *The Unexpected Return,* for the début of his son Charles, on 11 January.

The Young Milkmaid, for the début of the dancer Volange, on 20 April.

Carlos and Rosalba, or the Automaton-Lover, on 30 August, and *Theseus and Ariadne* on 22 November.

The Hungarian Hut, or The Famous Exiles, on 17 December.

In 1818: *Zéphire et Flore,* in the same form that Didelot remade it in Paris, on 3 February.

The Young Island Girl, or Leon and Tamaida, on 14 February.

The Shepherd and the Hamadryad, revived 25 July.

The Caliph of Baghdad, or The Youth of Harun-al-Rashid, on 30 August.

Adventures on the Hunt, on 2 October.

In 1819: *Raoul de Créqui, or The Return from the Crusades,* on 5 March.

Kensi and Tao, on 30 August. Besides this he revived in this year his earlier ballet, *Laura and Henry, or The Routing of the Moors,* on 3 November.

In 1820: *Charles and Lisbeth,* on 7 July.

Cora and Alonso, or The Daughter of the Sun, on 30 August.

Euthyme and Eucharis, or The Shade of Lybas, on 8 December.

In 1821: *Alceste, or The Descent of Hercules into Hell,* on 1 February.

Roland and Morgana, or The Destruction of the Enchanted Island (in five acts), on 30 August.

The Wooden Leg, on 14 November.

In 1823: *The Captive of the Caucasus,* on 15 February.

In 1825: *Phaedra,* on 24 September.

In 1827: *Dido, or The Destruction of Carthage,* on 24 October.

The Faun, The Offering of Love, The Mask, or The Spanish Evening, The Shattered Idol—these ballets were so hurriedly prepared that there was not even time to write them down. I am not recounting here the historical *divertissements* and the dances he conceived for various operas. In addition, he left several ballets in his portfolio: *Aeneas and Lavinia, Madcap Mind and Good Heart, The Father's Curse, Poverty and Mis-*

fortune, *The Pardon, Prosperity and Happiness, The Golden Braid, or the Youth of Medea, The Page's Pranks, The Noble Trait, Macbeth, Hamlet, The Enchanted Forest, Kamul and Zabara, Echo and Narcissus, Diana and Endymion*. He wrote the latter ballet at the age of thirteen or fourteen, in Sweden, when his imagination had only just begun to flower.

For a model I shall set forth one of Didelot's scenarios, which is already a bibliographic rarity.

EUTHYME ET EUCHARIS
or
THE VANQUISHED SHADE OF LYBAS
Tragico-heroic ballet in one act by Mr Didelot

CHARACTERS:

EUTHYME, a Greek warrior; EUCHARIS, a young girl of the city of Temessa; IPSIPILLA, her mother; the SHADE OF LYBAS, high priest, priests, soldiers, people, their children.

Mars, Victory, Venus, Cupid, Apollo, Hymen, Hebe, Bellona, Pallas, Mercury, the Graces, Zephyr, Flora, Thalia, Terpsichore, Clio, Euterpe, Polyhymnia, Melpomene, Erato, Urania, Calliope, cupids, glories, victories, gladiators and soldiers.

FOREWORD

I am not the first to have the honour to present this story to the public. Before me one of the great masters of the art of dance, Mr Noverre, made a ballet in three acts out of it, the programme of which is to be found in his works. I must arrange it into one act. In earlier times almost everything was new; costumes and theatrical effects were still not so well known, and thus it is not strange that during the time of that genius, the action could seem not too long even when it was in three acts, whereas now this would surely not happen. I have added the high priest and Eucharis' mother and have introduced the Greek deities at the end and given them episodic dances to avoid dryness; finally, I have tried to give the ballet a new course of action without departing from the [basic] story.

The action takes place in the city of Temessa at the time of Euthyme's return from the Pythian games; it is the very day when the unfortunate inhabitants of the city of Temessa must present a young maiden in sacrifice to the shade of Lybas.

The stage represents an ancient forest, the tree-tops of which are woven together and form a kind of arch, which all but prevents the penetration of the sun's rays. At the side the temple of Lybas is visible: a coffin stands inside it; around the temple are gloomy cypress trees, which give the place a particular despondency. On the other side lie remnants of ruined columns, steps, cornices, etc., from a structure that once stood there. On these ruins sit fathers and mothers with their daughters, despondent, the latter awaiting their fates.

At the back of the stage, cliffs and the sea are visible, the sea lapping up on them with its waves.

The urn in which lots have been cast is set upon a large stone, around which three old men—judges—are standing. Sunk in depression, the people crowd behind them. The young maidens named to the horrible fate slowly appear, accompanied by their mothers and relatives, from whom they are separated to stand before the judges. The parents of the innocent victims are seated on the other side, fearfully awaiting the judges' decision. Priests are standing on the steps of the temple, making preparations for the sacrifice.

Suddenly Ipsipilla appears, dragged in by soldiers; she is protecting her daughter, whom they want to tear from her arms. She runs and falls at the judges' feet, but in vain—they are unmoved! Already the incense is smoking and the sacrificial urn is ready. The maidens, trembling, draw lots, and seeing that the terrible fate has not befallen them, rush delighted into their parents' arms, and thank heaven for being spared. Finally it is Eucharis' turn; like her friends, she approaches the urn fearfully, and with trembling hand draws her fateful lot. She unfolds it, and seeing herself destined to die, falls unconscious in her mother's arms, who herself wants to be sacrificed in her daughter's place. She seizes the consecrated knife, but the priests, and Eucharis, coming to her senses, rush over to her and deflect the fatal blow. She reproaches them for preventing her from killing herself.

Seeing her mother's despair, Eucharis tries to console her, and hiding her own torment, says: 'It pleases heaven that I be the sacrifice to Lybas' shade; I submit to its will, but here, take this ring, give it to Euthyme; tell him that I shall love him to the grave and will remain true to him; and when he returns from the field of battle, then let him shed but one tear over my dust.' She bids farewell to her mother and her friends, and is ready to be sacrificed! But Ipsipilla again rushes at the priests and begs their mercy for her daughter.

A noise from beneath the earth suddenly makes all present tremble. Through thick smoke fiery words are visible on the coffin: *There shall be no compassion: she is my sacrifice.* They force Eucharis from her mother's arms, adorn her with flowers, and are about to carry out fate's destructive design when military sounds, repeated in an echo, are heard from a distance.

The people rush to the shore: it is Euthyme, returning from the Pythian games on a boat bedecked with the insignia of his victories. His weapons are arranged in the manner of trophies; he holds a wreath in his hands; palm branches cover the deck; he is surrounded by his friends, and a singer of the games, lyre in hand, hymns Euthyme's triumph. Ipsipilla rushes to meet him; Eucharis is already in his embrace. Euthyme learns of the horrible fate which awaits his bride, and swears he will not permit destiny's sentence to be carried out. In vain the priests attempt to tear Eucharis from his arms; they are afraid of Euthyme, who drives them furiously away, overturns the sacrificial altar, destroys the urn and calls Lybas to battle.

The earth begins to shake. The top of the coffin rises up in thick smoke; the fearful shade of Lybas appears in a white veil, which falls away and reveals a moving skeleton in a helmet. The people, horror-struck, scatter in all directions; only Euthyme is untouched by fright. The shade of Lybas demands from him the designated sacrifice; but Euthyme pays him no heed, and drawing his sword, challenges him to battle. The earth trembles; from it emerges a fiery sword and shield; Lybas' shade takes his weapons and begins to battle with Euthyme. The fight progresses evenly matched; at length they throw away their weapons and engage in hand-to-hand combat. Euthyme, trying to seize his opponent, embraces only a flame rising up out of the earth. Just then Lybas wants to abduct Eucharis; but Euthyme catches him and throws him to the ground. The shade quickly gets up and runs to the sepulchre; but Euthyme bars the way. The angry shade seizes his victim and rises with her into the air. A flame coming out of the earth dazzles Euthyme; but he disdains the horror, rushes into it and catches the rising shade by the foot, frees his beloved, and for the second time throws Lybas to the ground. The battle is renewed, and the shade, defeated, runs to the rocks bordering the sea, threatening Euthyme, who chases him. Lybas jumps into the waves, and his vanquisher jumps in after him, to seal his victory. Eucharis runs after her fiancé; but she has no sooner taken a few steps before the cliffs and sea disappear, and they find themselves transported to a temple of victory. Euthyme kneels before the goddess who wishes to reward his courage. Eucharis is in his arms. Bellona is also in the temple. Euthyme is quickly surrounded by other heroes, like himself, deserving immortality; winged victories crown him, Glory proclaims his deed. Cupid, accompanied by his mother and Hymen, join the lovers. Mars and Apollo also reward Euthyme. Jupiter himself, through Hebe, sends a crown of immortality and the gods' ambrosia. The young victims saved by him express their gratitude. Apollo, the Muses, Bellona and other deities of the ancients hasten to greet Eucharis. Terpsichore and Flora, gracing the celebration, sport with Zephyr; finally all come together to express their joy in general dances.

In this ballet we see an engaging story taken from Greek mythology—comprehensible, not long-winded, not too involved—and this conciseness is the ballet's chief merit. If long-windedness is unbearable in drama, it is all the more intolerable in a ballet, where there are fewer ways to explain the story. In this ballet an age of superstition is presented to the audience, when priests held sway over people plunged into coarse ignorance. In the temples they revealed to laymen the pretended will of the gods by means of oracles; you see gladiators who to amuse the idle public and the mob, thirsty for bloody spectacle, shed their blood in the circuses and whose numbers, in the latest period of the republic, increased more and more because the pleasure in blood spectacles grew and grew; you see a singer with his lyre glorifying the deeds of heroes crowned with laurel wreaths at the Pythian or Olympic games. A living picture of these legendary times unfolds before the spectators, and of the customs of the people, cultured and half-primitive at the same time.

The second change of decoration and subject is sharply distinguished from the first: in the first everything is dark, both content and locale. In the second, a most charming garden replaces the dense forest. In the middle of the garden stands the magnificent temple of Glory; in the interior of the temple a large sun rotates, from whose rays lights of various colours blaze on all sides. In and around the temple stand a number of burning sacrificial altars; next to it, on gold spheres, stand winged victories with laurel branches; next to Mars, the god of war, soldiers are stationed, waving banners over his head. In the middle of the stage is a charming grouping: the goddess of Glory stands in a gold chariot with a victory trumpet in her hand; she is surrounded by Amazons; their gold helmets with white fluttering feathers, glittering armour, shields, and lances impart a special triumphant aspect to the whole group. Victory descends from her chariot and crowns Euthyme with a laurel wreath for his heroic deed; Bellona with the Amazons adorns Eucharis with a helmet and armour. Bellona teaches Eucharis military evolutions, placing her in various heroic poses, showing her how to wield a lance, and Pallas—how to shoot a bow. They are joined by other Amazons and perform a majestic dance which is also pleasant to the eye. At that very moment cupids in various groups are visible in the air; they lie on baskets of flowers, supported by a single cloud moving about in the air; other [cupids] circle in the air, forming various groupings, holding multi-coloured veils, wreaths, and garlands. Past the temple fly genies with trumpets who glorify Euthyme's heroic deed; others hold red canopies in the air above the victor, which forms a magnificent picture.

With Didelot each group was intelligently thought out and excellently related to the content of the ballet. As he himself knew mechanics well, all his airborne groupings were realistic, and flying characters took true, not caricatured, poses. This ballet lacked only the flights of a flock of living doves, which he had used in the ballet *Psyché et l'Amour*. Didelot designed elastic corsets for the birds which were attached to them by a thin wire, and which allowed them to fly freely and to beat their wings around Amor and Psyche. 'Psyche' is the Greek word for 'soul' and 'Amor'—the Latin word for 'love'; what could better characterize the innocent soul and love than the fluttering of doves above them? I am amazed that [other] balletmasters have not borrowed this excellent ornament of the ballet from Didelot. And sylphides, I realize, now flutter about with Flora's wings—the very same little wings which were used in the same mechanism for the first time in 1815 in the ballet *Zéphire et Flore* on the stage at Paris.

In conclusion one must say that all true artists, painters, mechanics, and poets, looking at Didelot's ballets, took delight in his talent and marvelled at how he could join spirited invention with its beautiful realization so superbly in his ballets.

Now, at the conclusion of our article, we cannot omit a look at ballroom dancing and at the methods of teaching it during Didelot's time.

At the beginning of the 1800s in Petersburg there were first-class teachers of ballroom dances: Picq, Huard, and later Auguste, Didelot, Mlles Kolosova and Novitskaya, Messrs Dutaque and Eberhard.

The method of teaching at that time was as follows: for the most part children began to learn to dance at the age of eight or nine; at first they were required to do various kinds of battements; those who were by nature somewhat bow-legged, that is, whose toes almost touched, were improved through the performance of battements. Then they studied various *pas à terre*, that is, steps not involving jumps, to various measures of music for ballroom and different character dances. But the chief merit of the old dancing masters was that they kept their students for a long time at the *minuet à la reine*, because this dance straightens out the figure, accustoms one to bow adroitly, walk straight, extend the hand gracefully, in a word—makes all the movements and manners pleasant.

The minuet brought yet another virtue: if the student were stooped by nature, then through very frequently repeated orders to him by the teacher to hold himself straight, his young limbs straightened out and the stoop would disappear completely. If his head poked forward from habit, he was compelled to hold it straight, to straighten out his chest and not let his arms hang free. Thus the student attained a good posture in his entire body. Girls for the most part, on getting married, take leave of dances forever and abandon them completely. Hurriedly studying only waltzes, polkas, and other easy dances, they gain only temporary, wholly insignificant benefit. That was not true of those who studied according to the old teachers' method: women's figures remained straight and correct their whole lives, their gait was graceful, their movements pleasant. Even in advanced years in the Russian polonaise, in which one only walks and exchanges partners, middle-aged ladies and old men have an excellent *tournure*.*

Often one sees young people who did not study dance properly and at balls perform the French quadrille or waltz with their heads jutting out forward, concave chest and goose-like legs—*avec des pattes d'oie*—as the French call crooked legs in dances. This is of principal importance in dances: if a young girl who lacks a pleasant countenance dances gracefully and bears herself well, she becomes attractive to all because one sees nothing forced or caricatured in her. I have heard from excellent dancing masters that they proposed in several homes to teach children all the rules of the art of dance, explaining the value of battements, various *pas*, and the minuet. But they always received the answer that this was the old-fashioned way, that now nobody dances the minuet, that their children will not become theatre dancers, and that following these ancient rules of teaching they will throw money away to no purpose, and their children waste time in vain.

With this comes the usual request: that it might be possible to begin study with the polka, mazurka, *tremblant*, and waltz in two-tempi. There were teachers who agreed to this: they took the money and made caricatures out of fine children. These were teachers who themselves never studied dance, but, having been musicians to excellent dancing masters, went with them to lessons, noted the figures of the dances and certain *pas*, and then themselves took up teaching, like medical assistants who have observed hospital procedures and themselves sometimes have pretensions to the title of physician.

I shall relate one instance as an example. A dancing teacher was invited for the summer to a fine home on an estate in the steppe. An old woman landowner lived nearby; she invites him over and says: 'Is it possible, my dear sir, to teach my granddaughter to dance? Really, I am ashamed before the world that she has not begun to study. Of course, she is still a

---

* 'In 1815, when the allied armies were in Paris, this dance was introduced there also, and everyone loved it: persons of any age can take part in it. It represents the most suitable means to converse with a lady, and this is the main thing. One can say that the polonaise is invented for those who are able to conquer hearts only at a ball' (*Memoirs of the Duchess d'Abrantès, or Recollections of Napoleon* etc., part 16, p. 218, trans. K. Polevoy).—Ed.

child (she was about sixteen years old) but nevertheless it is time. Wherever one is invited everybody dances, and my Varenka can't take a step; only it is unfortunate that she is in very poor health: I am afraid of wearing her down (she was a plump and ruddy girl); can you not teach her the beginning positions and the leg exercises while she is sitting down? And later, in difficult steps, Feklushka and Ustyushka can hold her up by the arms: really, I worry so about her health! As this art is not complex, surely you will not take more than a rouble note per lesson.' The teacher, hearing the fee and the conditions being offered, answered: 'I do not teach sick people, for it is impossible: you would do better to call a doctor, and when your young lady is well, then I shall be at your service.' 'I see, Sir, that you have become conceited, though in other places I have heard how your kind taps on windows and cries, "Is there anyone who would like to learn to dance?"'

For the most part, parents teach their daughters to dance only to bring them out at balls and to provide them, by this means, with the opportunity to make an advantageous match, though nowadays almost all prospective bridegrooms are commercial people: give them California and all will be well, but short of that you won't get anywhere with them, least of all by jumping about. Once at a ball a young girl of the gentry was dancing, and a young man she knew said to another young man sitting next to him: 'I heard that you intend to marry,' and pointing out his acquaintance, added, 'There perhaps is a lovely bride for you: beautiful, very well educated, speaks several languages, sings and plays the piano well, but the main thing is that she has high morals.' The young man answered him: 'I very much agree, that "she is charming", but permit me to ask you about her "situation".' His friend answered: 'Of course, her parents are not rich, but to make up for it girl has high morals.' The eligible young man took his leave, and, seeing the daughter of a rich merchant, immediately engaged her in a galop.

At the time of which I am speaking the following ballroom dances were performed: the écossaise, waltz, cotillon, the French quadrille in eight figures, *Grossvater*, polonaise sautant, the *pergourdon* [= Périgordin?], the gavotte Vestris, and the mazurka in four pairs. Character dances were: the Russian Dance, *pas de châles*, fandango, matelot, Hungarian Dance, krakoviak, allemande, and the *pas de Cossaque*. I agree: why not teach children at the appropriate time these dances also, when they will already have formed their bodies through exercise and their limbs have acquired the necessary flexibility and posture? One often sees polka dancers who lack basic posture: instead of a graceful inclination of the head they place it in an unbecoming way, almost on their shoulder; instead of bending the body picturesquely, they shift the entire body to one side; instead of raising the arm naturally, they bend it at some mannered angle, imagining that this is grace. To dance the polka without the proper bearing is like ordering a student who cannot read music well to play a Liszt concerto, or assigning a child who doesn't know grammar to compose verses. Although the art of dance cannot be ranked with poetry, it has its first principles, and it can only be of benefit when these are thoroughly taught. The correct teaching of dance gives flexibility to the whole body and strengthens the muscles; it may be said that the student receives a completely new form as regards carriage, gait, and the positioning of the body. Some will think that dances exhaust a person; I shall say to the contrary that they improve health, and I offer as examples the following people, who are or were involved with dance, lived to a ripe old age, and continued their work almost to the moment of death: Dauberval, Gardel, Picq, Vestris, Didelot, Coulon, Lefebvre, Morchinsky, Solomoni, Munaretti, and Auguste; of those who are still alive: Duport, who is seventy-three, Taglioni, Titus, Jogel, Dutaque, Albert, Léon, and others. Some are teaching to this day and enjoy excellent health for their age. I am speaking only of those who led and are leading a life of moderation, and exclude those who take pleasure in earthly gifts. Quite a few die in the flower of their years, but here dances are not to blame, but wine. Dances, gymnastics, and fencing are closely related. Now in many aristocratic homes they teach children gymnastics, and in state educational institutions these arts constitute an indispensable part of a student's course precisely because they serve to strengthen muscles and to give a young person agility and quickness.

If I spoke of the virtues of the minuet, then it was not in the sense of advising that it be danced at balls, but in the sense that it is very useful at the outset of studies. In France during the last year of Napoleon's consulate the minuet *à la reine* and the gavotte Vestris were still in fashion: they were danced at balls. When Napoleon was at a ball at Mme Permon's, the mother of the Duchess d'Abrantès, her daughter danced it with Mr Lafitte several times in the course of the evening;* she was a student of the first teachers of ballroom dance, Dépréaux, St-Aman, and the well-known Parisian balletmaster Gardel, and was praised for her agility and grace; this dance much pleased Napoleon; he told the Duchess d'Abrantès that she danced it as well as Mlle Chameroy, who was the first danseuse of the Grand Opéra.

However old its origin, the French quadrille or contredanse still has widespread currency today, but has undergone significant changes: at one time it was danced, but now it is walked, and its figures are made in a way

---

* *Memoirs of the Duchess d'Abrantès*, part 3, p. 369.

comparable to how some people speak French, by mixing French and Russian words: 'Voulez vous aller diner at the merchants' club; il est là-bas, where it is pleasant to pass the time, je vous en prie, my dear sir, prêtez moi some roubles! . . .' They want to set the tone, and they think in French, but it comes out double-Dutch. The French quadrille these days is similar, some people never learning to dance it, but having observed only the figures, they walk calmly back and forth, as if promenading after taking mineral water. But anyway, now it is very excusable if in the French quadrille they walk and perform no *pas*, because at balls dress and manners in general have completely changed. I still remember a time when at balls, even in clubs, men could not make an appearance other than in dress coat, tails, and shoes, and now at formal nights they come in everyday coats and top boots with large heels. How can they execute a *pas* in such a get-up?! The mazurka with four pairs has now largely been abandoned; sometimes at a ball you will see one pair dancing the mazurka, who go around the hall in a circle, having banged their heels, turn—and enough. If the music played had not been a mazurka, one never could have recognized that this is what they were dancing. I think that many people to this day still remember the mazurka as it was before, when Ivan Ivanovich Sosnitsky or Saburov danced it—there was something to behold!

But in 1814, 1815, and 1816 the mazurka in four pairs was in great fashion: it was danced everywhere—on stage and in the salons of the best society. Sosnitsky danced it excellently, especially when he was in full dress uniform. His steps were simple, without any stamping, but he always carried himself nobly and elegantly. If he had to grasp his lady's waist in order to turn her, or to make a figure with her, all this was graceful in the highest degree. Dancing the mazurka he did not make any effort; it was all light, zephyr-like, and yet attractive. Sosnitsky danced the 'Sailor with the flag' just as well. If Sosnitsky was always so agile in his roles on stage, it was because he was helped by his acquaintance with high society—and his art in ballroom dancing opened the way for his entry there. At that time the very best homes in Petersburg vied with each other to have him teach the mazurka and other ballroom dances. In Moscow the actor of the Moscow theatre Alexandre Matveyevich Saburov also danced the mazurka well. He borrowed much from Sosnitsky, but he had more fire and life, although he lacked the noble *laisser aller* which characterized Sosnitsky's dances.

Speaking of Sosnitsky, one cannot help recalling his contemporary in the Imperial Petersburg Theatre, Nimfodora Semenovna Semenova, who also was the glory of the Russian dance. Her manner of dancing was captivating: during her dances she always had a pleasant smile, which enchanted the public, and in general her whole face gave the impression of some special languor and beauty; this gave her dance vitality. Her figure was *velichava* [that is, she moved in a stately way, slowly and majestically], all her movements were flowing, soft, light, unconstrained. She wore a Russian costume as no other; it was difficult to say which adorned which; the costume her, or she the costume. Doing perfect justice to the Russian dance, it must be said that at the same time she had a reputation for being the first beauty of the land and an excellent singer. For that reason many were the devotees of her talent, or, better said, her beauty; many opera glasses were directed at her charming head from the boxes and the stalls during her performance, at her eyes, full of fire and expression, and at her all-conquering smile. It seems to me that one can judge an actor in the following way. Look at the first scene of family life, where petty or strong human feelings are in play, check reality against the actor's performance, transferring the same passions on to the stage; if he is close to nature, then he is certainly a good actor. You could never reproach Mlle Semenova for a lack of naturalness. And in music, I think, sooner call a musician good who plays pleasantly than one who plays over-subtly, because music serves not for the playing of tricks but for the pleasure of the spirit. For that reason the musician who plays difficult pieces may please only those who take pleasure in rare things, but a musician who plays from the soul, with feeling and pleasantness, and effortlessly pours into every other soul a pleasant sensation—his playing is understood and pleases both the knowledgeable and those who are not. The same can be said of Semenova as a singer: as she did not have a [superb] voice, she did not engage in difficult passages, but made up for this with a charming manner and a special feeling.

Among the old teachers of ballroom dancing in Moscow Jogel is worthy of mention, who received an excellent education in his youth: he studied Russian, French, and music in the best possible way, and knew ballroom dances as no one else. Who in Moscow does not know of his talent? Through his hands three generations have passed! With art Jogel joins the inestimable virtues of society: he is resourceful, sharp, always merry and amiable. Being a teacher in aristocratic houses he can everywhere conduct himself as in the courtly era of Louis XIV, when civility was developed to its highest degree. Jogel was not a threatre dancer, but as an amateur he studied ballroom dances to the highest perfection. As painters are divided into several kinds—portrait, landscape, perspective, etc.—so one can divide dances into various types. Jogel can be numbered with the best dancing masters of *ballroom dances*. For the most part all his students danced and still dance excellently. The balls he gave in Moscow are referred to in the following way:

'Among the many balls, public and private, which, as usual, embellished this year's Christmas holidays (1849), we must pause with special pleasure on one: this was the ball of P. A. Jogel, venerable teacher and tutor of ours and probably of half of Moscow, veteran of the dance who placed on its feet and communicated lightness and grace to more than one generation, more than a thousand young men and women, and who for half a century has given lessons in ballroom dancing. The balls he hosts are the centrepiece of all, public and private. They combine the advantages of both while avoiding their inconveniences. They lack the inaccessibility of balls in private homes, to which not everyone can gain entry, as well as the aloofness of public balls. Simplicity and a traditional patriarchal atmosphere [in which a love of manners and customs is cultivated] hold sway at these balls, together with unfailing merriment. There is only one society present at them: the students of P. A. Jogel, who consider it both an imperative and wholly pleasurable responsibility to appear at his ball to show their indebtedness to their esteemed old teacher, having given him and themselves the pleasure of their presence at this ball.'

'And I think it is gratifying for the old teacher to see several generations of his students having a good time at his evenings. More than once I have seen a middle-aged lady walk up to P. A. Jogel, introduce herself as his student and present to him her daughter, granddaughter, and perhaps even her great granddaughter, all P. A. Jogel's students; and tears flowed unbidden from the old man's eyes. The distinctive feature of the balls which P. A. Jogel hosts is merriment. It infects everyone, forcing them to dance and to regret when the evening is over, that it ended so early, this enjoyment which the venerable host, himself merry to this day, is always able to breathe into his celebrations. On 28 December there were 900 people at the ball; everyone danced his feet off, and on leaving regretted that it was over. They wanted another, so much that many earnestly asked P. A. Jogel to give another ball, and the teacher, understanding his students' wish, promised to fulfil it.'

'In January 1800 P. A. Jogel began to give lessons for the first time. In the current year he will have completed fifty years, and the ball on 28 January will be like a half-century jubilee for the revered artist.'

In 1818, in Moscow on Rozhdestvenskaya Street opposite the Breach Gates, at No. 54, formerly the home of merchant Medyntsev and now that of the merchant Vishnyakov, Mr Munaretti conducted his dancing class, where students of various social strata, nations, and ages came every day because he gave lessons very cheaply. Often at his dance classes you could see the funniest collection of people among the participants of a quadrille, of which a most original picture could be drawn: of a child of about eight with a bun in his hand, who is jumping like a siskin;* of an old German man with his chin tied [to keep his teeth warm] who coughs and pants but dances until he is tired out, bends and makes kowtows; of an Armenian in national dress, with a black beard, serious face, muttering to himself that he does not understand the dancing master, who explains things poorly in Russian and substitutes a most amusing pantomime for words; of middle-aged foreign ladies from the stores in curlers, who with all the graces try to execute the *pas de zéphyr*, and glance at the rest as if to ask: 'And so, how do we look?'

As a professor of the art of dance, Munaretti always explained his lessons to his students; though he spoke many languages, he could not be understood in any of them: it was a Babylonian confusion of tongues. When his first group of students, having ended their hour of instruction, went off to their homes, new amateurs arrived at Munaretti's for dancing class, often including brides and grooms, learning dances so that they would not disgrace themselves at their weddings. These people, for the most part, paid him for a package deal: every student gave him 25 roubles in notes, and he took upon himself the obligation to teach them completely, in a prescribed time, all the dances which exist in the world. Many thought that for 25 roubles he would teach them verily unto death, saying that they had still not learned everything, a claim they made with an eye to further enjoying themselves in his dancing class.

But Munaretti was no simpleton: he grasped the situation and used a different kind of cunning in dealing with it: instead of the hour agreed upon, he taught them four hours running and ordered them to execute various *pas* without rest and to waltz to the point of fainting. His students were mostly petty merchants who sought profit in everything: they considered themselves winners, being taught four hours instead of one for 25 roubles, and did not suspect that this was a kind of trick. The dancing master so fatigued them that, being unaccustomed to dancing so long, they could not stand up the next day and their whole body ached. In a word, they felt the same way they would have had they ridden on horseback too long. At that point they cursed dances, renounced the money and wished only to recover quickly so they would not have to postpone their weddings.

For all this, Munaretti was also a fine Italian cook: he could make excellent macaroni, *stoffatti*, pâtés out of wild game, and other Italian dishes. Young people often came to him for quite another purpose, having in mind not the education of their legs in dancing class, but the education

---

* At Munaretti's during dance class one could have a snack. A poor old German gentleman lived with him who, in the event a partner was needed—cavalier or lady—had to join the quadrille: this was his job.

of their stomachs with champagne and macaroni. In this way Munaretti gained a double profit: from his hands and from his legs. He also put on masquerades; entrance was by pass. Lovers of amusing scenes came to them, for no masquerade passed without some interesting adventure. There would be something for the youth to laugh at, and at the same time Munaretti's dancing class had continuing success and was beneficial to society in its way. In the first decade of the century Munaretti served as a comic dancer in the Imperial Petersburg and Moscow theatres, but for various reasons he soon left the service. Munaretti was a good man though he was not very artful in his business; he was loved because he was funny and for his readiness to do anyone a good turn. Thus he danced through his life in jest, and danced through it not at all badly.

This era of the education of Russian legs in dancing can be defined by the following dances: écossaise, matradour, krakoviak, contredanse, cotillon, and German quadrille.

2
Libretto of *Raoul de Créqui, or The Return from the Crusades*
Charles Didelot

In his first sojourn in Russia (1801–12), Didelot had favoured mythological stories for his ballets, including *Apollo and Daphne*, *Zéphire et Flore* (Didelot's signature piece, which he produced at least six times in four cities), and *Psyché et l'Amour*. Upon his return to St Petersburg in 1816, and especially after the reopening in 1818 of the Bolshoy Theatre there, destroyed by fire in 1810, Didelot turned to dramatic pantomime. 'The public in those days demanded content in a ballet, interest—in a word, a play,' wrote theatre official and playwright Raphael Zotov.[1]

Raoul de Créqui is one of Didelot's most celebrated ballets and is typical of his dramatic pantomimes. He adapted the scenario from a number of earlier sources, including a historical account of the Crusades, opera libretti, and the stories of other ballets. The plot is similar to the closing episodes of *The Odyssey*, and the names of principal characters are taken from the *Nouvelles historiques* of D'Arnaud [a pseudonym of François Thomas Marie de Baculard], whom Didelot cites in his foreword. Didelot also cites the opera *Raoul sire de Créqui* of Dalayrac and Monvel, and may have been indebted to Grétry's *Richard Cœur de Lion* for the device of Raoul's song, which is similar in dramatic function to King Richard's song in Grétry's work.[2]

Opera, moreover, is today the best-known source of the stereotypical plot Didelot uses in *Raoul* and other ballets. The rescue opera, familiar from the example of Beethoven's *Fidelio*, is based on clearly prescribed characters and situations: a righteous leading character has been imprisoned by an unprincipled villain, but is freed by loyal followers in circumstances of great dramatic tension. A faithful spouse or lover usually makes the rescue, though in *Raoul* it is made by faithful retainers, as the hero's wife Adelaide (Adèle in D'Arnaud) is also incarcerated, and makes her own heroic escape from prison.

The element of virtue which runs through the story of *Raoul* might also be explained as an influence of rescue opera, but it is prominent elsewhere in Didelot's ballets regardless of their subject-matter or derivation. It is evidence of a strain of Enlightenment classicism which runs through all of Didelot's art.

---

[1] 'Memoirs of R. M. Zotov', *Istoricheskii vestnik* [Historical Messenger], vol. 66 (Oct.–Dec. 1896), p. 403.

[2] Abram Akimovich Gozenpud cites other operatic versions of *Raoul*, contends that Didelot borrowed ideas for this ballet from August von Schlegel and Salvatore Viganò, and discusses Catterino Cavos's music for Didelot's ballet (*Muzykal'nyi teatr v Rossii ot istokov do Glinki* [Musical Theatre in Russia from its Origins to Glinka] (Leningrad, 1959), pp. 500–15).

Whatever their circumstances, which are often melodramatic, his characters are linked to the Age of Reason by their poise, idealism, and a sense of civic duty.

In the production of *Raoul*, Didelot displayed his love of extravagant spectacle—in feats of stagecraft (the shipwreck, moving sets), blocking (two battles and a festival), and in tension-filled mimed tableaux, constructed along the lines of a living picture. The success of these devices was all the more extraordinary because Didelot produced them under stringent economies. Prince Pyotr Ivanovich Tyufyakin, Director of Theatres in St Petersburg during part of Didelot's career, 'was too excellent a proprietor to spend much money on ballet,' Zotov recalled.[3] To which Glushkovsky added, after pointing out that *Raoul* cost a mere 5,000 roubles to produce: 'The mounting of this ballet was got up so cheaply because at that time, according to theatre laws, large outlays were not made for the benefit performances of artists, and for their benefits they had to choose from decorations, costumes, etc., already in use, though what was lacking was made up new.'[4]

Raoul de Créqui enjoyed a huge success. Glushkovsky, who was present at the first performance, remembered:

> After the second and third acts Mr Didelot was called on to the stage by the entire audience; and many of the patrons from the boxes and stalls came to him in his dressing room; some extolled the story with praises, others the dances. Among them was Mr Orlovsky, a famous painter of the day who excelled in the representation of characteristic military figures, horses, etc. Turning to Mr Didelot, he said: 'Permit me in turn to pronounce my judgement; I have seen many pictures with a beautiful subject, but yours surpass everything seen by me in this genre: no master's hand could draw them as you do in ballet, because with you they are living, and animated by rich ideas.'[5]

In time *Raoul* sustained a hundred performances, always bringing in full houses.[6]

---

[3] *Istoricheskii vestnik*, vol. 66 (1896), p. 403.
[4] 'From Recollections of the Celebrated Choreographer Ch. L. Didelot', *Moskvityanin*, 1856, No. 4, p. 393. Didelot produced *Raoul* for his colleague Auguste, an important dancer and choreographer in the Petersburg company who took the part of Count Créqui (the Raoul of the ballet's title, though his elder son is the only character so referred to in the libretto).
[5] *Moskvityanin*, 1856, No. 4, p. 404.
[6] Zotov, 'Memoirs', *Istoricheskii vestnik*, vol. 66 (1896), p. 403.

RAOUL DE CRÉQUI
OR
THE RETURN
FROM THE CRUSADES

Grand pantomime ballet in five acts,

Composed by Mr Didelot

Music comp. by Mr Cavos and his student Zhuchkovsky; decorations by Messrs Kondratiev and Dranchet, the last [act] by Mr Canoppi; battles by Mr Gomburov; machines by Mr Bursey; costumes by Mr Babini

SAINT PETERSBURG

at the Typographer of the Imperial Theatres,

1819

Publication permitted:
St Petersburg, 30 April, 1819.
Secretary of the Censorship Division
Vasily Sots

To Mr Auguste
Gentle friend!
Our friendship does not require a surfeit of words. This ballet has been composed for your benefit performance, and I dedicate it to you. May it have success, that it might all the more deserve to be an expression of the sincere friendship and respect

Of your sincere friend,
C. Didelot

FOREWORD

My constant striving to bring the Public the greatest possible variety prompted me to set to work on an unusual genre, certain, however, that every genre is excellent except one which bores. And so if I manage to please and entertain the Public, I will have achieved my goal.

These events, drawn from the historical writings of D'Arnaud, were already put on stage in the form of an opera composed by Monvel. But my plan differs from his completely. I borrowed from him only one scene, namely, when the keys are stolen from the prison keeper; but even this scene is cast in another form altogether and has a different ending from that of the opera. With pleasure, however, I acknowledge the borrowing I made.

DRAMATIS PERSONAE

Count Créqui — Mr Auguste
Adelaide, *his wife* — Mlle Kolosova the elder

Alain, *young peasant, son of the prison keeper* — Mr Antonin
Raoul de Créqui, *elder son of the Count* — Mr Charles Didelot
Craon, *child of 7 years, younger son [of the Count]* — Mlle Rasova the younger

Baudouin, *former neighbouring sovereign, who in Créqui's absence has seized his dominions, and who is in love with the Countess* — Mr Shemaev the elder

Gentleman of the court — Mr Lyustikh
Two Ladies of the court — Mlle Novitskaya the elder, Mlle Ikonina

Young Raoul's lady — Mlle Osipova
Ladies-in-Waiting at Créqui's Court — Mlles Shemaeva, Pimenova, Dumond, Gornysheva, Azarova, Shcherbakova and Selezneva

Pierre, *a fisherman, retired soldier of Créqui* — Mr Eberhard
Marguerite, *his daughter* — Mlle Likhutina
Colas, *a gardener* — Mr Artemeyev
Nisette, *a gardener* — Mlle Ovosnikova
Simon, *the fisherman's son* — Mr Golts
Henri, *the fisherman's son* — Mr Striganov
Humbert, *Baudouin's counsellor* — Mr Baranov
Morlaques, *Créqui's arms-bearer* — Mr Kozlov
Gauvine, *Créqui's arms-bearer* — Mr Velichkin
Knight Renti, *who comes to Créqui's aid* — Mr Gomburov
Maturin, *eldest of Créqui's vassals, 105 years of age* — Mr André
Bertrand, *the prison keeper* — Mr Palnikov
Herald — Mr Bochenkov
Warrior and guard — Mr Didier

The progeny, of all ages, of a hundred-year-old man. Knights, pages, ladies, armsbearers, courtiers, hunters, peasants, gardeners, warriors, servants and vassals of Baudouin's court, which was previously Créqui's

ACT I

The stage represents a seashore rimmed with cliffs. On the left stands a fisherman's hut. Nets are stretched on stakes by the shore. It is dawn. A fierce storm is raging.

Fishermen, driven back by the storm, are returning to harbour. Simon and his father off-load the rich catch from their boat. Henri, less fortunate, docks on the other side. They are about to go home when suddenly, amidst flashes of lightning, they perceive a man contending with the ferocity of the waves. He was holding on to a tree branch floating out to sea, but the branch breaks and the unfortunate man sinks into the waves.

The fishermen rush to his aid. Marguerite is summoned, who with her brothers brings everything needed for the rescue; while they are saving the Count (for it is Créqui) the young peasant girl sees the wreckage of the aft part of a ship. A flag is still waving from it, and two people are still clinging to it. Rapid assistance also saves these two, who were about to perish; and how great the fishermen's astonishment and joy when they realize they have saved their lord and two of his arms-bearers.

Créqui regains his senses and embraces first his rescuers and then Morlaques and Gauvine, the faithful servants whom he thought had perished. The fishermen invite the Count into their hut. The storm quiets for a time, and good weather seems to return.

The Interior of a Fisherman's Hut

Créqui inquires about his family; a fisherman shows him portraits of the Countess and himself, objects of the inhabitants' adoration. He asks about Adelaide, and the peasants' sore distress frightens him and makes him fear misfortune; but they reassure him with recent stories of young Craon, who was still in the cradle at the time of Créqui's departure, and of Raoul, his elder son. Finally Pierre tells him the reason for their concern, which is that Baudouin wants to force the Countess into marriage, and that this tyrant has seized all of Créqui's domains. What is he to do in this sad situation?

Surrounded by dangers, having lost his troops, having been tossed about by the storm, knowing Baudouin's power, and wanting assurance of his friends' and vassals' devotion, Créqui decides to go into hiding for the time being, and asks for help. He hastens to write to his supporters, and Pierre names all the people still loyal to him. Meanwhile Gauvine has disguised himself as a poor shepherd, and Créqui entrusts him to deliver the letter despite the storm, which has again begun to rage. For his part, the Count also plans a disguise, and dresses up as a pilgrim who has returned from Palestine. That he might better attest to Baudouin that he has died in the crusades, Créqui wraps his own banner and arms in black crêpe, the emblem of death; he writes a letter to his wife, and entrusts Morlaques to hand it to the Countess in Baudouin's presence; he proposes to go to her himself in his new guise and to assure her of the contrary.

These preparations have hardly been completed when Marguerite runs in with frightening news that Baudouin, caught in the storm during his hunt, is coming to the cottage with his entire retinue. Créqui, Morlaques, and the fishermen hasten to leave; only Marguerite remains to receive Baudouin, who soon appears, accompanied by Humbert, warriors, and courtiers. His dark and brutal mien reveals his character. 'Are you here alone, young lady?' 'I shall run and fetch my father.' 'No! Stay.' Baudouin marvels at the young peasant girl's innocence and beauty, and amuses himself for a time at her embarrassment and fright; but presently, as he remembers the lovely Countess, he falls into a dark reverie again.

At this moment Pierre appears with Créqui, already dressed as a pilgrim and carrying a harp on his shoulder belt. They arrive as if from the road, and Pierre scolds Marguerite for not coming out to greet him. Baudouin interrupts him with a question: 'Who is this person with you?' 'I am a lowly pilgrim; I was hunting for shelter from the storm, and this good fisherman took me in.'

'Old man, you have come at the right time for my wedding,' says Marguerite, 'you have a harp, and we shall dance.' And in joyful expectation she dances before Baudouin while Créqui accompanies her on the harp. Soon Baudouin finds even this boring: he stands up and walks around the cottage; his gaze suddenly meets the picture of Adelaide, and his movements express a fervent passion; but just as suddenly he sees the portrait of Créqui, and fury flares up in his heart. He turns to Pierre: 'How dare you, wretch! . . .' 'No, it is I,' says Marguerite, 'I found this portrait and hung it here.' Pierre, wishing to allay Baudouin's suspicion, rips down the portrait and wants to tear it up; but Baudouin stops him.

Créqui takes advantage of the moment to persuade Baudouin that he, Créqui, is dead. He pretends to laugh: 'Ah, my God! To go to so much trouble over a dead man!' 'What do you mean, a dead man?' 'Exactly that! I myself saw him dying of wounds in Palestine, whence I have recently come.' Love is gullible, and Baudouin promises a rich reward to the pilgrim if he will repeat this news in the Countess's presence—to which Créqui agrees. Only Humbert, it seems, distrusts the false pilgrim, and follows all his movements.

Just then Alain, Marguerite's fiancé, arrives for his bride with many villagers; he runs around the room looking for her, and great indeed is his surprise when he runs into Baudouin instead. Marguerite brings in the meal but all are in a quandary, for the presence of a lord inhibits them, since the storm has still not passed. Finally Créqui proposes that Baudouin amuse himself with the peasants' dancing, the music of his hunters, and that of Créqui's own harp. Baudouin, momentarily less gloomy and nurturing flattering ambitions towards Marguerite, agrees; the dances begin, and he, it seems, takes part in them. But Humbert reawakens suspicions about Créqui in his master's heart, and with him devises a way to lure the pilgrim to the castle.

Finally Baudouin stands up, throws a purse to the bride and groom, and commands the entire wedding party to go to the castle; the pilgrim is also invited, and is entrusted to Humbert's surveillance. Baudouin departs, leaving Humbert. The peasants go into the church for the wedding rites, and all exit. At that time Morlaques and Henri appear, who had been hiding up to then. Simon shows the latter the shortest and best way to the castle, a road on which he will encounter nobody.

ACT II

The exterior of a castle, of which one part takes up the entire left side of the Theatre. Further away hills are seen, and a meadow, a river with a bridge, and in the distance the sea, which washes up on the other side of the castle.

Adelaide is sitting on a balcony, playing on her harp a romance which Créqui had composed for her. Her tears flow at the recollection of her beloved. Next to her sits little Craon, who tries to console her with caresses. Presently her elder son Raoul enters, and in his wake comes the entire wedding party. Pierre, Alain, Marguerite, and the vassals ask the Countess to favour them with her presence, and she, accompanied by the young Raoul, leaves with them.

All rush to meet her in a delight of sincere devotion, and among them Créqui, whom Humbert has not let out of his sight, takes pleasure in seeing his devoted wife. The celebration grows; people of all ages, even a hundred-year-old man who is brought in on a litter, take part in it. The old man asks to see the Countess once again before he dies, as there is no longer any hope of ever seeing his good master. Moved, Adelaide gives him a medallion with Créqui's likeness, and mingles her tears with his.

The celebration continues, and young Raoul chooses to take part in these village diversions. Knights of the court arrive and propose to Raoul a mock single combat in honour of the ladies. He accepts the challenge and vanquishes all competitors, laying down the wreath of victory at his mother's feet. Other games follow, in which the disguised elder Créqui takes part.

The young Craon arms himself in the manner of his brother, and wants to imitate him. Créqui's cheerful disposition attracts the young Craon, who demonstrates his skills before him. The delighted father takes a wreath, crowns his son, and raising his arms to heaven, asks the Countess for permission to kiss the boy. Tears flow from Créqui's eyes, but one look from Humbert dries them up and serves to warn the incautious father. The boy then admires the sounds of the harp, and Créqui hastens to satisfy his curiosity. Among the tokens they exchange, Craon recognizes the mark of a crusader. 'Mama, mama! This old man must have been in the East!' Man and wife are approaching each other, and everything is about to be explained, when Baudouin's arrival interrupts their conversation, which has hardly begun.

Humbert goes to greet him, and walking back has already poured the poison of doubt into Baudouin's breast. 'Who is this old man, Countess?' Baudouin asks her, 'and what did he tell you?' The Countess explains, and Baudouin questions the pilgrim about the fate of Créqui. The false pilgrim informs the Countess of Créqui's death, and Baudouin's joy betrays him. The Countess is incredulous, claims that the old man has been bribed, and reproaches Baudouin for such a low deed.

Suddenly a page enters, announcing that a stranger requests permission to see the Countess. Baudouin commands that he be allowed in . . . It is Morlaques! The Countess recognizes him; everyone crowds around him, asking him about his master; but his gloomy mien, the black crêpe covering his weapons and the feigned tears announce to all the news he brings. Morlaques gives her Créqui's letter; in vain the young Raoul tries to prevent her from reading it, but she has already done so, sensed everything, and faints. Créqui's situation is even more horrible now: he must watch his wife's torments yet cannot assure her of the truth. Having regained consciousness, Adelaide despairs. Baudouin reads the letter and cannot conceal his joy, thinking himself freed of a hated rival; his suspicions regarding the pilgrim vanish, replaced by trust.

Créqui takes advantage of this turn of events and assures Baudouin that his presence just then is a burden on the Countess which he hopes to lighten with the aid of music. Baudouin agrees, despite Humbert's persistent misgivings, and exits. Créqui, also fearing his son's impetuosity, tries to send away the young Raoul; but the latter will on no condition agree to leave his mother in such sorrow. 'Good youth,' says Créqui pressing him to his heart . . . then, suddenly remembering his role: 'Forgive me, sir!' 'Do not worry, my friend, I venerate age, and am pleased to embrace such an esteemed man.' Créqui repeats his request that the young Raoul withdraw, and the latter, sensing a special affinity between himself and the old man, at last obeys.

Créqui sends away the others and then, taking up his harp, begins to play the romance that he once composed for the Countess, and which she herself had been playing on the balcony not long before. The first strains of the music have hardly reached her ear before she revives, looks about, smiles, takes the harp in turn, and repeats the romance. With sympathy and curiosity she gazes at the pilgrim: she tries to recognize his features, and Créqui, drawn by an irresistible feeling of love, shows her the ring and bracelet she gave to him at the time they parted. This enables her to identify him. Who can describe the feeling of a spouse's delight and joy? She now recognizes Créqui, and thrilled by the bliss of the meeting, they are at the point of forgetting all... when suddenly Baudouin appears with Humbert, and Créqui, making a sign to his wife, reverts to his disguise.

Encouraged by Humbert, Baudouin returns with new suspicions in his soul. 'Do you think, Humbert,' Baudouin says to him, 'that this old man could be Créqui?' 'He could very well be!' and Baudouin trembles with rage. Humbert restrains him, saying that first one should be sure. Baudouin walks up to the Countess and smiling praises the pilgrim's virtue, that he had so quickly cheered her up. 'And now then, Countess, as you ought not to have any further doubt about Créqui's death, nothing stands in the way of our union, and continued resistance will be in vain!'

The young Raoul responds in his mother's stead with pride and scorn. Baudouin rushes at him in a fury, but Adelaide throws herself between them. Créqui makes a sign to the Countess and she orders her son not to irritate Baudouin, who is delighted with the flattering hope of reciprocated affection. He throws himself at her feet, begs her to take off her veil of mourning and replace it with a wedding veil. Emboldened by her husband's signs she pretends to agree, and permits her veil to be removed.

In a fury Raoul wrests it from Baudouin's hands, rushes to a sword hanging among the war trophies, and charges the tyrant; but Créqui wrests the sword from his hands, fearing for his son's life, and the disarmed Raoul weeps over his father's trophies.

On Créqui's advice, Adelaide entrusts her son to Baudouin's protection. The latter delights in the change which has come over her, and returns her veil, while the false pilgrim proposes that the young villagers' marriage celebration be resumed. Baudouin agrees, and the scene becomes merry again. Only Humbert keeps an eye on Créqui and his spouse. During this time Gauvine steals in among the dancers to Créqui with letters from friends coming to his aid; he leads Créqui to one side, conveys this joyous news and gives him the letters; but that instant they are seized by the vigilant Humbert, who has seen all and who has crept up on them. General confusion interrupts the celebration. Baudouin begins to read the letters; his fury increases with each one. 'I am betrayed,' he cries, 'this is Créqui!'

Seeing that dissimulation is futile, Créqui rushes into his wife's embrace; young Raoul, informed by Morlaques, rushes to his father's feet, while young Craon hugs his father around the neck. The vassals encircle this touching scene, and serve as a defence against the violence of evil; Morlaques unfurls Créqui's standard over the entire group.

Baudouin is enraged. The Countess, inspired by love and danger, uses every means at her disposal to induce the troops to come over to Créqui's side, but in vain: the soldiers know only Baudouin, and only a few officers take sides with their former leader. A warrior runs in announcing that the army of Chevalier Renti has invaded Baudouin's territory; hope inspires Créqui's party. A battle begins; but what can a handful of courageous people do against superior numbers and strength? There is no reasonable hope for rescue when three soldiers make a special attempt, despite the odds, to capture Créqui, the young Raoul and the Countess.

Two of them manage at length to take the men; but the third tries in vain to get near the Countess. Seeing it impossible, he runs off when the battle is over. The two soldiers present their prisoners to Baudouin, who, delighted, gives them captured weapons as a reward and charges them to guard the prisoners. Faithful Morlaques, wounded and dying, kisses Créqui's standard for the last time, hands the banner to the young Raoul, and expires.

A herald is announced; he has come from Chevalier Renti, demanding that Créqui and his family be delivered to him. Baudouin arrogantly refuses and dispatches the messenger with his proud answer. Then he entrusts Humbert to keep close watch on the prisoners, and the young Raoul is delivered under guard to one of Humbert's followers and the soldier who had captured him. Créqui and his wife are also led to the dungeon, and reach out in sorrow to their son, who consoles them and assures them of heavenly assistance. Tearing himself away from his guards, he flies to his parents, embraces them as he kneels, and receives their blessing. But he is quickly taken away and led to a boat moored at the shore, which soon gets underway and disappears.

Trumpets are heard in the distance, summoning Baudouin to battle, and he withdraws with his soldiers.

ACT III

The interior of the tower in which the Countess and her younger son Craon are incarcerated.

The child is sleeping at his mother's knees; she looks tenderly at her beloved son. Suddenly, frightened by a horrible dream, Craon awakens and rushes to embrace his mother. Calmed by her, he asks her if she too would fall asleep, and she, as if agreeing to the child's request, sinks into a profound reverie. Craon thinks that she has fallen asleep, kneels down, imploring the Almighty to grant her peace of mind. Presently hunger begins to work at him, and finding a piece of black bread and a glass of water he decides, whatever his disgust at the unpalatability of this discovery, to use it to his advantage.

Just then he hears his mother crying, and runs over to her. An arrow suddenly flies through the grating of the tower window and falls at the Countess's feet. Attached to it is a message: 'Is the Countess imprisoned here?' Adelaide tears off half of the note, and in accordance with its instructions attaches her answer to the line to which the arrow is attached, pulls it sharply, and the arrow flies back whence it came.

Someone is heard entering the dungeon: it is Baudouin, coming to his victims; he has vanquished Créqui's allies. In one hand he holds the enemy's standard; he offers the other to the Countess. Infuriated by her refusal, he runs to her and tears Créqui's wedding ring from her hand, when suddenly the same arrow, shot for a second time along the same course, flies in, striking and wounding Baudouin, who is not wearing armour.

The Countess runs to him to extract the arrow and deliberately tarries with precautions to gain time to read the message attached to it. Then she tears off the line attached to the arrow and hurries to bind Baudouin's wound. He runs around the dungeon hunting for the villain who attacked him. Finally his gaze fixes upon the fated window—and everything is revealed. The Countess tries to inform her friends about the danger they face; but Baudouin seizes Craon, pushes the boy's mother away from the window, and calls for a soldier. The soldier enters. Baudouin orders him not to let the Countess near the window; at her slightest attempt to do so he is to throw her son, whom Baudouin has handed over, out of the window and down the side of the tower. Then he leaves.

With entreaties, cries, and promises, Adelaide tries to arouse the soldier's sympathy, but in vain; he is unmoved. She falls at his feet, kisses his hand, invokes all that is holy and precious to him on this earth, but he coarsely rejects her. In despair the Countess grabs her son and cries out: 'Save yourselves!' The soldier wrests Craon back and repulses her with such force that she falls down. In her fall she sees the standard Baudouin left behind; despair inspires boldness in her as the barbarous soldier is at the point of casting down her son, and like an enraged lioness she charges the cruel soldier and strikes him in the chest with the spear point of the standard. The villain lets fall the child, but with wavering steps he runs after the Countess and her son as they flee him, still trying to strike them down with his dying hand. Finally his strength abandons him, and gasping, he dies.

Raising thanks to heaven, the Countess seizes the signal line sent by her friends, and soon they lower to her by turns a thick rope, a saw, an axe, and a rope ladder; she hurriedly saws through the grating, secures the ladder, and is at the point of escaping when her son's fright stops her. How will she lower down the timid child, whose slightest cry will betray her effort? How is this problem to be solved, how is he to be exposed to the dangers attendant on their escape? . . . The moments are precious, and maternal tenderness provides the answer: she rips away the standard, and wrapping her child in it, binds him to her. Craon, fearless, even encourages his mother with his kisses, and finally, amidst the glint of lightning and thunderclaps which announce the storm for a second time, they put this horrible place behind them.

The outside of the prison tower, washed by the sea; a forest and cliffs make up the environs; near the tower the remains of a long-ruined bridge are visible.

The storm waxes stronger; a fishing boat next to the tower awaits the Countess, who has already descended the rope ladder. A fisherman helps her embark, unfurls the sail, and quickened by the use of oars, the boat speeds off. They are barely underway before Baudouin arrives. His rage is indescribable. Seeing the impossibility of appeasing it, he nevertheless orders his men to take aim at the fugitives; but a strong wind diverts them from the tyrant's revenge. Créqui, however, remains within his power, and Baudouin decides to vent his wrath on him; ordering that Adelaide be pursued, he hastens to Créqui.

No sooner have Baudouin and his soldiers left than another boat appears, bearing young Raoul and the two soldiers. The barbarian had ordered the young Créqui brought to this tower to die; one of the soldiers accompanying him has already raised his sword to strike when suddenly the other, the one who took Raoul prisoner in the battle that morning, averts the killing by plunging his own sword into the would-be murderer's breast. Then with the young Count's help they throw the dead man's body into the sea. His liberator lifts the visor of his helmet, and Raoul recognizes the faithful Simon. He rushes to embrace him, gives thanks to Providence, and they hurry to leave this unhappy shore; they think they see a boat in the distance on which the Countess is escaping, and they make haste to follow it.

ACT IV

Créqui's dungeon. The stage is divided into two parts: in one, on the left side, Créqui lies; a guard stands in the other.

Créqui lies on a bed of straw; his arms, legs and waist are bound to the wall by chains. The soldier who took him prisoner stands guard in the other compartment. Bertrand, Marguerite, and Alain bring Créqui his dinner. The young newly-weds try to mollify their father and persuade him to free the Count. Bertrand would like this too, but the guard . . . 'Promise him money, papa; with money everything is possible.'

'. . . What's that you say?' cries the guard, 'I heard everything, and now you must pay me . . .' They fall at his feet, beg mercy—and he lifts his visor. Who can describe their joy and astonishment when they recognize Henri? They embrace him, and Créqui is at the point of being freed when suddenly Baudouin's voice is heard. He appears with Humbert and Simon (who as such is still unknown to Baudouin).

They open the door to Créqui's dungeon, and Baudouin takes pleasure in the Count's torments. 'Villain! You have come to revile me. Give me my freedom, my sword—and on the field of battle come challenge me over who reigns in Adelaide's heart and in these dominions.' 'Madman!' Baudouin answers, 'Adelaide is already my wife! Behold your ring—do you not recognize it?' 'It cannot be! I know my wife, and I also know you, villain!' And to show his confidence and bravery, Créqui takes up his harp, playing and singing of his love for Adelaide.

The enraged tyrant breaks the instrument. 'Then do you not at least recognize these clothes?' 'O God! These are my son's! Villain, my son is no more!' He charges Baudouin, but the chains restrain him and he falls back, helpless. Midnight tolls. 'You have two hours to prepare for death,' Baudouin says, giving secret orders to Humbert. By this time Simon has managed to signal Créqui of his identity, and assures him that his son is still alive and that the blood on the clothes, apparently Craon's, is Simon's. Hope again revives the despondent father.

Baudouin, taking pleasure in his enemy's miseries, is at the point of leaving when Créqui stops him, and with the firmness appropriate to virtue and courage, reproaches him for all his evil-doing, calls down upon him the vengeance of the righteous heavens, and threatens him with the punishment he deserves. For the first time Baudouin's soul is troubled by some vague disquiet; for the first time he trembles before God's judgement, and for a moment indecision possesses his cruel heart.

Pacing quietly and deep in thought he walks around the dungeon while Simon again informs Créqui that Adelaide and Craon are saved, then disappears with lightning speed. Finally brutality wins over Baudouin's soul, and the least spark of compassion disappears: the time of Créqui's death is fixed. He orders the prison keeper to give Humbert the keys, and the trembling Bertrand obeys, looking sadly at Henri. Baudouin exits, taking Bertrand with him. Humbert closes the doors behind them, preventing the young newly-weds from leaving, though they beg him to let them go. 'I am very tired,' Humbert tells them, 'I want very much to go to sleep, but you two with your new status in life will surely keep me awake.' He amuses himself at their expense, and seated among them orders them to dance, each separately.

Henri meanwhile asks Humbert about Créqui's fate, and seeing that he is getting sleepier with every word, proposes that he rest, assuring Humbert that he, Henri, can be relied upon to guard the prisoner. Content with this arrangement, Humbert thanks Henri, entrusts his prisoner to the vigilant sentry, and is about to give in completely to his sleepiness. As a precaution he places his table and chair across the door of Créqui's cell. Henri asks him for the key, so he can let the young people go, and Humbert is at the point of giving it to him but suddenly stops, changes his mind, ponders the matter—and falls asleep.

The time has come to save the Count, but how . . . Alain advises that Humbert be killed, but Henri can only fight, not kill. Anyway, the matter before them is how to rescue the hero, and all resolve to act according to the circumstances. Marguerite wrests the keys away from Humbert very stealthily, and Alain relieves him of his sword. After many anxious moments, this and other deeds accomplished, they open the dungeon door and remove Créqui's chains; his delight is indescribable.

All is ready for the escape, but as they leave Marguerite—alas!—suddenly lets fall the key ring, too heavy for her, which wakens Humbert when it hits the floor. Before Humbert collects himself Henri throws him to the ground, and holding a sword to his neck threatens him with death for the slightest movement or sound. Créqui takes Humbert's sword and Alain, who knows all the exits from the dungeon, opens another unguarded door through which they pass. All of them flee, and Henri hurries behind.

But Humbert rushes after them at lightning speed, stops them and seizes the door as if to close it. Henri strikes him in the arm with his sabre, pushes him away and runs out, locking the door behind him. Humbert's rage and despair increase at hearing Baudouin's voice demanding that the door be opened. He cries out that he cannot, whereupon the impatient Baudouin breaks it down and enters. With Baudouin's first question: 'Where is Créqui?' Humbert falls at his feet, explains what happened, shows him his wound—and the livid Baudouin orders him executed in Créqui's place. Then he orders the door through which the fugitives escaped to be broken down. The door is just broken when bells sound the alarm, announcing an enemy attack. Baudouin rushes off to do battle, speeding on in Créqui's tracks.

ACT V

The interior of a dwelling constructed in the ruins of an old castle, which serves as a storage place where fishermen fold their nets. Many underground passageways, known only to them, lead to this place.

In the distance the sounds of battle can be heard, gradually coming closer. In horrifying anxiety the Countess marks, it seems, each minute as it passes; next to her the young Craon, the Chevalier Renti who came to Créqui's assistance, young Raoul and Simon are all trying to console her and assure her of the battle's success and her husband's fate.

Every minute messengers appear with favourable dispatches, but Créqui himself has not yet come. Pierre goes to find out what has happened to him. He has barely left when a knock is heard: 'Open, open up! It is we!' First Alain and Marguerite come in, and behind them Créqui himself. Who can describe the delight of man and wife'! . . . Suddenly Pierre runs back, breathless, and says that they are being pursued. 'But do not be afraid, there is an underground passage unknown to them. Save yourself and fly to the field of glory! I shall remain here with my son and these warriors to defend the Countess.'

With the soldiers' help he lifts the heavy stone which conceals the passageway; Créqui, Renti, young Raoul, and several soldiers hasten to go out. But Baudouin's soldiers have already burst through the door with torches. The soldiers holding the stone at the entrance to the underground passage put it down to take up the defence. As the battle begins, the Countess, unable to follow her husband, escapes with her younger son on to the gallery, on the other side of which Simon hopes to rescue her. They are pursued and almost caught when suddenly Pierre, with an axe, cuts down the support linking the gallery with the staircase—just as Simon is defending the latter from destruction.

The gallery falls with a crash, and the brave Simon with it, holding on to a piece of wreckage. With the collapse of the staircase the entire hut is destroyed. Entering at this moment, Baudouin orders that Créqui be found among the ruins. Suddenly he sees the Countess, and wishing to salvage at least one victim for his lust for vengeance, he orders the gallery on to which she has escaped, impossible for anyone to reach, to be burned.

Créqui appears with his son and his friends, and stands between them

Raoul de Créqui 65

and his wife. The rivals begin a fearsome battle. As they fight, the gallery falls in ruins; Pierre takes Craon and jumps down with him; the Countess falls from the ruins, but at this very moment young Raoul, having disposed of the enemy standard, sees his mother's danger, flies to her aid and takes her into his arms, covering her with his shield and waving his father's standard over her. Baudouin, struck at last by a mortal blow, falls almost at the Countess's feet. The army shouts cries of victory; all rush forward in triumph with congratulations. Créqui, encircled by his family and friends, stands utterly happy at the centre of the most joyful grouping—and with this scene the ballet ends.

In reminiscences published in the journal *Moskvityanin* in 1856 (No. 4, pp. [385]–412), Glushkovsky recounted the story of *Raoul de Créqui*, recalled its first performance, and described various scenic effects of the original production. He was concerned with technical devices and with conveying artistic impressions in greater detail than Didelot was in preparing the libretto.

Glushkovsky glosses the libretto. Describing the dungeon scene with Adelaide and Craon, for example, he recounts the action vividly:

> The countess seizes the line of her friends' signal, and quickly draws in a thick rope, saw, crowbar, and rope ladder. She begins to saw the window grating, her son helping her, but it seems to her that things are moving slowly; she takes the crowbar, knocks out some bricks around the grating; from these efforts she falls exhausted and invokes God's help. Her son gives her courage: with his weak little hands he casts away the half-crumbled bricks of the grating. The countess, her strength renewed, begins again to knock away the bricks around the grating with the crowbar. The bricks quickly fall away and the grating comes down . . .' (pp. 398–9).

Describing the *changement à vue* which immediately followed this, Glushkovsky continued:

> The full view transformation of the tower was extremely effective as it relates to the ensemble of this scene: the countess with her son had just managed to be saved in the window as the spectator, within seconds, saw another scene with the two of them coming down the rope ladder to the waiting boat. This quick change of action, from the interior of the castle to the exterior, made a powerful impression on the audience . . . To make such a quick change of scene they lowered different people, who took the same roles, down the ladder in the second set; the same height and costume, the distance of the castle, night-time on stage—all this so masked the change itself that the spectator could not help guessing that other artists were involved' (p. 400).

In a ballet so full of pantomime, Glushkovsky was concerned that the dances not be slighted, and provided a description of the celebration given for the countess in Act II:

A beautiful tournament was given at this celebration, at which the countess's older son competed with the knights in jousting; it was especially effective when from the knights' heavy blows their shields and lances broke to pieces. [He adds, in a footnote, that 'shields and lances were mechanical: under pressure of a spring they came apart'.]

... After the tournament the ladies, standing in an oblique line, dance, and delightedly throw laurel wreaths in the air to the victor; young Créqui, dancing, runs through the line, artfully catches each wreath in the air with his lance, then mixes in with the general dancing. The ladies, dancing, form two lines; among them soldiers are grouped; the right line of ladies throws wreaths to the left, and the left responds in kind, so that for a time a large winged avenue of wreaths is formed in the air, which makes a pleasant picture for the eye; at this moment the soldiers move to the centre between the lines of ladies, striking various military poses, and catch the wreaths in the air with their lances as they dance. By means of frequent rehearsals all this was so well co-ordinated that rare was the knight who let his wreath fall; in the event of such misfortune, Didelot had arranged for pages dancing around them and playing on lyres to pick up the wreaths unnoticed.

After this, dancing vassals with palm branches formed a grouping with the ladies; the ladies attached multi-coloured veils to the branches and, standing up on little stools, raised their shawls aloft. All this formed fantastic triumphal gates with hints of country simplicity, beneath which pass the knights, hand-in-hand with the ladies, and the pages with their badges in the manner of flags. The whole of this riotous movement takes place with dances, and at the end a large general grouping is formed from it' (pp. 401–2).

In the libretto Didelot indicates simply that after Humbert falls asleep in the dungeon scene Henri, Marguerite and Alain resolve to save Créqui and do so 'after many anxious moments, this and other deeds accomplished'. Glushkovsky offers a more detailed picture of Créqui's rescue, which turns out to be the kind of moving tableau which was a specialty of Didelot's. 'The public was so taken with this scene that a profound silence reigned in the boxes and the stalls as it was played; all were afraid of waking Humbert' (p. 409).

After Baudoin's henchman falls asleep, Marguerite crawls under the table to take the keys hanging from his scarf, then:

Henri gives Alain his lance, bends down on one knee and orders Alain to stand on the other, making it easier for him to crawl on to the table and move aside the countersunk bolt at the top of the door; Alain follows orders and crawls on to the table; Marguerite, in a picturesque pose, from beneath the table stretches out her hand with the keys and gives them to Henri. Alain, seeing the realization of his undertaking, thanks God for its success. After this Alain, standing on the table, and Marguerite beneath it in a pleasant pose, move aside the bolts at the top and bottom of Créqui's door. Henri gives Alain the keys. Alain unlocks the dungeon door, pushes it quietly with the point of his lance, it opens wide and he jumps from the table into the dungeon. Marguerite cautiously crawls out from beneath the table and walks into Créqui's cell behind Humbert's chair; for the remainder of this scene Henri poses in a grouping with his sword above Humbert's chest. After many fears and worries, each played out, the cell is opened and the chains removed from Créqui, whose delight is indescribable' (pp. 407–8).

Music completed the effect. 'It was composed for [string] quartet,' Glushkovsky recalled, 'and to make it quieter mutes were placed on the strings.' It was co-ordinated with machines in the wings which supplied sounds appropriate to the bolts and the door lock as they were moved. These details produced no small effect, Glushkovsky wrote, 'any concern on the part of Créqui's rescuers frightened them, for it could disrupt their plan to free the condemned man; this circumstance provided the opportunity to create several pictures in various graceful poses with facial expressions of joy and fear' (p. 409).

3
Libretto of *The Captive of the Caucasus, or The Bride's Shade*
Charles Didelot

At the end of a chapter describing the accomplishments of Russian ballet artists around the year 1800, Soviet historian Vera Krasovskaya concludes:

> And so, the Petersburg ballet theatre at the frontier between two centuries had achieved a high level of its own creative development. It possessed an excellent, ever-improving troupe, a magnificent school, and an experienced balletmaster, Ivan Ivanovich Valberkh.[1]

How Russian was it? Adam Glushkovsky listed four balletmasters in the Petersburg company at about the same period; besides Valberkh there were the Frenchmen Didelot, Duport, and Auguste.[2] Writing about the ballet in 1817, Raphael Zotov clarified the relationship between Russian and foreign members of the company, pointing out the extraordinary salary disparities between them (Louis Duport had accumulated 160,000 roubles by 1812, Maria Danilova, a Russian ballerina reputedly driven to an early grave in 1810 from excessive rehearsal, earned 2,500 roubles a year). According to Zotov, with Didelot this situation began to change:

> The celebrated Didelot arrived, and our ballet came alive. . . . Up to that time the management always tried to bolster the ballet troupe with expensive foreign dancers. Didelot announced that he would make first-class European talents out of Russian students, and he kept his word. Istomina, hymned by Pushkin, Novitskaya, Nikitina—were at the head of the first wave of our superior women dancers, and several years later came Zubova, Teleshova, Azarevicheva, [the men] Shelikhov, Golts, and many others, who brought our ballet to the first rank of European fame.[3]

Didelot's teaching caused an important metamorphosis in the Petersburg company. For more than two decades he applied superior pedagogical skills to Russian students (stemming from his own work with the greatest masters, including Dauberval and Noverre), making new foreign talent unnecessary by the early 1820s. In 1825 the number of students in the theatre school had grown to

---

[1] V[era Mikhailovna] Krasovskaya. *Russkii baletnyi teatr ot vozniknoveniya do serediny XIX veka* [Russian Ballet Theatre from its Origins to the Middle of the 19th Century] (Moscow and Leningrad: 'Iskusstvo', 1958), p. 100.
[2] *Moskvityanin*, 1856, No. 4, p. 389.
[3] 'Memoirs of R. M. Zotov', *Istoricheskii vestnik*, vol. 65 (1896), p. 304; vol. 66 (1896), pp. 30, 401.

120—more than the schools of Paris or Milan—from some two dozen at the turn of the century.[4] In his hands, the Petersburg ballet became more dominantly Russian.

He made parallel advances in repertoire. *Divertissements* and ballets on Russian topics had been produced in the eighteenth century and were especially prominent, spurred by patriotism, during the time of Napoleon's invasion. But subject-matter alone was not enough to give stature to these works. Like Russian dancers, a Russian ballet needed the surpassing mastery of a Didelot. His most important contribution to this repertoire was *The Captive of the Caucasus*.

Based on Alexandre Pushkin's poem of the same name, then recently finished, Didelot's ballet has been hailed since its first production as a landmark of Russian ballet. Without disputing the significance of that perception, *The Captive* in retrospect seems more Russian in the eyes of its beholders than in the light of any rational assessment. As would be true again forty years later with Arthur St-Léon's *The Little Humpbacked Horse*, the story and the sheer audacity of adapting it for ballet made it Russian.

The conception and design of *The Captive of the Caucasus* conform to the same principles which had served Didelot well in earlier ballets. In *The Captive* as in *Raoul*, a high-minded hero held prisoner is freed by a loyal friend. The balletmaster takes every opportunity to display the exotic locale and the mores of its people, and the adventure ends with virtue being rewarded. To this extent, Didelot made over Pushkin's poem in the image of his other ballets, inventing the final scene in which the Circassian khan takes Russian citizenship. In addition, Didelot gave names to Pushkin's anonymous characters.

Such extraordinary adaptation was called for in part by the nature of the original. The Russian captive in Pushkin is not high-minded, but a disillusioned romantic whose 'fiery youth | He proudly began without a care,' but who, with a stormy life, destroyed

> Hope, joy and all desires,
> And the memory of better days
> He kept within his withered heart.
> People and the world he had come to know,
> And knew the value of a specious life.
> In the hearts of friends he found but treachery,
> In hopes of love a mindless dream.

The poem is also romantic in the connotations of the word 'captive', for the Russian is both a physical prisoner of the Circassians and a captive to the beauties of the place, held in thrall by the natural surroundings and the people:

> But it was this storied nation that attracted
> The European's constant attention.

---

[4] Aleksandr Grigor'evich Movshenson and Yurii Iosifovich Slonimskii, 'Novoe o poslednikh godakh deyatel'nosti Didlo' [New Information on the Last Years of Didelot's Work], *Teatral'noe nasledstvo* [Theatrical Heritage] (Moscow, 1956), p. 68.

> Among the mountain people the captive observed
> Their belief, customs, upbringing,
> He loved the simplicity of their life,
> Their hospitality, their thirst for battle,
> The speed of their unbridled movements,
> Their fleetness of foot, their strength of hand.

Pushkin's captive is far removed from the civic virtues of Didelot's heroes. He is a Byronesque adventurer with distinctive Promethean features that Pushkin may have meant to be autobiographical. Like Prometheus, the Russian lies banished in chains, his head inclined upon a rock, in a locale where Pushkin himself was also banished.

The effect of Didelot's changes was to make the ballet scenario more active and extroverted than the poem, a transformation required by the medium. While Didelot routinely altered his sources to make them fit his criteria for a ballet, we cannot assume in this case that he was blind to the romantic elements in the story. On one hand Gorislava, who at first bolsters Rostislav's sense of conjugal responsibility and later, as a shade, releases him from it, is guided by the same sense of virtue found in Raoul and Adelaide. On the other, the device of a female spirit intervening in a mortal man's affairs of the heart is fundamental to the kind of ballet the Taglionis would bring to Russia in the 1830s. Such a clear break with the balletic conventions of the eighteenth century in anticipation of romantic ballet is an important claim *The Captive* makes to our attentions.

The ballet is also important in the context of Russian national identity. Didelot's adaptation made *The Captive* a European ballet set in an exotic locale. For most Russians of the time the place which Pushkin describes with such ravishing effect was still exotic, but secondary in importance to the world view of the captive and the beauty of the language, which could not help but stir a sense of sympathy and of pride in being Russian. Didelot could not convey the language in his medium, and that he changed the psychology of the hero must have been noticed and excused by audiences who accepted the ballet as homage to their national poet. Nor by 1823 could Didelot's efforts on behalf of Russian artists have gone unnoticed. *The Captive* was produced at a time when Russian talent was flourishing in the Petersburg company, in which it is possible to sense a change, attributable to Didelot, from the mere participation of Russian artists, which for decades had been passive and neutral, to more active involvement.

THE CAPTIVE OF
THE CAUCASUS
OR
THE BRIDE'S SHADE

Grand pantomime ballet in four acts, comp. by Mr Didelot, the story of which is taken from the famous Poem of Mr Pushkin; music comp. by Mr Cavos, arranged for orchestra by his student Mr Zhuchkovsky; new decorations by Mr Kondratiev; costumes by Mr Babini; machines by Mr Natier.

Presented for the first time at the Bolshoy Theatre for the benefit performance of the Dancer Mr Auguste, 15 January, 1823.

SAINT PETERSBURG
at the Typographer of the Imperial Theatres,
1823

Publication permitted:
Saint Petersburg, 13 January 1823
Secretary of the Censor[ship] Committee of the
Min[istry] of Int[ernal] Aff[airs]
Colleg[iate] As[sessor] and Cav[alier] V[asily] Sots

DRAMATIS PERSONAE

Kzelkaya, *young Circassian girl, orphan, brought up by Sunchulei, the Circassian khan, who took her captive as a child from another horde*	Mlle Istomina
Rostislav, *young Slavonic Prince, son of Branislav, taken prisoner by the Circassians*	Mr Golts
Gorislava, *young Slavonic Princess, fiancée of Rostislav*	Mlle Velichkina the younger
Branislav, *Slavonic Prince, living in the city of Terki on the shore of the Terek*	Mr Auguste
Sveneld, *young Slav, friend of Rostislav*	Mr Charles Didelot
Sunchulei, *Circassian khan, who has taken up camp with his horde along the other side of the river Terek*	Mr Eberhard
Tamar, *wife of Mangu, commander of the Circassians*	Mlle Azlova
Tugorkan, *Tartar chieftain*	Mr Castillon
Relatives of the Circassian khan (and then Slavonic girls):	
Konkacha	Mlle Teleshova the younger
Zyuyunbeka	Mlle Zubova the younger
Ksugeriya	Mlle Azarevicheva the elder
Zalika	Mlle Selezneva the younger
Mangu, *Circassian chieftain*	Mr Artemieyev
Akhmat, *Circassian chieftain*	Mr Pix-Durnisel
Girei, *Tartar khan, friend of Sunchulei*	Mr Butkevich
Oktai, *Tartar chieftain*	Mr Didier
Timur, *Tartar chieftain*	Mr Trifonov
Asmud, *Slavonic chieftain*	Mr Marcel

Radim, *Slavonic chieftain*   Mr Shelikhov the elder
Pretich, *Slavonic chieftain*   Mr Gomburov
Gusli players: Messrs Palnikov, Didier, Dembrovskoy and Dmitriev, Mlles Karachintsova, Didier and Yanysheva.
Circassian men and women, Tartars, Slavs of both sexes, Boyars, Gusli players, slaves, etc.

FOREWORD

This ballet, being presented by me to the Public, is taken from the beautiful Poem: *The Captive of the Caucasus*, written by Mr Pushkin. All literary authorities praise this excellent work of Russian Poetry. I asked that a short excerpt of it be translated for me, and I found the subject-matter most interesting. Of course it would be much better if I could read the original, to sense all the beauties of the style and the richness of the ideas, but, unfortunately, not reading Russian, I had to be contented with translated extracts. It would have been untoward in a ballet to see an ending such as occurs in the poem, because the spectator in spite of himself would be sad over the fate of the tender, passionate Circassian girl, whereas at the same time, the Russian prisoner could not return her love without betraying the vows which tie him to his fiancée in his homeland. (I proposed that she still be his fiancée, otherwise I could not make clear and quickly express in pantomime the reasons why Rostislav refuses the love of his charming rescuer. Our art with difficulty expresses past or future action; here it was impossible.)

Avoiding one problem I encountered another. Rostislav is in love with two women! Which of the two should he sacrifice? Having moved the action back into mythical times, and following the story of the Poem, I composed the above-mentioned ending. In the year 1594, Tsar Fyodor Ivanovich sent Prince Khvorostinin to spread his conquests into Circassian lands, where their Prince Sunchulei built a little city on the other side of the Terek (according to Vsevolozhsky's *Hist[orical] and Geog[raphical] Dictionary of Russia*). If the action of the ballet occurred in these times, then it also could occur in about the year 980 (Virgil himself permitted himself a similar anachronism). Of course I could leave behind the Circassian who died in the waves of the river and bring forth her shade after her—but the moral sense would be broken because the Circassians were still idol worshippers at that time. On the other hand, the shade of Gorislava would not produce any untoward impression. From the moment of her appearance on stage her life already seems doomed, from all the enemies' calamities and tyranny which she has endured. The spectator would be conditioned to think that she cannot survive all her sufferings. That is my plan! Whether it succeeds—on that the Public will pronounce its verdict. Even otherwise it will see my efforts and struggles as it pleases.

My colleague and friend Auguste assisted me in composing the Russian dances of the last act, and who better than he has investigated thoroughly the true character of national dances? Who more pleasantly than he to enliven them with his art? And therefore in the third act, from the scene of the denouement, I yielded my place to Mr Auguste. I would of course have asked him to make dances in the first act also, if they were not so closely allied to the principal action. Among the games of the Circassians I introduced the lassos they use to catch horses and prisoners, which I blended with the dances of the first act. May our joint efforts warrant the Public's favour.

ACT I

The stage represents two adjoining camps, Tartar and Circassian. People are dancing, playing various instruments, and smoking. Women spin yarn and rock their children. Young Circassians carry in an eagle they have caught in a net. They amuse themselves with the bird. Tartars and Circassians are preparing for games, their favourite pastime. Divided into two groups, the Tartars and Circassians vie with each other. The Circassians use their lassos, the Tartars various weapons; the latter are finally vanquished. The khan rewards those who excelled; the victors celebrate their triumph.

By the end of the festivities, the captured eagle has managed to break through its fetters, and as if craving vengeance and spoils, flies to a child in its cradle, lifts it into the air and bears it up to its nest on the peak of a distant cliff. The entire horde is aghast; the despairing mother cries out in vain; the father lets go an arrow, but it falls short. The mother runs along the chasms and cliffs to the mountain where the eagle set down; the predatory bird has hardly touched down when a true shot fells it, but the mother has already reached its nest, taken her child, and in her joy ignoring all dangers, runs back to her delighted family. Celebration attends this event. Meanwhile a young Circassian, having watched the fall of the stricken bird into the river, has swum over to it, taken it to his comrades, and handed it over to the children for revenge.

All this is interrupted by the arrival of a Slavonic prisoner, who is dragged in covered with blood and dust; his armour is ruined, his weapons broken; guile and superior numbers have overcome him. The entire horde rush over to him and curse the unfortunate man; his very life is in danger when suddenly the young Kzelkaya, the Khan's ward, taken by the handsome Slavonic Prince, stays the furious crowd, implores the Khan to grant him mercy, and with effort secures his agreement... 'If my weapon had not broken... if you had fought me openly' (says the Russian)... His courage pleases the Khan; Kzelkaya suggests to the Khan that a rich ransom can be had for him—a thought which curbs the Khan's brutality. Kzelkaya leads Rostislav to the Khan's feet to thank him for his mercy; but the proud Slav walks up to him courageously and says: '... Khan! Give me my weapon, command them to fight with me before you, and you will see if I warrant your esteem!'

The Khan, convinced of his intrepidity, orders that a weapon be given him, and that he fight the men who took him prisoner. The fierce Rostislav lays them low immediately! 'Do you not see, Khan, that not weapons and courage, but pernicious lassos won out over me!' Much of the crowd rushes towards him; but the Khan stops them, favours the prisoner and invites him to his table. Kzelkaya is delighted with Rostislav's courage and victories; she followed his every movement; her tender, fervent heart had mistaken a growing passion for pity. The fatal fire already flows in her blood—by turns pangs of jealousy and delights of joy press her spirit. Wishing to please Rostislav, she treats the captive to food and drink...

Just then a Tartar comes in and announces to the Khan that crowds of Russians are roaming the surrounding areas in search of Rostislav. The Khan immediately orders all to take up arms and to hide in the mountains among the rocks for a surprise attack on the Russians. Everyone is about to leave, but first as a precaution they put the Russian prisoner in chains. Kzelkaya's entreaties are in vain; the Khan threatens her, and all hasten to new spoils. The women also disperse to their huts; only the Circassian girl stays with Rostislav.

The tender concern and sympathy she constantly expresses betray her ardent secret. Rostislav perceives her passionate love; sadly he looks at his betrothal ring and sighs, recalling the vows which bind him to his homeland. 'What is the matter, Russian? You have become so sad.' 'Leave me, sweet Circassian girl, leave me to myself, I do not deserve your tender caresses, your love.' 'Can I leave you? No, never! To live and die with you is my destiny, my desire!' Rostislav insists that she leave, and a grieving, despairing Kzelkaya walks away slowly, with measured steps, to hide her torments.

Overcome by emotion, Rostislav calls to his benefactress; she rushes back to him, he reaches out to embrace her—and Kzelkaya falls into his arms. The noble youth, however, cannot deceive the young Circassian: he remembers his obligation... but how to reveal to the passionate girl that he cannot love her, that he cannot reciprocate her passions? Kissing her hands, he moistens them with his tears, but is resigned to silence.

Kzelkaya, delighted, takes Rostislav's perplexity as caused by love. Carried away by outbursts of feeling, she describes her ardent passion for him, offers him hand, heart, freedom, and escape to distant lands . . . Rostislav remains silent, trembles . . . 'Sweet Circassian,' he says to her, 'pity me, the unfortunate one; my heart . . . ah! it is no longer mine! . . . Look at this ring: it is a mark of eternal union with my fiancée!' . . . Rostislav has hardly finished when Kzelkaya, suddenly horror-struck, falls unconscious in his arms. Just then the wild cries of the horde, returning victorious with numerous prisoners, are heard. Kzelkaya regains consciousness, looks down at Rostislav's ring, and in horror pushes him away.

Meanwhile the prisoners being led in recognize Rostislav and rush to fall at his feet. Behind them a woman appears, worn out with fatigue and sufferings, dragged in by two Tartars; in her pale features life is struggling with death; her lacerated feet, her torn clothes and dishevelled hair show how much she has suffered from the barbarians. Almost unconscious she falls on the stone, uttering a heavy moan. Rostislav has no sooner looked at her than he recognizes Gorislava, his fiancée, rushes to her, and roughly throws back the Tartar who wanted to drag her further. To the Khan's question: 'What does this behaviour mean?' he explains that this is his bride, his wife-to-be—and presses her to his heart.

Kzelkaya trembles at the sight of another in Rostislav's embrace; jealousy and fury torment her breast; with a fierce hand she seizes a dagger to strike down her rival . . . But feelings of magnanimity overwhelm this outburst of passion: she throws down the dagger. 'Let me alone suffer, let me alone perish, why should they face more unhappiness?' Kzelkaya runs to Gorislava, waits on her, consoles her, begs the Khan to let her watch over both prisoners, pays Gorislava's ransom to the Tartar who captured her—in a word she expends the most tender attentions on the unhappy couple.

Meanwhile the Tartars separate the prisoners; night falls; all retire to their huts.

ACT II

It has grown calm. Kzelkaya alone stays awake in the night silence in front of the tent in which both prisoners are held. Cautiously she brings a saw and other instruments which enable her to break Rostislav's fetters. She stops, listens; a moment of jealousy possesses her heart: she draws back the curtain, looks at the sorrowful pair—and is ashamed of her feelings. 'I have brought you freedom; be happy; flee from here; here is a saw! Quickly! Every minute is precious.'

They set to work; the smallest noise frightens them. Half-dead, Gorislava grows stronger in spirit, revived by the hope of freedom. Rostislav finally breaks the chains. 'Quickly, quickly to the river,' Kzelkaya tells them, 'Swim across, to your homeland on the other side. Farewell, I wish you happiness.' Rostislav and Gorislava thank their rescuer, beg her to come with them, try to drag her along—but she breaks away and bids them farewell, embraces them and tearfully hurries them on. Rostislav ties the frail Gorislava to his waist, rushes into the river and swims off, inspired by the joyous hope of imminent rescue.

Kzelkaya follows them with her gaze until the cliff obscures her sight of the freed pair; then the press of her emotions saps her strength, and she falls in a swoon. Revived, she looks about, despairing. Everything is quiet and deathlike around her. Dark thoughts fill her mind; she runs to the river bank, sees Rostislav and Gorislava, who are saved and thank her in gestures from afar. She finds the garment thrown to her by Rostislav, presses it to her heart, and . . . throws herself into the waves.

Rostislav sees this horrible spectacle, and despite his fatigue, despite the storm which has come up during his escape, he jumps into the river again and swims off to save the woman who rescued him. The storm is raging at its height, lightning rends the dark clouds, thunder shakes the earth. Rostislav reaches a rock which juts out in the middle of the river, climbs up on it, and looks around him. He spots Kzelkaya when lightning flashes, and jumps into the depths again.

Meanwhile Branislav, Rostislav's father, upon learning that his son is a prisoner, has gathered an army, and taking advantage of the night darkness has approached the horde of unwary Circassians. By the riverbank Rostislav's garment is the first thing which meets his gaze. In despair Branislav thinks that the barbarians have killed his son. Vengeance seethes in his breast. 'Warriors! To your shields! To the enemy! Find my son or avenge his death!' The battle begins. Circassians and Tartars run about everywhere in fear and disarray; everywhere they encounter Russians and everywhere are vanquished. Russian prisoners, taking advantage of the confusion, break their chains, take up arms and rush to attack the enemy. After a cruel battle and despite the efforts of the valiant Khan, the horde is defeated, the Khan is taken prisoner, and the Russians celebrate victory. The captives are taken away in carts, and the victorious army sets out on its return trip by torchlight. All rejoice; only Branislav weeps, over the loss of his beloved son.

ACT III

In the distance on the other side of the river the ruins of the devastated horde are visible, that is: the decoration of the first act is transferred to the opposite shore, and the action takes place on this side, which was formerly the opposite shore.

Despairing Gorislava waits for Rostislav, who has still not returned. In her sorrowful imagination she pictures her betrothed perished in the waves, and the dark presentiment of death drains her strength. 'What will become of me?' she says, 'death is my only hope!' Exhausted, she is drawn again to the river-bank, looks out on to the water, and sees nothing but the blue waves. Sensing death to be near she kneels, raises up her weak arms in a last prayer to heaven, and expires, pronouncing Rostislav's name.

Meanwhile the storm has passed, and the ever calmer waves wash Rostislav and Kzelkaya ashore. 'Cruel man! Why did you save me, why did you not let me die?' she says to him. Rostislav runs to look for Gorislava, as does Kzelkaya; at last he finds her, dead, at his feet. His cry draws the Circassian girl, and they cannot believe what they see. They try to revive her, to restore her to life—all in vain—she is no more! A profound despair seizes them, and they kneel in silence before Gorislava's body. In their despair they do not hear the celestial harmony which is suddenly audible; Gorislava's shade quietly rises up from the earth, lingering for a moment in front of the lovers, who look at her in fright and reverence. Gorislava takes the betrothal ring from her finger, gives it to Rostislav and indicates that he give it to Kzelkaya. Rostislav rushes to embrace Gorislava—and grasps naught but empty space. He and Kzelkaya kneel before the shade, which floats away from them.

For a long time they look off in her wake, when suddenly martial sounds awaken them from their reverie. It is Branislav's army. The shade returns for a moment and stops before the earthly souls; she shows Rostislav his father, and in a trice the young Prince is at Branislav's breast, pointing sadly at his bride's shade. Branislav inquires directly of the shade, and receives from Gorislava the command to unite his son with Kzelkaya. He promises to obey, after which the shade continues her journey to the heavens, bidding farewell to her earthly friends.

ACT IV

The stage represents Branislav's palace.

A numerous gathering of Boyars and Courtiers is celebrating the happy events of that day. Pages, the royal bodyguards and slaves serve at table. Through the columns of the chamber magnificently illuminated gardens are seen, where the people also take part in the festivities of their Prince. Even the Tartar and Circassian prisoners are enjoying their conquerers' feast.

Everything is joyful. Branislav presents Kzelkaya to his kinsfolk and courtiers, tells of her delivering Rostislav from captivity, and of his intention to unite them by ties of marriage. All approve his choice. The captive Khan willingly becomes a Russian subject and kneels before his sovereign. A general celebration ends this happy day.

II
Marie and Filippo Taglioni in Russia

INTRODUCTION

On 31 October 1829, at a performance of Didelot's *Thésée et Ariadne* at the Bolshoy Theatre, there was an unpleasant exchange between the balletmaster and the newly appointed Director of Imperial Theatres, Prince Sergei Sergeyevich Gagarin. Raphael Zotov, who claimed to be present, gave the following account of what happened:

> The dancers on one occasion were changing costume slowly, and Prince Gagarin ordered the performance to begin at once. Didelot responded indifferently to this command. The prince was enraged, Didelot no less so, and it ended with the prince turning to me and ordering him placed under arrest. 'A man such as Didelot is not to be arrested,' the balletmaster said, and yet I had to obey the command, and Didelot submitted. Of course, the next day Didelot sent in his resignation and was released from service. . . . I of course sympathized with the retirement of the famous Didelot, but it was difficult to reconcile myself with these two capricious people, of whom one had all the power on his side, and the other—talent.[1]

This display of pique was the latest in a series of controversies, provoked by both sides, which had marred Didelot's relations with the theatre management since 1824.[2] The resignation permitted Gagarin openly to pursue his wish to hire another balletmaster. Alexis Blache and Antoine Titus came to St Petersburg for this purpose from 1832 to 1838 and from 1832 to 1850, respectively, but it soon became clear that to hire other balletmasters was not to replace Didelot.

As choreographers, Blache suffered repeated failures and Titus, except for *Kia-King* (1832) and *Julius Caesar in Egypt* (1835), contributed little of his own to the repertoire, devoting his efforts instead to the revival of others' works and to pedagogy. These men lacked their predecessor's creative gift and his leadership, especially as Didelot had displayed it in his advocacy of Russian artists. The national identity which Didelot inspired in the company would repeatedly be tested, in the remainder of the

---

[1] R[afail] M[ikhailovich] Zotov, 'The Memoirs of Rafail Mikhailovich Zotov', *Istoricheskii vestnik* [The Historical Messenger], vol. 64 (Apr.–June 1896), pp. [762]–797; vol. 65 (July–Sept. 1896), pp. [26]–50, [301]–321, [593]–616; vol. 66 (Oct.–Dec. 1896), pp. [27]–53, [400]–427, [765]–796: vol. 66, p. 771, vol. 65, p. 310. Aleksandr Grigor'evch Movshenson and Yurij Iosifovich Slonimskii, citing official theatre records, indicate that Didelot uttered an 'inexcusable impertinence' to the Prince ('Novoe o poslednikh godakh deyatel'nosti Didlo' [New Information on the Last Years of Didelot's Work], *Teatral'noe nasledstvo* [Theatrical Heritage] (Moscow: 'Iskusstvo', 1956), p. 90). In an anecdotal account closely related to Zotov's 'Memoirs', Pierre d'Alheim claims that Didelot was arrested for disregarding the Prince's repeated demands to shorten the intervals of a performance (*Sur les pointes* [Paris: Société du Mercure de France, 1897], p. 184; recounted in Mary Grace Swift, *A Loftier Flight: The Life and Achievements of Charles-Louis Didelot, Balletmaster* (Middletown and London, 1974], p. 184).

[2] These controversies are examined in detail by Movshenson and Slonimskii, 'Novoe o poslednikh godakh deyatel'nosti Didlo', and reviewed in Swift, *A Loftier Flight*, pp. [177]–184.

century, by the management's policy of inviting foreign stars to dance in Russia. These artists often brought fresh repertoire and a new balletmaster with them, and reinvigorated faltering public interest in ballet. Didelot's coming to Russia had itself been the result of this policy, and in the middle of the 1830s, the management invoked it again.

Marie and Filippo Taglioni came to St Petersburg after enjoying extraordinary success in other European cities.[3] She danced her first performance at the Bolshoy Theatre, in *La Sylphide*, on 6 September 1837 and her last, in *Gerta*, on 3 March 1842, the second day of Orthodox Lent, when performances were still being given 'for persons of a different creed'. For five years between September and Lent she performed all her celebrated roles, including that of Héléna in Meyerbeer's *Robert le Diable* and in the entire range of ballets created for her by her father. Though she appears never to have met Didelot, Marie Taglioni brought to St Petersburg stronger links of continuity with his repertoire than did either Blache or Titus. She had danced in a revival of *Zéphire et Flore* at the Paris Opéra in 1831, and had made her London début in that ballet in 1830. For his part, Filippo was at one point to follow Didelot's lead by adapting a poem of Alexandre Pushkin for ballet: *The Fountain at Bakhchisarai*, which was the companion piece of *The Captive of the Caucasus* among that master's so called 'southern poems'.[4]

Writers differ in their assessments of the Taglionis' long-term impact on the Petersburg ballet, though on other points there is no dispute. As a great artist appearing before a new audience, Marie galvanized public interest in ballet. Within three weeks of her début, local merchants had responded to public enthusiasm for her: portraits were available, 'large and small, some which look like her and some which don't, most of which fall into the latter category,' as well as a pastry bearing her name 'with her likeness in all the roles she has performed,' so that 'those who have yet to see the real Taglioni will receive a most favourable impression of her from this sugary representation, and those who have seen her will have the sweetest recollection'.[5]

In Marie Taglioni's third season, enthusiasm for her was still running high. Journalist Fyodor Koni wrote:

---

[3] On the Taglionis see, among other sources, V[adim Moiseyevich] Gaevskii, *Divertisment; Sud' by klassicheskogo baleta* [*Divertissement;* The Fate of Classical Ballet] (Moscow, 1981), pp. 8–42; Ivor Guest, *The Romantic Ballet in Paris* (London, 1966), pp. 73–99, 103–5, 109–17, 122–30; Guest, *The Romantic Ballet in England; Its development, fulfilment and decline* (Middletown, 1972), ch. 8; Krasovskaya, *Russkii baletnyi teatr ot vozniknoveniya do serediny XIX veka* (Leningrad and Moscow, 1958), pp. 213–23; André Levinson, *Marie Taglioni* (Paris, 1929); Natalia [Petrovna] Roslavleva, *Era of the Russian Ballet* (New York, 1966) pp. 58–60.

[4] Though nothing came of it, this intention was referred to in 'The Chronicle of the Russian Theatre for the 1837 Theatre Year', *Severnaya pchela* [The Northern Bee], 1838, No. 108 [16 May], p. 431.

[5] 'Miscellany', *Severnaya pchela* 1837, No. 216 [27 Sept.], pp. 863–4.

Look at our opera house; what a change! Its normally empty walls are lined with spectators; the aisles are full; even the gilt is brighter! Rossini's music overflows with the most splendid melodies—but no one listens! Singers strain their voices—but no one hears them! The orchestra thunders and roars—the talk and noise of the audience cover its sounds! But a well-built little foot appears, the wings of ether flutter, a familiar garland begins to whiten [the stage], and everything is hushed, everyone holds his breath so as not to let pass the slightest rustle of this little foot, not to miss the imperceptible breath of a little wing. And in fact, Taglioni cannot create: she perfects when it is impossible for perfection to conceive anything higher. Where the imagination of the most ardent poet gives out, only there does all the might of her immense talent begin.[6]

Marie also gave criticism new life. The difference she made is easily measured by comparing notices of her father's compositions before and after her participation in them. P. Yurkevich's review of *The Revolt in the Seraglio* in Titus's production illustrates a tendency of pre-Taglioni criticism of the 1830s to devote the bulk of an article to recounting the story or to some digression, leaving the assessment of the production and its performers to a brief comment at the end. Yurkevich's tone is at worst sullen and mocking, at best polite:

Surely nobody will argue with us when we say that in our day choreographers and balletmasters have fallen on bad times. Whence to derive stories for their ballets? Whence the picturesque and original situations which could awaken the drowsy attention and the blunted taste of a peevish, business-oriented, boring and particular century? . . . Mythology, with all its zephyrs, cupids, graces and shepherds, has become antiquated; the *gods have abandoned Olympus*, and continue to live on the plafonds of theatres and academic halls. . . .

After this, nothing is left to any respectable balletmaster except to collapse in despair on the floor with some spectre of a fashionable opera, and sit beneath the stage, in the darkness, until the time when a bright, luminous thought lights up his creative brain!

Mr Taglioni, balletmaster of the Grand Opéra in Paris, first of all balletmasters in the world, found himself in precisely this situation, because nobody ever had such a daughter as the young Taglioni, and no one had composed a ballet like *The Revolt in the Seraglio*.[7]

When Yurkevich reviewed the work again, with Marie Taglioni in the leading role, his complaint about the story was blunted by an apostrophe to the dancer:

Our incomparable sylphide, with one wave of her little foot, rends asunder all the heavy theories of encyclopeadic construction! With Taglioni every ballet is above

[6] F. —ni, 'Miscellany. Taglioni in Petersburg!' *Severnaya pchela*, 1839, No. 204 [12 Sept.] p. [813].
[7] P.M., 'The Petersburg Theatre. First Performance of the ballet, *The Revolt in the Seraglio*, on Wednesday, 18 December, at the Alexandrinsky Theatre', *Severnaya pchela*, 1836, No. 2 [3 Jan.], pp. [5]–6.

perfection. See! There she is, beautiful and unattainable, like a dream! She flies, she dances—and how she dances! She sings like Paganini's violin, she draws like Raphael—and all is said! After this explain ballet and choreography, idea and dance! There is no ballet, no balletmaster, no dance, no idea! There is only Taglioni, and this is enough for us![8]

Marie produced a similar reaction when she first danced in *La Sylphide*, the other ballet which preceded her to St Petersburg in a production by Titus:

In the second act Mlle Taglioni dances. Is it really possible that this dance entails difficulties, really possible that these *pas* are studied and learned by heart? A magnificent poem is concealed in each gesture, in the slightest movement of the body. Thoughts passionate and tender, which destroy and elevate the soul before us, wordlessly flash in her expressive face, from her burning eyes. A winged, fascinating dream, elusive and undefined, plays and sports before your eyes; the dream is pellucid and tender, like the wings of the Sylphide, capricious, like a woman, the aspect of which it assumes, ravishing, like Taglioni . . .[9]

In time, critics attempted to distinguish the artist from the art. The choreographic composition of Filippo's 'white' acts was new, as Vera Krasovskaya explains:

Marie Taglioni's ethereal dance existed in her father's ballets not for itself, but was included in a defined system of images, and required a corresponding ensemble. In the *grand pas classique* from *La Sylphide* the heroine appears within a swarm of similarly ethereal sisters. The dance of the mass is not raised to the virtuoso height of the dance of the ballerina, but was its analogy in plastique, sounded in the same poetical key, reduced to a kind of abstract monotony. The corps de ballet as an organized unit was especially important in the system of expressivenesss of F. Taglioni's romantic spectacle.

. . . New stylistic tasks demanded from the corps de ballet a new expressiveness as well; one of its most important features was the regulated synchronization of the dance. A single position of the body of each participant, the same line of the arms, raised in a measured movement, the uninterrupted design of the group arabesque, the simultaneity of the upward sweep of a jump—all this accompanied the dance of the ballerina or seconded it harmoniously. In some cases groups of corps de ballet dancers—now in the same, now in a growing, now in a diminishing number of participants—one after the other repeated and varied the cycle of assigned movements, in the manner of a fugue setting out the statements of the theme by each voice in turn.[10]

---

[8] P.M., 'Bolshoy Theatre. *The Revolt in the Seraglio*. Sixth début of Mlle Taglioni', *Severnaya pchela*, 1837, No. 228 [9 Oct.] p. [909].

[9] E., 'Bolshoy Theatre. *La Sylphide*. Début of Mlle Taglioni', *Severnaya pchela*, 1837, No. 209 [18 Sept.], p. 834.

[10] *Russkii baletnyi teatr ot vozniknoveniya do serediny XIX veka*, p. 219.

In addition, meaning as we find it in the ballets of Didelot—the virtuous moral—was transformed in the Taglionis' repertoire. 'Taglioni was enigmatic,' writes Oleg Petrov, a historian of ballet criticism,

The nature of Taglioni's dance was revealed through abstract, universal explanations. Emotionally staggering, Taglioni compelled one (whether or not the Russian public and criticism wanted) to understand and accept her art and thereby brought about a significant change in the 'old' awareness, 'drawing' it into new artistic-aesthetic stands.[11]

In the earliest Russian reviews of Marie's dancing one senses a new emphasis on abstraction. 'The ideal of grace, the ideal of dance, the ideal of pantomime—that is Taglioni,' declared a reviewer of *La Sylphide*.[12] 'Our choreographer Mr Taglioni created one truly poetical work: *La Sylphide*,' added Fyodor Koni,

When his daughter represents the Sylphide, it seems that she is the poet's thought personified, or his corporeal child turned into a visible spectre of imagination. It is impossible to suspect the ordinary in her, for everything that she does approximates to magic.[13]

Meaning in the white ballet is the first point on which the Taglionis' stay in St Petersburg has been evaluated. In the perspective of later developments, Soviet aesthetician Vadim Gaevsky concludes:

White ballet was bequeathed to us by the nineteenth century. It is, perhaps, the principal discovery of the ballet theatre in the last 200 years. Like many other discoveries in ballet, it is impossible to date precisely, impossible to attribute to an author—to a solitary genius. There are several authors: Filippo Taglioni, partly Perrot, in large part to Petipa, Lev Ivanov, and Fokine. We are dealing with a discovery which was extended and developed, which realized itself.

... In all times it was understood as a professional absolute. From the balletmaster, from the performers an ideal mastery is demanded, a professional perfection. Both the technique and the school of the artists must be infallible, the line of the arabesque must be infallible.

The arabesque is the principal participant in this play. The romantic arabesque is a graphic formula of the absolute. The subject-matter of white ballet is the movement of the action towards the arabesque. On this path the ballet spectacle was freed from representational tasks, and all the more from tasks of being 'characteristic'. For the first time in the history of ballet a spectacle was built on the unalloyed imagery of pure classical dance. In a certain sense the white ballet

---

[11] O[leg Alekseevich] Petrov, *Russkaya baletnaya kritika kontsa XVIII–pervoi poloviny XIX veka* [Russian Ballet Criticism of the End of the 18th and the First Half of the 19th Centuries] (Moscow, 1982), p. 77.
[12] E., 'Bolshoy Theatre . . .', *Severnaya pchela* 1837, No. 209 [18 Sept.] p. [833].
[13] F. —ni, 'Miscellany . . .', *Severnaya pchela*, 1839, No. 204 [12 Sept.], pp. [813]–14.

is a *tabula rasa*, untarnished by naturalistic illusions, unburdened by concrete associations—either from the living world or that of inanimate nature.[14]

Filippo Taglioni's white ballets were his most significant contribution to the repertoire. As for the rest—*La Gitana*, *Gerta*, and the others—it would seem that the low level of his choreographic invention was redeemed only by Marie's gift as a performer. 'Taglioni the old man,' wrote Fyodor Koni five years after father and daughter had left Russia,

> having given the world an incomparable dancer-mime, began to grope about in the depth of his soul and called forth therefrom several of those lunatic visions, with which only dreams and wine strike and coddle people . . . Only Marie Taglioni could give life and enchantment to these fantastic images, because their versimilitude resided in her individuality, in her talent lay their charm. Marie Taglioni disappeared and a whole swarm of sweet visions dispersed, the spectator fell asleep and felt the whole oppressive stupidity of the ballet.[15]

The difficulty of sustaining interest in her father's compositions may be responsible for Marie's extraordinary exertions during their last weeks in Russia. She took a benefit on 26 January 1842, and then danced twenty-one more performances in the next thirty-four days, an effort which also speaks to her physical stamina at the age of thirty-eight.

A second question, which still awaits concensus, is the degree to which the Taglionis were an enchanting interpolation in the history of the Petersburg ballet, too self-contained to interact with the company and leave a lasting benefit. If this is true, their stay was a palliative, temporary relief from Blache and Titus which left the company in worse condition from years of benign neglect. Even Zotov, who supported theatre policies and was not especially well disposed towards the ballet, implies this:

> When we will speak of the era of Marie Taglioni we shall say, of course, that the ballet was raised yet higher, but already then it was not ours. It was a meteor which dazzled for a while, to leave us afterwards in the darkness for a long time, and indeed which cost us too much.[16]

Marie was given a lavish farewell, reported with warmth and sadness in the press. There is none of Zotov's sentiment in the long apostrophe published in *The Northern Bee*:

> We have never before seen in Petersburg such an enthusiastic public in the theatre as at the last performance of *Gerta*! In every spectator lay a supernatural, electric force which broke to the surface in unanimous applause and unquenchable cries of 'Bravo!' Yes and how charming she was on this evening, when for the last time

---

[14] V. Gaevskii, *Divertisment*, pp. 10–11.
[15] 'Talisman', *Repertuar i panteon* [Repertoire and Pantheon], 1847, vol. ii, p. 31; quoted in Krasovskaya, *Russkii baletnyi teatr ot vozniknoveniya do serediny XIX veka*, p. 222.
[16] 'Memoirs', *Istoricheskii vestnik*, vol. 66 (Oct.–Dec. 1896), p. 404.

Petersburg gazed upon her sporting about the stage in her gauzy attire as Elfrida, on her sharply profiled Aragonese dances, on her elegant dance in the magnificent enchanted gardens! One could not have recognized our public on that evening: the thought that this perfection, this ideal it was seeing for the last time, made them bold to request a repetition of everything distinguished by special charm among all that was elegant; on the least pretence calls for her followed one after another—after every dance, after every scene. Meanwhile, in one of the intervals, after Mlle Taglioni had been called out three times and the hall had quieted, a child's meek little voice was suddenly heard: 'Taglioni!' The delighted public, seething like red-hot lava, picked up the call and expressed again its gratitude to the fleeting Sylphide at her farewell.

In the course of the ballet she was called out fourteen times on this evening, and at the end of the ballet another sixteen times, and so *thirty times* in all! Amidst these enthusiastic and touching farewells, she, as on the preceding Sunday, came to the front of the stage and said, in German: 'Accept my most heartfelt thanks! I hope that I shall come back.'

Meanwhile in the wings, we were told, a most touching scene took place. Our young women dancers, whom Mlle Taglioni managed to win over by her modesty and tenderness, during the last act decorated her offstage dressing-room with fragrant fresh-cut flowers, even strewing them along the staircase leading up to it. Mlle Taglioni was touched by this to the depths of her soul, and bade farewell here to her modest friends who remain in the north and to the personnel of our theatre.

Mlle Taglioni, in fact, is a rare example of good nature, gentleness, and courtesy in manners, and brought all who but knew her to love and esteem her. It is difficult to find a woman more cultured than her and more modest besides, especially among visiting artists, normally gifted with many more pretensions than talents, and who thereby imagine that all their colleagues are plotting against them. Everyone sincerely and decidedly regrets the departure of Mlle Taglioni, on the contrary, because regardless of her first-class talent, she is a good and kind woman, and for artists, a blood sister. Leaving with our public the sweetest remembrance, she takes with her its unanimous love.[17]

La Fille du Danube was the first ballet by Filippo Taglioni not already produced in St Petersburg before he and his daughter arrived, and one of two (the other was *La Sylphide*) that survived their departure, being revived in 1880 by Marius Petipa at the express wish of Alexandre II. Marie performed it in Russia for the first time at her benefit on 20 December 1837, after which it was repeated on each of the dozen ballet nights which followed in the next month. She danced in it for the last time within a week of her final performance in St Petersburg in 1842.

The reception given to La *La Fille du Danube* was characteristically enthusiastic:

[17] 'Miscellany. Farewell to Taglioni', *Severnaya pchela*, 1842, No. 51 [6 Mar.] p. 202.

The ballet presented yesterday, *La Fille du Danube*, had a marvellous success. The balletmaster Mr Taglioni, the creator of the ballet, who produced it on our stage, was called for after the first and second acts. Mlle Taglioni was never so captivating as on this evening. The calls for her were endless; we lost count of them.[18]

By the sixth performance a French quadrille on motifs from the ballet had been arranged and published with a dedication to Marie, and after the tenth performance: 'The closeness and crowding in Mlle Taglioni's performances are not decreasing: it seems that the walls of the Bolshoy Theatre have grown closer together since the time they started giving *La Fille du Danube*.'[19]

In his extended review of the new ballet Faddei Bulgarin, who cast his article in the pre-Taglioni format long on digression and short on evaluation, also found something nice to say. After rambling through the history of Greek and Roman dance and degrading ballet in a comparison with drama, he made a connection with his stated topic by claiming that Marie Taglioni 'revived our memories of ancient elegance in plastique, of perfection . . . At the sight of Mlle Taglioni the elegant, artistic Greece is animated in our mind . . . Taglioni alone reconciles us with ballet.'[20] In a passing remark, Bulgarin also gave us some hint of stage designer Andrei Roller's treatment of the underwater scene: 'Finally we saw Mr Roller's charming decoration, the subterranean grotto of the deity of the Danube and this deity itself in the form of a raving giant with hair and clothes of running water. The tableaux vivants made up of Danube nymphs and their graceful dances one can call truly poetical, and the creator of the ballet, Mr Taglioni, was called out amidst loud applause.'[21]

The leading female characters of *La Sylphide* and *La Fille du Danube* are half-woman, half-spirit, who differ in the elements they inhabit: the Sylphide rises into the air, the Daughter of the Danube descends into the waters. The latter is similar to the legendary Undine, immortalized in de la Motte Fouqué's fable and its Russian translation by Vasily Zhukovsky into a poem. Undine is also a strange girl of unknown origin discovered on the bank of a river who will later be courted by a nobleman. But Undine's motivations, both earthly and spiritual, are clearly stated: she wants to marry to gain a soul. Taglioni's heroines express no interest in marriage

---

[18] 'Miscellany', *Severnaya pchela*, 1837, No. 291 [22 Dec.], p. 1162.
[19] 'Miscellany', *Severnaya pchela*, 1838, No. 17 [21 Jan.], p. 67.
[20] 'Bolshoy Theatre. *La Fille du Danube*, by the balletmaster of the Imperial Theatres Mr Taglioni, given for the first time for the benefit performance of Mlle Taglioni, 20 December 1837. First Article', *Severnaya pchela*, 1838, No. 1 [1 Jan.], p. 2.
[21] '*La Fille du Danube*, ballet in two acts by the balletmaster of the Imperial Theatres, Mr Taglioni. Second Article', *Severnaya pchela*, 1838, No. 2 [3 Jan.], p. 6.

and are enigmatic in general, though the Daughter of the Danube, like Undine, is protected by the ruling spirit of the waters.

In the following translation, designations of act, scene, and tableau are placed in the centre of the page; in the Russian edition of 1837 (and the Parisian edition of 1836, where they were less numerous) they were placed in the margins.

4
Libretto of *La Fille du Danube*
Filippo Taglioni

LA FILLE DU DANUBE
Pantomime Ballet
In Two Acts
and
Four Scenes

By Mr Taglioni
Music by Mr Adolphe Adam

Saint Petersburg
Printed at the Typographers I. Glazunov and Co.
1837

Decoration of the 1st scene
by Mr Fyodorov
Decoration of the 2nd scene
by Mr Sabat
Decoration of the 3rd and 4th scenes
by Mr Andrei Roller
Machines by, and machinist:
Andrei Roller
Costumes: Messrs Baltier and Matier

FOREWORD
to the French edition

The legend on which the story of this ballet is based has nothing except its name in common with the famous German play *Donauweibchen*, which is well known to almost all who inhabit the shores of the Danube. It was the inspiration of first-class German poets.

We decided to present it to the public in full, despite the fact that the tale borrowed from it for the ballet, the most absorbing episode in the whole story, is its denouement.

The notes which the reader will find in this programme are neccesary for a clear understanding of the ballet.

You must not think that the choreographer intends to develop this story in full detail, which is marvellously embroidered with superstition;

La Fille du Danube 91

the most demanding critic will not notice any literary pretensions whatever in our brief tale; we attempted to preserve the chaste style of this German legend. Our esteem for the old German tale is such that we did not even divide it into parts in accordance with its theatrical presentation. Wishing to transmit it to the reader in its complete and original form, we limited ourselves to simple indications in the margins, which reveal to the reader all the necessary linkings, divisions of the episode into acts, tableaux, and scenes, which is true to the original in all respects, even down to fine details.

Perhaps the huge success of our libretto must be attributed precisely to this reverential esteem: in Paris and London eleven thousand copies were sold.

Pantomime

ACT I

SCENE I

Messrs Golts, Eberhard, Emile Gredlu
Mlles Taglioni, Savitskaya, Ovoshnikova, Andreanova II

Nymphs
Mlles Spiridonova, Rasova, Kapylova, Paradizova

Peasant girls
Mlles Didier, Efremova, Godovikova, Butorina, Monroit, Dyukova, Amosova, Sokolova, Dolbilova, Lyadova, Danilova, Selezneva, Semenova, Bekker, Terikhova, Vaghina, Katlerovskaya, Labazina

Peasant men
Messrs Shambursky, Pimenov, Bogdanov, Gorinovsky, Grigoriev the younger, Morozov, Godovikov, Vasiliev, Meister, Rodek, Zhigachev, Grigoriev the elder, Spiridonov the elder, Semenov, Katlerovsky, Seleznev, Rokendorf, Politaev

Peasant children
Boy students: Manokhin, Mukhin, Ramazanov, Auguste, Tanashev, Smirnov II, Lukin, Lyustikh
Girl students: Yakovleva, Ryumina, Lukasheva, Kasterovskaya, Zakaspiiskaya II, Goshishok, Levkeyeva and Klinina

Two heralds

Pas de trois
Mr Emile Gredlu, Mlles Taglioni and Andreanova II

SCENE 2

Pantomime
Messrs Lachouc, Golts and Eberhard; Mlles Taglioni, Andreanova II and Ovoshnikova

Pages
Girl students: Danilova, Gorina, Polyakova, Beloutovtseva, Gering, Kostina, Ushakova, Sysoeva, Kolosova, Ivanova, Stepanova and Golovanova

Gentry
Boy students: Voronov, Johanson, Gorin, Yablochkin, Petrov I, Fedosov, Petrov II Gorkago, Fomin, Sergeyev, Zakharov and Gusev

Noblewomen who do not dance
Girl students: Volkova II, Bormotova I, Greneva, Kapylova, Volkova I, Kashirina, Semenova I, Shtofert, Volkova III, Mironova, Kostina and Ushakova

Noblewomen who dance
Mlles Spiridonova, Rasova, Kapylova, Paradizova, Gammer [a student], Bormotova II, Erash and Zakaspiiskaya

Rosiers

Girls from the Valley of Flowers
Mlles Vasilieva, Grigorieva, Gundurova, Bystrova, Shiryaeva, Ivanova II, Savinova, Chernoyarova, Avdiukova, Antipova [a student] Fedorova and Karostinskaya

Dancers of the Allemande
Mr Golts, Mlles Taglioni, Andreanova II, Vasilieva, Grigorieva, Gundurova, Bystrova, and [students] Fedorova and Karostinskaya

Pas de cinq
Mr Theodore-Gerino, Mlles Volkova, Smirnova, Alexis, and Teleshova

ACT II

SCENE 3

Messrs Lachouc, Golts and Eberhard
Mlles Taglioni, Andreanova II and Savitskaya

Gentry
Noble women
Peasant men and women
Warriors and honour guard

SCENE 4

Pantomime and Dances

Nymphs
Mlles Taglioni, Savitskaya, and the danseuses of the corps de ballet.

Finale
Messrs Lachouc, Eberhard and the others who are taking part in the pantomime and the dances.

DRAMATIS PERSONAE:

The Daughter of the Danube	Mlle Taglioni
Baron Willibald	Mr Lachouc
Rudolf, *his sword-bearer*	Mr Golts
Irmengarda	Mlle Ovoshnikova
Nymph of the Danube	Mlle Savitskaya
A young girl	Mlle Andreanova II
The Baron's officer	Mr Eberhard
The Danube	Artemiev

Girls of the Valley of the Flowers
Undines and nymphs
Courtly ladies
Knights
Pages
Heralds
Peasant men and women of the fourth scene
Honour guard
Warriors
Musicians

Enchanted valley of the Donaueschingen! The mountains of Ferenbach which shield it on one side, and the meadow running along the banks of the Danube to Neiding, somewhat lower than the point where the Brigach flows into this river, make the valley of the Donaueschingen one of the most charming places in the world.

The rarest flowers, the most luxurious vegetation cover this valley. Sunlight in this wild country, sultry in July, is blocked for nine months of the year by mists which come down from the mountains out of which the Danube flows.

Nympheae water lilies and forget-me-nots, raising their white, rose, and blue flowers above the meadow grasses, follow the course of the river in all its windings, as if to form a garland.

This enchanted valley, which for good reason is called the Valley of Flowers, was for a long time a dependency of the Brunlingen sovereigns; from 1365 it belonged to the Baron von Neiding, and now is part of the Fürstenburg princedom.

Baron Willibald, Count of Meringen, sovereign of Neiding, Balding, Donaueschingen, and other places, was left the full and sole owner of all these estates in 1420, on the death of his father and older brother who were killed on 11 June in the service of Emperor Sigismund beneath the walls of Prague, which the celebrated leader of the Hussites—Ziska—was defending.

Willibald's older brother (his name is not known), wearied by the misfortunes suffered in three tragic marriages within five years, sought death, following the emperor everywhere.

His first wife, daughter of the sovereign of Brunlingen, died after the first month of marriage; the cause of her sudden death was an unexplained mystery; the widow of Baron Willingen, whom he married two years later, died a week after the wedding, at the same hour on the same day as his first wife—indeed, at the very moment when she uttered the name of her predecessor. These two horrifying misfortunes, stamped with some hellish secret, acted powerfully on the mind of the poor husband; it seemed to him that he was causing the disasters to everything dear to his heart. He was near suicide, but on the advice of the Pope, to whom he dispatched ambassadors concerning this matter, he decided on a third marriage.

With great effort he won the hand of the daughter of the ruler of Ferenbach; ambition forced the father to sacrifice his only daughter. This third marriage was celebrated with the greatest magnificence: the public reading of a papal letter was to have put an end to the persecutions of evil spirits; the newly-weds retired to the marriage bed, having reverently kissed the words written in His Holiness's hand. Futile rite, idle precautions: in two hours the young bride, in the arms of the unfortunate baron, had already breathed her last.

Thrice widowed, death remained the poor young man's only bride; he went in search of it, and found it on the field of battle.

Baron Willibald, thus left at twenty-five the sole possessor of his predecessors' estates, received this sorrowful legacy with trepidation. Recollections of his brother inspired in his heart a superstitious if wholly excusable thought, especially in those times when neither exalted nor lowborn was free of prejudice.

The young baron, blessed with all the virtues of soul and body, could not resolve to marry despite a passionate wish to do so; he feared (doubtless much more than he feared his brother's fate) the refusal of the nearest sovereign houses, who kept their distance, supposing that a sad destiny threatened his family. Doomed to solitude, he would happily have sacrificed half his domains, and would keep from his vast legacy only the Valley of Flowers, if he could but share this meagre leaving with a beloved spouse. Only one thought consoled him: that perhaps the evil fate reigned *not in his* family; that perhaps his brother's *wives* brought such a fatal destiny to his house.

Fully committed to this idea, he swore eternal hatred to all women who could put the particle *von* before their surnames, or embroider coats-of-arms on their velvet dresses. Every noblewoman, baroness, or countess, raised the thought of an inequitable marriage and damnation; yet so much did he want to marry, thus to prove that the curse of heaven was not hanging over his house, that he decided to use every possible means to reveal the mystery and to make its purpose well known.

Therefore he did not want to turn to the barons, to the sovereigns of Brunlingen, Willingen, Ferenbach, and other places; he decided to marry a simple girl, a poor vassal, innocent, without distinguished forebears, without aspirations, without a past or a future.

He decided to offer his hand to a daughter of nature and to share with her the possessions which heaven or hell had given him; this was the only means of avoiding obstacles to the realization of his plans.

At that point he directed his gaze at the Valley of Flowers. Willibald remembered that as a child he ran around these enchanted places, filled with the scent of the most beautiful flowers, which announce with a murmur the first waters of the Danube; that from the height of the mountains the empty silence shielded the joyous meadows; that in childhood he would count these cottages, these little roofs hanging on to the cliffs. He recalled how he marvelled at the beauty of the girls, now robust, now pale, like the sun in those lands, their white brows the same colour as the *nympheae*, their eyes tender, like the valley plants, and moist like the Danube mists.

There, in this flowered valley, among the simple-hearted maidens, he decided to find the companion of his life, though all the baronies and all the counties in the world clamoured against a marriage of such unequal rank.

Superstition, which reigned among the rulers of the nearby localities, took root there and held sway in the highest possible degree; it distressed proud barons, noble counts, and spread horror among them. But here, in this ravishing valley, it resided in poetical dreams and innocent fantasies. Strange rumours circulated about one of the valley girls, rumours which even reached as far as the castle of Neiding.

One morning an old woman, Irmengarda, found a young girl, whom she called her daughter from that time on. The girl was kneeling, surrounded by forget-me-nots which grew in front of the grotto from which the source of the Danube flows. Her friends called her Flower of the Field, Feldblume, Fleur-des-champs, to preserve the memory of her mysterious origin. No one for ten miles around knew her before this time; a mute, she could not give any information either about her birth or about her life; no wanderer had ever seen her before. Just a few hours before she was found at the river's edge the Danube, without any storm or subterranean sound, overflowed its banks along the meadows: but soon the waters were running within their channels again and resumed their normal course. That is all they would say in connection with the girl, wanting to make her appearance strange.

After this, much good was said of her: loving the old Irmengarda as a mother, she was the most quiet and modest of all the valley girls. She played with others of her age with such charm, innocence, and kind-heartedness, that there was not so much as the shadow of envy in her friends.

Every morning, they would say, she knelt at the place where she was seen for the first time; she brought an offering of flowers to the Danube, or prayed to the rising sun. Such behaviour was an unfathomable mystery. Old Irmengarda, however, fearing to lose her heaven-sent daughter, did not leave her for a minute, swore on her soul that she once saw the shade of Father Danube in the river, and in general told many tales on her account, invented by superstition, because everything about this girl was mysterious and enigmatic.

ACT I SCENE I

Dawn lights the Valley of Flowers, along which the first waves of the Danube flow; the sun, already pouring its light over the cottages scattered in the hills and valleys, illuminates two huge cliffs which form an entrance to the main source of the river.

Tableau 1

The Daughter of the Danube comes secretively out of her cottage; her mother is still resting and therefore cannot observe her; she brings an offering of flowers to the Danube, her customary morning prayer, and having received her father's blessing, goes back into the cottage.

Tableau 2

Enter young Rudolf, Baron Willibald's sword-bearer. He has given his heart and future to the mysterious girl. The joy that he finds so close to the object of his love is quickly marred by the thought that his noble ruler will never agree to let him marry the mysterious girl. But the time of their meeting has come, and his sad forebodings disappear.

Tableau 3

The doors of the cottage open quietly and the girl comes out cautiously, fearing to wake her mother. Rudolf runs to meet her and tries to reassure her. They comfort each other; innocent, like their love, they amuse themselves picking flowers which are scattered about the banks. Then, weary of games, they sit down on a bench of turf. On this occasion Rudolf is more affectionate than usual. The girl listens to him trustingly and puts her head on his knee; her brow is resting in her beloved's hands; he places a garland of forget-me-nots on her head.

Tableau 4

Meanwhile the nymph to whom Father Danube has entrusted his daughter, without taking her gaze from the couple, comes out of the grotto and puts them into a deep, magic sleep. Their hearts pound as they dream. The nymph approaches, puts a ring on the hand of each, and they begin to breathe as one.

Tableau 5

The nymph disappears. The two wake up; their feelings are confused; they rush to each other's embrace, as if wanting to realize their dream. Rudolf is now the mysterious girl's fiancé.

Tableau 6

Just then old Irmengarda comes out; ruthlessly she drives off the young seducer, whose whole estate is naught but love, and whose rights reside in nobility of heart. The old woman dreams of an incomparably better future for her daughter: some mighty sovereign, a baron, perhaps the Holy Roman Emperor himself are the only ones worthy, in her opinion, to possess such a treasure. The poor sword-bearer despairs, but the girl's gaze consoles him; he exits.

Tableau 7

As the girl is about to go back into the cottage, youths and girls of the valley appear, and invite her to take part in games. Refusing at first, she quickly yields to her friends' urgings.

Tableau 8

The dances end. Trumpets are heard; all come forward in amazement. It is Baron Willibald's heralds. The baron has heard a great deal about the mysterious girl; he did not want to lure her by his power—to the contrary, he orders that all the girls of the valley be invited to his castle; from among them he wants to choose his life companion: the gentlest and most beautiful must receive the title of his wife—baroness. The inhabitants of the valley surround the heralds and shower them with questions; the reason for his coming is still unknown; but a joyous presentiment fills the girls' hearts. One of the heralds unfurls a banner with the following inscription:

'Baron Willibald wishes to choose a bride from among the girls of the

Valley of Flowers, and invites them to a celebration today at the castle of Neiding.' Shouts of joy go up from all sides; in anticipation, mothers are enjoying their daughters' happiness and glory; old Irmengarda is the most delighted.

Tableau 9

Alone, the Daughter of the Danube is struck by the news. The honour that everyone else wants so passionately scares her.

Table 10

The news also troubles poor Rudolf; he despairs . . . All his hopes are dashed: the girl's sorrow does not fully convince him that she returns his love; her attractiveness clearly tells him that she will be chosen; in vain the girl consoles him. If only she could become ugly, the more to allay Rudolf's concern! She comforts him with the promise to disappoint the baron by feigned awkwardness; finally she reminds him of the dream which united them, and of their nymph-protectress.

Rudolf calms down; the girl, satisfied with her clever plan, runs off to her mother, who must dress her in her best clothes.

Tableau 11

The youth of the Valley of the Flowers, hearing of the celebration being readied, crowd around Rudolf and ask him to lead them to the castle. The sword-bearer agrees, and all rush after him.

SCENE 2
Tableau 1

Meanwhile in the castle of Neiding the celebration is being prepared; the baron's guards come into the broad entrance hall of the castle, arrange themselves in formation and render honours to the noblewomen and the famous knights who are arriving from all quarters.

The Danube, which flows beneath this very terrace, reflects the brilliance of the soldiers' gold and steel armour in its deep transparent waves. The magnificence, the richness of costumes, everything is brought together in this festival, the reason for which is still a mystery to the noble personages.

Tableau 2

Surrounded by his retinue, Baron Willibald comes into the hall. Pages enter first; the honour guard concludes the procession. He affectionately greets the crowd which is mingling around him. A secret thought troubles him: will his orders be executed precisely? Will the girls of the Valley of the Flowers come at his invitation? These thoughts distract him amidst the respects he is being paid.

Tableau 3

A herald enters and announces that the girls of the Valley are arriving.

Tableau 4

They come in; at their arrival the baron smiles; he is already enjoying the amazement, anger, and indignation of the noblewomen who thought that he, whom they had cast off, was fated to eternal bachelorhood.

The girls are decked out in their best clothes: their white dresses adorned with simple flowers seem all the fresher among the fancy costumes of the noble ladies. A garland of *nympheae* and a bouquet of forget-me-nots are the Daughter of the Danube's sole accesories.

Baron Willibald, whom this contrast pleases as much as it offends the noble ladies' pride, invites the valley girls to join the celebration. They dance, waltz, mingle . . . The Daughter of the Danube remembers her promise: sorrow darkens her features; she affects a deliberate clumsiness. Irmengarda conceals her indignation with difficulty, and Rudolf his joy; but through this feigned awkwardness, a thousand times enchanting compared with the studied movements of her rivals, the young baron's heart divines the real Daughter of the Danube. She is chosen, she will take the title of baroness. Irmengarda's fury turns to joy, Rudolf's joy to sorrow.

The dances end. The ladies do not hide their indignation; they want to leave, but the baron has yet to satisfy his vengeance; he has yet offended them too little. He needs one more gibe; he showers the ladies with cold greetings, entreats them to stay and be witness to his happiness. Each affectionate word directed at them is a bitter mockery; the joy on his face wounds their hearts; they mask their feelings with sardonic smiles.

The baron walks up to the girl of his choice and speaks to her of marriage, glory, and prosperity. The general astonishment and indignation grow: the girl declines his proposal. Willibald insists, begs, places at her feet his exalted station, riches, and pride.

Rudolf runs between them, reminding the girl of her love and the baron of his mercy; he begs, he threatens; but Willibald rejects him angrily and takes the girl's hand; she tears herself away and runs to the balcony, which overhangs the Danube.

Everyone is horrified. The girl expresses her aversion to the marriage and declares her love for Rudolf; she curses the baron, tosses her bouquet of forget-me-nots to the young sword-bearer, and throws herself into the Danube.

Tableau 5

It is too late to assist. Her friends' cries, the ladies' evil joy, the baron's grief, Rudolf's despair—all supplement this sad picture.

ACT II SCENE 3
Tableau 1

This sad event has clouded the young sword-bearer's reason; he runs along the avenues of the park, far from the castle, which is visible on that side of the Danube. Servants, peasants, and soldiers hurry after him, wishing to prevent a new misfortune.

Tableau 2

But they soon lose sight of him. Rudolf returns, despair etched on his face. He walks along the bank, begging the waves to return his beloved. Nothing can dispel his grief: neither Irmengarda's consolation nor her tears, nor the exhortations of his sovereign.

The despair of friends means nothing to him, nor do the honours which await him if he abandons his perished love. Glory no longer attracts him; his happiness lies beneath the river, and there is where he hopes to find it. This thought consoles Rudolf, and gives him momentary calm which he thinks is true happiness. Wishing to share it with his love, he takes the bouquet she threw to him in farewell, which he has kept next to his heart, and covers with kisses all which remains to him of beauty and love.

Before him is the river which conceals all his desires! He rushes towards it to find again what he lost or to perish . . .

Tableau 3

Suddenly he hears a quiet, mysterious harmony. The Nymph of the Danube appears, surrounded by undines; next to her is Rudolf's beloved. He cannot believe his eyes, and takes this as a sign of his deranged imagination. He falls to his knees, begging the girl to come to him and the nymph to return his bride.

Tableau 4

The girl comes ashore, but Rudolf's efforts to embrace her are in vain: she is only a spectre.

Father Danube, having called his daughter back, does not want to return her to an unworthy world. The Daughter of the Danube can belong only to him who will take her from her father's arms.

The girl evades pursuit; Rudolf's frenzy scares her, but he tells her, 'In parting with you I have lost my reason; in your presence I feel only love's distraction.'

The sound of footsteps diverts Rudolf's attention from his beloved.

Tableau 5

When he turns to her again, the spectre has disappeared. Rudolf falls into utter despair, and in a frenzy turns to the river responsible for his misfortunes. The murmur of its waters is his only response.

Tableau 6

The footsteps were those of the baron and his servants, who were following Rudolf. Willibald sees Rudolf; he calls some of the girls, who quickly encircle him. A happy thought occurs to the baron, one which should restore his favourite groom's reason. He wants one of the girls, dressed the same as Forget-me-not but covered with a veil, to appear before Rudolf.

Recognizing his sovereign, Rudolf tries to hide but a guard bars his way; he seizes a sword from one of them and wants to attack the baron; the girl with the veil rushes between them.

Tableau 7

Rudolf stops . . .

He thinks that it is she: the sword falls from his hands. Willibald and his retinue withdraw to the side to see what the deception will bring.

Rudolf falls at her feet, embracing her passionately; with tears in his eyes he begs her to remove her veil, and dispirited by grief, his head sinks on to his chest.

Tableau 8

Seeing that Rudolf has grown calm, and thinking that the deception has succeeded, the baron entrusts the girl to take advantage of Rudolf's error and bring him to the castle. Then the baron leaves.

Tableau 9

Just then Rudolf reminds her of those happy moments which they spent together, of games in the Valley of Flowers; he shows her the ring which joins their hearts; he speaks of the pure, unselfish love of which they dreamed. But the girl does not understand him; instead of answering she asks him to follow her. Rudolf is about to agree but her veil comes off at that moment and he realizes his error. He pushes her away angrily, takes the bouquet of forget-me-nots, presses it to his heart, and then throws himself into the river. The baron and his retinue arrive too late to prevent this tragedy.

Suddenly the river becomes turbulent and overflows its banks; a peal of thunder is heard. A great mystery occurs: the Danube embraces its daughter's spouse.

SCENE 4
Tableau 1

Rudolf is unconscious; a crowd of nymphs supports him as he descends into Father Danube's grotto. The nymph whom he saw in his dream in the Valley of the Flowers approaches him and returns him to life and reason.

One trial remains. The first time, when Baron Willibald presented Rudolf with the young girl covered with a veil, his transport deceived him; he must now conquer such deceptions.

Tableau 2

Nymphs covered with veils surround him and entice him with the charms of delusion: his eyes having deceived him once, let now his heart recognize the one he loves; let him touch her with his hands, and the girl will be his.

He rejects the first nymphs; he is not drawn to them. Others come up to him, ethereal, like the Daughter of the Danube, and like her tender and charming; he hesitates ... and then refuses. He does not love them, he will not swear his love to them forever. Then all encircle him, offering him shells and the rarest plants of the underwater realm, describing the pleasures of love and happiness.

But the bouquet of forget-me-nots in his hands serves as a talisman against their delusions and treachery. No one touches this sweet bouquet; none among them dared. One only, walking past him, touched his hand ... It is the Daughter of the Danube, it is she! She wants to hide from him, but Rudolf pursues and finally catches her; he is in his beloved's embrace.

Tableau 3

They kneel before the Nymph of the Danube and ask her to return them to the world, which can never separate them again.

Tableau 4

Just then a seashell rises from the depths of the river and floats to the surface, flowing with the current.

Tableau 5

The undines and the entire subterranean court surround the Nymph, who points the way; the undines rise in groups above the water; Father

Danube appears in all his magnificence; he blesses this wonderful union and returns the young pair to the world.

Tableau 6

The baron, old Irmengarda and the court run to the shore, bow down before the river, then embrace the happy lovers.

This chronicle ends with the touching account of the death of Baron Willibald, Count Meringen, sovereign of Neiding, Balding, Donaueschingen, and other places.

Willibald, unhappy in love as his brother in marriage, did not want to spoil the well-being he could not hope for himself. He knew that God alone has the power to grant true happiness to people, and retired to a monastery. Wishing to augment heaven's good deed, he gave the happy couple the estate of Donaueschingen as dowry. Ten years later, on 21 December 1430, Baron Willibald died in the Augustine monastery at Bologna; in memory of his benefactor, Rudolf built a chapel, the ruins of which can be seen to this day on the road from Donaueschingen to Ferenbach.

III
The Life of a Russian Artist

INTRODUCTION

Watching the parade of visiting celebrities pass through Russian ballet in the nineteenth century makes it easy to forget the large number of native artists who were seldom famous but served faithfully amidst the comings and goings of foreign stars. Timofei Stukolkin and Anna Natarova were two such artists. In their memoirs they describe their student days, and Stukolkin his advance into the first stages of fame. They tell of the tribulations of making a reputation at home, and offer a special perspective on the ballet as they knew it by placing important events in the history of the Petersburg company within the context of everyday existence, full of anecdotes and lore.

Timofei Alexeyevich Stukolkin (1829–94) was one of a handful of Russian-trained male dancers in the nineteenth century who came fully into his own over a long career (others are Nikolai Golts, Lev Ivanov, and Pavel Gerdt).[1] He specialized in acrobatics and caricature, and was a celebrated Doctor Coppélius and Don Quixote. His most famous role was Drosselmeyer in *The Nutcracker*, which he created. An artist who danced in revivals of Didelot's *The Hungarian Hut* and *The Captive of the Caucasus*, who performed on the same stage with Taglioni, Elssler, Perrot, and St-Léon, and then went on to star in Tchaikovsky's last ballet, Stukolkin exemplifies the continuity provided by the permanent cadre of Russian artists.

But Stukolkin does not refer to many of his most important achievements in the 'Recollections'. He acknowledges in the text that he never kept diaries or notes on his life, and it is likely that he began work on the 'Recollections' not long before his death in 1894, of a heart attack which struck him down between Acts II and III of a performance of *Coppélia*. 'He died like a soldier at his post,' one eulogist wrote, 'serving the art of choreography, which he loved passionately and never betrayed, to his last minute.'[2]

The full title of Stukolkin's memoirs is: 'Recollections of T. A. Stukolkin, Artist of the Imperial Theatres. Copied from his account by A. Valberg'. They were published in the journal *Artist*, No. 45 (January 1895), pp. 126–33 [Chapters I–III], and No. 46 (February 1895), pp. 117–25 [Chapters IV–VIII]. Like his life, Stukolkin's 'Reminiscences' ended unexpectedly, with a promise 'to be continued' which was never fulfilled.

---

[1] On Stukolkin see also V[era Mikhailovna] Krasovskaya, *Russkii baletnyi teatr vtoroi poloviny XIX veka* [Russian Ballet Theatre of the Last Half of the 19th Century] (Leningrad and Moscow, 1963), pp. 443–5; M. Borisoglebskii, compiler, *Proshloe baletnogo otdeleniya peterburgskogo teatral'nogo uchilishcha, nyne Leningradskogo Gosurdarstvennogo Khoregraficheskogo Uchilishcha. Materialy po istorii rueskogo baleta* [The Past of the Ballet Division of the Petersburg Theatre School, now the Leningrad State Choreographic School. Materials for the History of Russian Ballet], 2 vols. (Leningrad, 1938–9), i, 143, 160, 183, 361–2.

[2] 'T. A. Stukolkin†', *Vsemirnaya Illyustratsiya* [World Illustration], No. 1340 (1 Oct. 1894), p. 272.

5
'Recollections of T. A. Stukolkin, Artist of the Imperial Theatres'

> There Didelot was crowned with glory.
> There, there in the protection of the wings,
> My youthful days were swept away.
> (Pushkin)

I

Very long ago it was. In 1836, my now deceased father, thanks to the patronage and co-operation of A. L. Nevakhovich, supervisor of the repertoire at that time, had me, a seven-year-old tot, appointed to the Petersburg Theatre School, which at that time had just been transferred from its old building facing Ofitserskaya Street and the Ekaterininsky Canal to its new site on Theatre Street, where it still is. At first I was enrolled as an external student, but in a few months, after I demonstrated significant progress, I was made an internal student, a boarder, and enrolled at state expense.

I quickly got used to the new situation and was drawn into school life, which in those days was regulated as follows: we got up at 7.00 a.m., drank *sbiten* [a hot drink of honey and spices] (we still dared not think of proper tea then). From 8.00 to 10.00 classes in academic subjects were given, from 10.00 to 1.00 in the afternoon classes in the arts—that is, singing, dances, music, drama, fencing, and the like. At 2.00 p.m. we had lunch, and from 3.00 to 5.00 we continued our academic classes. In addition, very often at 12.00 noon we would have to leave for some rehearsal, which lasted until 3.00 or 4.00, and sometimes to 5.00 in the afternoon. Those who took part in rehearsals, of course, lunched separately from the others, later. Almost every day in the evenings we went to the theatres, as students would sometimes have to perform.

Academic studies, of course, suffered in this way of life—all the more since the teachers of academic classes were in most cases beneath all criticism. The study of art was incomparably more successful, which is readily explained by the felicitous choice of teachers and the brilliant models which we could see on stage, and likewise by the traditions of the school. According to regulations of that day, every male student was obliged to study singing, dancing, and drama, and how to play some musical instrument. Thanks to this [policy] absolutely every student, however untalented, found work in the theatre on leaving school, work which could be done more or less successfully—that is, if he lacked dramatic abilities he went into the ballet, if not ballet then into opera, if not suitable for opera then as a musician, and that failing, as a prompter or a supernumerary or a figurant or as a machinist, etc. The main thing was that he went into the field for which he had been prepared in school since early childhood.

Our management reasoned absolutely correctly that once a child has spent eight to nine years preparing for the theatre, he is not suited for any other profession, and consequently the theatre administration is obliged to employ him in an occupation native, intimate, and known to him. For some reason, however, this view has now changed. Thus, for example, it is sometimes impossible to place a youth, supported in school to the age of eighteen to twenty, on the official stage. Preparing him several years for theatrical service, in the end it is observed that he lacks aptitude for this line of work, and forgotten that he lacks training for any other! Or consider this: before, as I mentioned, our school provided the theatre with personnel for all the necessary professions of the stage, from premier danseur to souffleur. Nearly thirty years have passed since it produced one first dancer (the last such was Pavel Gerdt); it graduates only coryphées, corps de ballet, and figurants, and to compensate for a deficiency of male dancers foreigners must be hired, such as Mr Enrico Cecchetti. Cecchetti, however, was additionally useful because his talent, and the success he immediately won in Petersburg awakened a sense of competition in our own balletic youth, who began, as they say, to watch him, to learn, and be perfected.

I return to the old days. At the time I am describing, the administration encouraged and extensively developed a sincere and passionate love for theatre among the students of our school. Independent of any participation in [official] school performances, which were given twice a week in Lent, we put on our own, improvising costumes and decorations, and even plays. Blankets, sheets, benches, beds went into this effort—in short, everything around us in the school environment. The quality of these improvised pieces was, of course, directly dependent on the age and talents of the actor-students involved. Thus, for example, scenes were sometimes given in which robbers raged about with improvised pistols in their hands, firing shots produced by the clapping of desk tops. Other stage effects were devised, and our pieces were always given with a full ensemble. To round out the similarity of this theatre with the real thing, we even built a buffet for the sale of bread and butter and delicacies. These performances were approved by our superiors and were of great benefit to us, gradually training us for the stage, providing us with methods and skills which the actor not born to the art acquires with such difficulty, and which a spectator at amateur performances notices the absence of so clearly.

Our administration's approach to the study of theatrical arts was, I repeat, as correct as our training in academic subjects was poor. Although we were taken through languages and mathematics, geography, and history with mythology and archaeology (!), the results, alas, were most deplorable. The antique, bankrupt system of instruction, our way of life, and the choice of tutors and instructors—unsuccessful to the point of mirth—hindered our progress. Among them, and of the administrative personnel of the school of the 1830s and 40s generally, I can remember the following persons and names: tutors Mikhelson and Knapich; the school doctor Marochetti, who treated all illnesses with the same remedy—the laxative herb jalappa, which, it seems, contemporary pharmacopoeia does not even acknowledge; teachers A. P. Olimpiev in history, Tyazhelov in geography (which for some reason we could not tolerate), and Nikolai Ivanovich Kulikov in drama.*

Unfortunately only the last-named, with his knowledge and gift, could bring the slightest benefit to us through teaching, as the others were not just bad pedagogues—it is incomprehensible that they could be tolerated in an educational institution. A case in point was the tutor Panov, who came to this post only because he was a graduate of our school and was totally unsuitable for the stage. He passed the time, a wooden ruler in his hand, fencing with the wall, seeing some opponent there invisible to the rest of us. Another teacher—Solich—a man of nearly eighty and a paralytic, always carried a section of rope in his pocket, with which he beat students mercilessly. Yes, and the rest were likely of the same ilk...

Only such a system could have produced a student like Andrei Martynov (brother of the celebrated actor Alexandre Astafievich, a born talent). Arithmetic was a special problem for him. Multiplication tables were a mine of wisdom, and normally after questioning—how much is 5×5 or 7×7—to which he answered unimaginable rubbish, the teacher unceremoniously asked him which he preferred—a ruler across the hands or to go without lunch—and Martynov usually chose the first, received a helping of blows across the hands and returned to his seat.

Indeed, many of his colleagues would have made the same choice, because, although they didn't feed us well at all, it was difficult to sit through the afternoon without lunch. Our young appetite was the best

* Kulikov died in St Petersburg in 1890 at the age of 79. He was in general a talented person, having begun his career as a cellist in the Moscow theatres, and later transferred to the drama company; then he was transferred to Petersburg as an actor, and subsequently was made régisseur. He wrote or translated more than 130 plays for the Russian stage. His memoirs, which are not without interest, were published in *Russkaya starina* last year [= 1892]—footnote of A. V.

seasoning of our scanty menu. I well recall those slices of meat, thin as paper, with a rather bad potato sauce—which was considered a special dish. Appetite sometimes compelled bolder students to make raids on the kitchen with pillage in mind. Among the school legends passed from generation to generation is one about a student (whom in fact I don't recall) making such a raid. He chatted up the cook, as they say, and taking advantage of his distraction, stole from a dish several sausages being prepared for lunch. Not realizing that the links weren't cut apart, he began to drag a whole garland of sausages behind him as he made his escape from the kitchen. This circumstance spelled his end. Fearing he would be relieved of his booty, he was reluctant to let go the sausage, was overtaken by the cook, and, of course, he paid cruelly for his bold foray.

Of my colleagues, these were my closest friends: Pyotr Ramázanov,* Isakov, Lavrov, Volkov, Svishchev, and the aforementioned Martynov. In this company, it seems, I was punished for the first time by being deprived of a holiday: having received permission to have tea in the bathroom (we were especially flattered by the fact that we could have a party by ourselves, even in the bathroom), we took it into our heads to greet the new year there and meticulously slaked our thirst with a contraband bottle of rum. The noise and uproar with which we came into the dormitory enabled Mikhelson, the tutor on duty, to observe our abnormal condition. This scared us frightfully, as we feared that he would report us to inspector Aubel—the terror of the boy students. But we managed to mollify the tutor, who, instead of reporting our misdemeanour to the director, deprived us of our holiday on his own authority. Lavrov and I still felt the effects of our binge the next day. We both had to sing at liturgy, during which Lavrov felt so poorly that he fainted and had to be carried from the church.

II

Speaking of my vocal studies, I am obliged to point out that I cannot read music to this day, although the whole time I was at school I was a choirboy in the school church choir and even sang solo parts. My musical ear, excellently developed by nature, consistently helped me. Thanks to it, as I shall relate presently, I even sang on the professional stage.

At the time of my appointment to the school, Lachouc had already begun to teach me dancing, then Frédéric Malaverne, and finally the famous Emile Gredlu. I consider myself much indebted to Gredlu, as he

---

* Son of A. N. Ramazanov, who was graduated on to the stage in 1813, a talented actor in servants' roles in vaudeville and comic opera, an excellent Figaro, and who died in 1828. Besides Pyotr he had a son Nikolai, a professor of sculpture, who was born in 1815 and died in 1867; he was creator of the bas-reliefs on the memorial to Tsar Nikolai and author of the book, *Materials for the History of Art in Russia*—Footnote of A. V.

was first to see the comic dancer in me, and gave me the proper direction. But I must say nevertheless that at times my power of observation and ability to imitate also promoted my success on stage. Seeing what others do I tried to do the same, and practising, soon copied them, sometimes quite successfully. For example, it being my 'day off', I saw in the Mikhailovsky Theatre during a performance of the vaudeville *Les trois bals*, how smartly French artists danced the cancan. After that, during a Christmas masquerade at school, I started to cancan, with agility and good technique, during the dances. Such an unusual dance for a school evening quickly attracted a lot of attention. As I was in costume and had a pasted-on nose, no one recognized me, then one of our supervisors, having called me over and recognizing who I was, asked me with some curiosity where I learned *that?*—to which I responded directly, where and under what circumstances. The authorities were quite satisfied with my answer. At that time, however, I was still so young and inexperienced that I didn't realize I was doing something reprehensible and unbecoming by dancing the cancan. Copying the artists I had seen and practising, I developed my agility to acrobatic perfection. For example, I could tap my head with my foot. My teacher of classical dances, the aforementioned Gredlu, always praised this litheness in me. It was all the more important since I soon had to use my abilities on the stage, to apply them in my day-to-day work, as I was to appear before the public very young. Not long after entering school I took part in the ballet *La Gitana*, in which Marie Taglioni danced, a first-class celebrity of the day, and daughter of the ballet's author.* I do not, however, retain a clear and distinctive recollection of this performance. I remember only that the harlequinade in this ballet, produced on stage, represented a small *balagan* theatre of the fairs, on the balcony of which I played Cassandre, Colombine's father, in a pantomime. After that, when I was eight or nine years old, I played (but only at rehearsal, replacing someone who was ill) in *The Captive of the Caucasus*, Didelot's beautiful ballet, written after Pushkin.†

I first performed a speaking role at the age of ten, that of the merchant's son in the vaudeville, *The Excursion to Tsarkoe Selo*. The road to Tsarskoe Selo was still new then and very fashionable, and this play was, so to speak, apropos. My role in Kulikov's vaudeville, *Actors Among Themselves*,

---

* Marie Taglioni was born in Stockholm in 1807. From 1827 she danced in Paris, from 1832 in Berlin, and the end of the 1830s in Petersburg with a huge success. Having married Count Gilbert de Voisins she left the stage forever—footnote of A. V.

† Charles Frantsevich Didelot was balletmaster of the Petersburg Imperial Theatres and a teacher in the theatre school. His talent and system were very beneficial to our ballet. He was born in 1767 and died in 1837. He composed many ballets: *Alceste, The Caliph of Baghdad, Phaedra, Laura and Henry, Cora and Alonzo*, etc., etc. In the notes to *Evgenii Onegin* Pushkin writes of him: 'Didelot's ballets were filled with lively imagination and unusual charm.' See in *Onegin* Stanzas 18 and 21, Chapter 1, about Didelot—footnote of A. V.

from about the same time, also had spoken lines. Kulikov himself staged this play, as too the following one, *The Cossack Poet* (by Shakhovskoy, with music by Cavos), the performers of which were all boys and girls of the theatre school. We learned our singing parts in this work by ear, with Kulikov at the violin. Both vaudevilles were first produced on the stage of the school, and then *The Cossack* was passed for production at the Alexandrinsky Theatre, where it succeeded well enough to be given several times, and was even requested by benefit artists for performances of the French and German drama troupes. According to regulations of that time, a benefit artist was allowed to vary a performance as desired, to make it more original and more interesting.

In one divertissement from the period being described, an 'English Dance' was produced in which the celebrated artists of the time, Fleury and Didier, performed.*

My teacher Gredlu produced this dance for a school performance in which I had to impersonate a lady, and for the first time I warranted the approbation of my patron Nevakhovich. Some time later, on the same stage, I performed the comic dance 'Polichinelle', again with great success. Dressed in a multi-coloured costume and with humps back and front, I imitated a polichinelle, my movements seemingly caused by little ropes attached to my arms and legs. I produced such an effect with this dance that I was called for several times despite rules governing school performances which prohibit this. At that time I also participated in a French performance, as one of the students in the play translated into Russian by Pyotr Karatygin under the title *The Schoolteacher*, subsequently performed on the Russian stage.

III

On 26 April 1845 I turned sixteen. With this important day for me, I was, as is noted in my record of service, 'placed into the ballet troupe as a dancer'; and from this day is counted, according to theatre regulations, my active service to the theatre. In the second half of this year Gredlu produced for students on the stage of the school theatre the one-act comic ballet *The Millers*, in which I was given the principal role of Sautinet (on the large stage Fleury performed this role). In this role I was not so much to dance as to combine comic pantomime with dizzying acrobatics. Our performance of *The Millers* so pleased Director of Theatres Alexandre Mikhailovich Gedeonov that he decided to repeat it on the large stage, and on 4 October *The Millers* was given in the Mikhailovsky Theatre. This performance was my professional début in a responsible role. I am

---
* Pyotr Ivanovich Didier, a dancer-comic, French by origin, was born in 1799. Fleury, a dancer-comic as well, also French, was in Petersburg from 1831—footnote of A. V.

not my own judge, but according to some data it could be declared that the public liked my performance. In one of the newspapers there was even a remark, in the notice of this ballet, concerning my agility: 'Stukolkin has no bones.'

Speaking of pieces performed at that time exclusively by young people and which pleased the Petersburgers, I cannot omit mention of the small but merry ballet *The Simpletons* (*Les trois innocences*), in which three ladies and three cavaliers danced, I among them. Besides these named roles, at the end of the same year I danced again in the Alexandrinsky Theatre, an 'English Dance' with the student Reinshausen.

The following year (1846), it seems, the state acquired the Petersburg circus of Leonard and Paul Cuzent, and the management of the Imperial Theatres, in order to provide its own circus artists, proposed to students of the theatre school who wished, to train to appear in the state's circus arena.*

I decided to test my abilities in this new field, and prepared so earnestly for it that under the supervision of Paul Cuzent I was soon riding while standing on horseback and performing various *finesses* of the circus art. But after once falling from a horse and striking the barrier hard, and having lain two weeks in the infirmary covered with compresses of cantherides, I flatly rejected horsemanship. Cuzent was very saddened by my decision, since he thought I had shown great progress. I remember to this day the name of the horse on which I often rode. His name was 'Fritz', and belonged to the celebrated rider of that time, Laura Bassin, who made a good match, marrying into one of the most aristocratic families.

Paul Cuzent repeatedly tried to persuade me to stay with the circus, assuring me that my build was suited to the profession and that my legs were long enough for the 'post'—a routine calling for the rider to stand on several horses running at once. Before I fell off the horse, it was going so well that after one of the trial performances of the circus (one without a paying public), the Minister of the Imperial Court, Prince Volkonsky, was so pleased that he requested the performance be repeated before a regular audience, and it came off with great success. Our trick-riding was especially liked.

Others worked in the state circus at the same time as I did: Anna Petrovna Natarova (subsequently an esteemed artist of the Alexandrinsky Theatre, who retired about two years ago), Alexandra Matveyevna Chitau,

---

* The state circus was located in the building of the theatre-circus constructed in 1847 by the court architect A. K. Cavos. In 1859 this building burned down, and, after reconstruction by the same architect, was reopened in 1860 and renamed the Maryinsky Theatre. In later times, from 1885, a number of alterations and improvements were made by the architect V. A. Schroeter—footnote of A. V.

Strekalova, Fyodorova and E. N. Vasilieva, who also later worked successfully on the dramatic stage, and others. Natarova, I recall, suffered a misfortune similar to mine. She was released from the company and placed in the state theatre after twice falling from a horse and twice breaking her arm.

Students training for the circus enjoyed the right of free admission to it. One fine evening I happened to be there under these auspices, and with the other spectators was saddened that a public favourite, the artist L. Viol, could not perform in the second part of the programme due to a sudden indisposition. Having returned to the school and sitting down to my modest dinner, I was suddenly informed that Gedeonov required my presence immediately at the Alexandrinsky Theatre, where the evening performance was still going on. As the Director of Theatres was a very elevated personage compared to me, a lowly student of the school, I left for the theatre not a little anxious at the impending meeting with the chief. I had just walked into the theatre when I met the actor Maximov I,* who begged and prevailed on me to play, the next day for his benefit performance, the role of Jocko (the monkey) in the play *The Monkey and the Suitor*— as a stand-in for Viol, who, as it turned out, had broken his collar-bone during the performance at the circus. Despite all my excuses, Alexei Mikhailovich insisted that I honour his request. I must add that his benefit performance had been postponed several times due to unforeseen circumstances, for which reason Maximov was very upset by the accident which would prevent Viol from appearing in the role of the monkey. At last Gedeonov, who was present, categorically declared that the next day I would be performing the role of the monkey. At that point I risked telling him that I was positively terrified to appear in such a role after the talented and famous gymnast Viol, who had performed Jocko many times in the German Theatre. To this I was told that Viol as Jocko did not really act like a monkey, 'and you will do it better, in a more natural way'.... In that 'good old' time the entire theatre management used the subordinate [second person] 'you', and no one was shocked by this.

Then they gave me a translation of the play (specially translated from German for Maximov's benefit performance by a certain Grigorovich), ordered me to learn it, and to make notes on it. On my way out of the theatre Nevakhovich stopped me, gave me a little money, and told me to come the next day to the Zam menagerie, then located at the corner of Bolshaya Morskaya and Kirpichny Alleyway, to observe the monkeys and see how they play their roles, in their cages if not on stage. Returning

---

* Alexei Mikhailovich Maximov I was born in 1813; the son of a Petersburg artisan, he was educated in the local theatre school. He was noted for his talent as an artist of first roles in the dramatic theatre. In 1861 he died of consumption—footnote of A. V.

to the school I practised a bit, jumping over beds, and then, unable to sleep, read through the play and made the necessary notes for my début in a completely new role. The next day at rehearsal I played the part of the monkey, and apparently satisfied the management; then I went to the menagerie and after that visited Viol, whom I found lying in a bed all bandaged and tied up. He received me very graciously, and since he could not move, he instructed his student Pacifico to show me a number of devices useful in performing the role.

The benefit performance took place that evening. As a play, *The Monkey and the Suitor* did not have any particular merit, and it interested the regular public of the Alexandrinka only because Viol, fashionable at the time, would appear in it, because it was being performed on the Russian stage for the first time, and finally and chiefly because it was Maximov's benefit, and he was a Petersburg favourite. Actually Viol played the monkey masterfully, as a consequence of which I, his inexperienced stand-in, was in a terrible fright waiting for the curtain to go up. I must add that before my entrance I requested it be announced that because of Viol's indispostion his role would be performed by 'student Stukolkin'. But Gedeonov flatly refused to do this, and Viol's name was left on the affiche.

The curtain went up. I went on stage dressed in a monkey suit, excellently made up to boot. In Act I Jocko must leap down from a rather high hillock, a distance of one *sazhen* [2.13 metres]. They did not teach us such *tours de force* in school, and I, as it were, stopped short . . . But applause burst out, and thus approved, crossing myself, I jumped perfectly well and then, scratching with all the grimaces of a real monkey, ran on all fours around the stage. A burst of applause encouraged me, and completely self-possessed I entered into the role and played it through with complete success. At the end of the performance the public left the theatre quite satisfied that it had seen the talented Viol, and the next day the newspapers lavished praise on the *Italian* artist. Only in the little world of the theatre was it known that Viol had been a counterfeit. After two or three performances the play was taken out of the repertoire of the Russian company, and this little story was forgotten.

In the same year I often 'was taken up', that is, I 'played', in theatrical parlance—the following roles: the evil genie in Titus's ballet *The Two Sorceresses*, my previous part in the familiar old *The Millers*, the Scottish peasant Gurn in Taglioni's beautiful and poetical *La Sylphide*, produced for the artists of the Moscow Theatre Sankovskaya and Montague. Finally, I danced in divertissements at the French and German theatres, as well as a 'comic *pas*' and '*pas chinois*', performed several times with my regular partner, the student Reinshausen.

Stukolkin's Recollections

In 1847, a new role was the gnome in the *féerie*, or, as the affiche of the time announced it, in the Grand Fantastic Operetta-Vaudeville: *Some Great Pills; Every Time I Take One, Thanks but No Thanks*. This piece was also produced for Maximov's benefit performance.

At the beginning of the next year (1848), I was made, for distinction, a *state boarder*, that is, one having room and board in the state quarters of the school. I took my own clothes, enjoyed complete freedom and received 10 roubles a month. In March I was finally graduated, 'set at liberty' with a salary of 240 roubles in notes per year, which was less than a small remuneration for such laborious work as one had to perform at the time being described. Thus, irrespective of *Pills* and *The Millers*, in which I played ten times in the course of the year, sometimes I even had to take part even in tableaux vivants, which in those days were produced during the Lenten concerts given by the Imperial Theatres. Some of the tableaux were very successful and effective, thanks to the gifted decorator Roller's participation in their production and construction.* The measure of his talent is clearly reflected in the fact that the management is still using his works. Of those I shall name, for example, the remarkably beautiful background decoration to Act III of the ballet *Catarina*, which represents a view of Rome with a view of St Peter's basilica during carnival illumination.

After that I also danced in many divertissements and in Dauberval's small but graceful, poetic and comprehensible ballet *La Fille mal gardée*, or, as we more often call it by the names of its characters, *Lise and Colin*, in which I performed the comic role of Nicaise while the part of Lise was taken by the European celebrity, Fanny Elssler,† who made her début before our balletomanes on 10 October 1848 in the ballet *Giselle*.

La Fille mal gardée was particularly memorable to me for an extraordinary occurrence which happened several years later and could have affected my career very sadly. Once Donizetti's opera *La Fille du Régiment* was being given, to be followed by *La Fille mal gardée*. I do not remember where or why, but I had had a little too much at dinner, and feeling somewhat *au bon courage*, going to the theatre I prayed Lord God that

---

* The Bavarian Andrei Adamovich Roller was born in 1805, died and was buried in St Petersburg in 1891. In 1834 he entered the service of the Imperial Theatres as a decorator and machinist; in 1839 he was awarded the title of academician by the Academy of Art; in 1856 he was made professor of perspective painting, and he retired in 1879. He drew decorations, incidentally, for the following operas: *Robert and Bertram*, *A Life for the Tsar* (the moving groups of cardboard people along the rear wing in the last act of this opera are his invention and work), *Ruslan and Lyudmila*, *The Bronze Horse*, *L'Africaine*, *La Juive*, *Charles the Bold* [i.e, *William Tell*], *I Puritani*, etc., and for the following ballets: *La Sylphide*, *The Daughter of the Danube*, *La Gitana*, *Giselle*, *Paquita*, *Satanilla*, *Faust*, *Armide*, *Trilby*, *Le Corsaire*, *The Golden Fish*, and many others—footnote of A. V.

† F. Elssler, a star of the first magnitude in the choreographic heavens. She was born in 1810. She had a sister, Thérèse, also a dancer, but less well known for her art than her morganatic marriage to Prince Adalbert of Prussia, which gave her the title and surname Baroness Barnim—footnote of A. V.

Emperor Nikolai Pavlovich would not be present. Frightened at what might happen, I consoled myself that as long as the opera was going on I could go to the buffet, have a cup of black coffee, and put myself somewhat to rights. As luck would have it, I met a hussar I knew in the buffet, with whom, unable to restrain myself, I drank another glass or two of wine, which did precious little to improve my condition. Then, getting ready in the dressing-room, to my horror I heard one of my colleagues announce that the tsar had arrived at the theatre! Be that as it may, when my time came I went on stage and performed, but I won't tell about how I felt. Appearing backstage after the interval from my dressing-room, I realized to my great despair that the emperor was on stage; he complimented me (my comic manner always entertained him), and learning that I was in my dressing-room, asked Didier, who was playing the part of the old woman, if I were not his student, to which Didier answered that I was not his, but Gredlu's. At that point the emperor noted that I was in especially good form this evening. By then I had become quite sober, out of fear and joy.

The fashion for Fanny Elssler caused the appearance on our stage in that year of the original *vaudeville à propos*, *The False Fanny Elssler, or the Ball and the Concert*, in which I took part on several occasions. On 21 December Elssler's benefit performance was given, for which she chose Perrot's beautiful ballet *Esmeralda*, then new.* Elssler took an active part in the production of this ballet, demonstrating to practically every artist how to perform his or her role. Elssler's greatness as a mime was proved by the fact that she—a beautiful young woman—when showing Didier how to play Quasimodo (a deformed cripple, don't forget), brought those present to tears. Impersonating him, she kisses the footprints of Esmeralda whom he passionately loves, and conscious of his ugliness, he conceals his burning passion, suffering and despairing unbearably. Her every movement and gesture was full of truth and the most realistic life and drama. . . . How marvellously she acted! The author of the ballet, making his début, played the principal male role of Pierre Gringoire. I played one of the beggars. The assignment of such an insignificant, inconspicuous role to me, who at that time was already enjoying some celebrity—yes, and my colleagues as well—showed that our new balletmaster Perrot did not like Russian artists in general, and outstanding ones in particular, and tried, as they say, to keep them down, not to give them an opportunity. Only after eight years (in 1856) was I given a more responsible role in

---

* The balletmaster Jules Perrot was born in 1800 [sic] and died in 1892 in Paris, where from 1828 he was balletmaster at the Grand Opéra. Having spent 5 years in Petersburg as balletmaster, he produced several of his ballets: *Esmeralda*, *Délire d'un artist*, *Faust*, *Caterina*, *Le Corsaire*, *The Naiad and the Fisherman*, and others. Perrot was married to the celebrated ballerina Carlotta Grisi—footnote of A. V.

Esmeralda—that of Clopin Trouillefou, the leader of the beggars. At first Christian Johanson was Gringoire, but when Perrot arrived he took this role for himself and played it in his début before the Petersburg public. Several years after he left I asked Johanson, to whom in all fairness the role in question should be returned, to let me take it, which he did in a comradely way. Not wishing to go on about myself, I will say only that my performance in this role pleased the public; that I did not spoil the part of Gringoire, to which it may be added that I had to perform it many times later and with several ballerinas. Thus in *Esmeralda* I performed three roles in all. Several years after that I performed in this ballet again. P. S. Fyodorov was in charge of repertoire at that time.* In the interval, coming on stage, he reprimanded me for not having shaven, saying that I was showing 'disrespect for the public' . . . Only with difficulty did I explain to the supervisor that my appearance was deliberate, that I was made up according to the concept of my part. Gringoire was a poor student, dressed in rags [in Act I], but in the second act I would be made up differently and enter shaven.

'Ah, so this was make-up!' was the only thing the stern supervisor could think of to say.†

At about this time a completely unexpected storm burst over me. During the production of *Les Huguenots*, at that time new, I and my friend Alexandre Nikolaevich Picheau were put into the corps de ballet. Both of us, as artists who performed responsible roles, considered this beneath our station, for which reason we reported ourselves ill, did not appear in the opera, and went instead to the Bolshoy Theatre where a ballet was being given. There, as fate would have it, we ran into Gedeonov, who, remembering that we were reported ill and seeing us on stage, got frightfully angry, reprimanded us in the strongest terms, called us 'puerile children', and ordered the régisseur Marcel, as a punishment, to place us in the opera corps de ballet. We did this for some time, and danced in the corps de ballet in *Les Huguenots*, but then the management softened its stance and began to use us in accordance with our talents and artistry. By

---

* Pavel Stepanovich Fyodorov was born in 1800 and died in 1879. He was supervisor in charge of repertoire and the theatre school. He wrote and translated many vaudevilles, which were mostly produced on the stage of the Alexandrinsky Theatre.

† The story of Victor Hugo's novel, *Notre Dame de Paris*, full of profound meaning, poetry and drama, was used by composers as well as balletmasters like Perrot. Thus on 18 November 1859 on the stage of the Alexandrinsky Theatre, for the benefit performance of A. A. Bulakhova (born Lavrova), Dargomyzhsky's 4-act opera *Esmeralda* was given for the first time, in which the benefit artist performed the title-role. Phoebus was played by Bulakhov, the syndic by Petrov, Quasimodo by Vasiliev I, Fleur-de-Lys (Phoebus's fiancée) by Lileyeva, Chevrèz (his friend) by Gumbik and the leader of the gypsies by Matveyev. Ten years later on the Italian stage the 4-act opera of the same name by Campano was given. The singer Trebelli-Bettini used it for her benefit performance on 18 December 1869, with the following personnel: Phoebus—Bettini, Claude Frollo (archdeacon)—Graziani, Tristan Hermit—Meo, Esmeralda—Volpini, and Estella—the benefit artist—footnote of A. V.

the way, concerning Marcel. Despite his foreign surname he was in language and character purely Russian, having begun his career as a coryphée or a dancer in the corps de ballet, and later having been made ballet régisseur. He spent a very long time in this post, which made him equally sympathetic to senior and junior artists. Not blessed with brilliant gifts as a dancer, he was most beneficial as a régisseur. Having learned the affairs of our ballet from years of practice, he personified the exceptional ballet régisseur. In addition, because he had an excellent memory, he could always show how a ballet was performed or staged, or some *pas* or variation, etc., and in case the balletmaster was absent he could always substitute for him successfully. In short, as régisseur Marcel was positively invaluable, and I attribute this solely to the fact that he was a ballet artist himself. In later times we have had as ballet régisseur such anomalies as actors of the drama theatre—and worse yet, of the German theatre. I think any layman, someone outside the theatre milieu, would understand as much as such a régisseur about what is useful to our ballet. One cannot help adding to the list of the late Marcel's important traits his ability not just to be a good colleague, but also to unify, to rally the community of artists—to ensure freedom from that dissension always injurious to people and affairs.

IV

On 4 February of the following year (1849) Perrot's ballet *Catarina, Daughter of the Robbers* (music by Pugni) was produced for the first time, for Fanny Elssler. She danced in the title-role, as usual with huge success, and I performed two insignificant roles: a bandit and a soldier.

Speaking of this ballet, I consider it relevant to say a few words about its composer. Born in 1812, Italian by birth, Cesare Pugni was a talented and productive composer who wrote music for nearly 300 ballets. He worked for the management at a salary, not receiving an honorarium for each work, as 'establishment' composers of the theatre do now. He had to write music for a number of ballets each year, as specified by the management. Several of them are quite beautiful and melodious. Despite his industry, Pugni died in 1869 [that is, 1870] in utter poverty, without having provided for his family, for whom a mixed benefit performance was given in May of 1870, in which I participated.

Besides *Catarina*, *La Fille mal gardée*, and *Esmeralda*, which brought good houses and for that reason were often performed in the period I am describing, I was again employed, as before, in tableaux vivants produced at Lenten concerts. In those days they were very beautiful, as it seems I have already mentioned, and even our major stars appeared in them. Thus, for example, in a concert given by Roller on 27 December [*sic*],

Fanny Elssler and the balletmaster Perrot were shown off. After that I danced several times again in divertissements and in benefit performances: the 'Sabotière', the 'Redowa' with the German artiste Graff, the 'Marie polka' with the French artiste Esther, a Cossack Dance with Pavlova, etc.

I seem to recall it was the beginning of 1851 when Emperor Nikolai Pavlovich, as a surprise for the Empress Alexandra Feodorovna, commanded the five best pairs of dancers from the Warsaw ballet to come and perform the so-called 'blue mazurka', which took its name from the colour of its performers' costumes. These dancers were: Kvitovsky, Poppel, Menier, Maevsky, and Gillert; of the ladies I remember the surnames of only two: Kotlyarovskaya and Dams. At the performance of some Italian opera—I don't remember which one—the curtain was raised during an interval and, unexpected by the public (nothing of this was indicated on the affiche), our guests made their début with great success. In fact they performed their national dance in a masterful way. The blue mazurka so pleased the emperor that he ordered it repeated as follows: with four Polish men to dance with Russian ladies, and four Polish women to dance with Russian cavaliers. The mazurka's success in this mixed staffing exceeded all expectations. Our performance of this remarkably beautiful and effective dance so pleased and electrified the public that clapping broke out before our entrance, when still from the wings the metallic clink and rattle of our *skobki* [spurs] were heard. When pair after pair of us flew out on stage, the roar of applause swelled to a fortissimo. Of the Russian artists besides myself who danced in this mazurka, I remember Alexandre Picheau and Shamburgsky.

After that the Poles performed a small ballet by Steffani, *The Peasant Wedding*, the corps de ballet of which, consisting of seven cavaliers, was Russian. This ballet also enjoyed great success after the Poles left and all the artists were Russian. The Poles showed us that the mazurka must be performed in [two] completely different ways: either nobly, elegantly, in the manner of Polish gentry, with softer, more elegant movements, or with fire, stamping in metal-trimmed boots and throwing hats—that is, in a peasant-like manner. Unfortunately these distinctions in the performance of the mazurka are rarely observed now, even among specialists . . .

Subsequently, at Gedeonov's order, I performed, besides the mazurka in *The Peasant Wedding*, the role of the suitor, opposite Picheau as the matchmaker. Then Gedeonov passed on to us what the emperor said: 'Our dancers perform it no worse than the Poles'. . . after which we both received a per-performance stipend of 3 roubles, assigned by the Director of Theatres himself without our requesting it. In general my situation on the stage became noticeably better in a material sense at this time. My per-performance fee was later increased to 5 roubles. At that time pay rises were quite insignificant, and our salary itself, as I said already, was incomparably more miserly than now, for which reason we set great store on these small increases and were indescribably happy with them.

A few more words about the blue mazurka. The following story, sharply etched on my memory, shows how much it pleased Nikolai Pavlovich. Once in Peterhof, I don't remember the occasion, a gala performance was given on Olga Island in the open-air summer theatre. *The Naiad and the Fisherman* was being performed. In the interval the emperor paid a call, and most unexpectedly asked that we dance his favourite mazurka. When he learned that we had not brought the costumes for this dance with us, he commanded us to dance it in the costumes we had. Thus we performed a Polish dance in the costumes of Neapolitan fishermen. Meanwhile I had foreseen his request and amid the preparations for the train ride to Peterhof had said, 'Should we not take our Polish costumes just in case?' but my words were like a voice crying in the wilderness. Fortunately our conductor, Lyadov,* brought along the music. At the end of the year *The Peasant Wedding* was produced and performed in Moscow, for which purpose Shamburgsky and I were sent there. Having successfully fulfilled this commission, we quickly headed for home.

V

On 25 January 1852 I married the daughter of the revered ballet artist N. O. Golts.† Several days after the wedding, at a performance of *Catarina*, the emperor visited the stage. It must be noted that he was not only interested in theatre and was well versed in our business, but also knew many of us by sight and by name, which is explained by the fact that Emperor Nikolai was a great theatre-goer, and rare was the evening he did not spend in the theatre.

---

* Alexandre Nikolaevich Lyadov was a celebrated ballet Kapellmeister; his ballroom orchestra was in great fashion in his time. Alexandre's brother, Konstantin Nikolaevich, born in 1820, studied in the theatre school under Prof Solivier and conducted the orchestra of the Russian opera; in 1868 he turned over his baton to Napravnik, and died the same year.

† Nikolai Osipovich Golts was born in 1800, graduated on to the stage from the Petersburg theatre school in 1822; he died in 1880. A student of Didelot and Prince Shakhovskoy, Golts was a talented dancer, mime, and dancing master. He took part with great success in the ballets *The Captive of the Caucasus*, *The Caliph of Baghdad*, and *The Hungarian Hut* (Didelot), *Medea and Jason* (Noverre), *Raoul Bluebeard* (Valberkh), *The Demented Woman* (Auguste), *Le Déserteur* (Dauberval), *Revolt in the Seraglio* and *The Daughter of the Danube* (Taglioni), *Paquita* (Mazilier), *Satanilla* (St-Georges [and Mazilier]), *Fiametta* and *The Humpbacked Horse* (St-Léon), and many others. Golts staged the dances in the operas *Ruslan and Lyudmila*, *Rusalka*, *A Life for the Tsar*, *Mazepa*, (of Baron Fitinhof), etc. Nikolai Osipovich enjoyed a brilliant reputation as a teacher. Among his students one may name the late Tsarevich Nikolai and the composer Glinka (see *The Memoirs of M. I. Glinka*, publ. 1887 by Suvorin). Golt was married to N. I. Valberkh, the younger daughter of the first, indeed practically the only Russian balletmaster, Ivan Ivanovich Valberkh (born in 1766, died in 1820)—footnote of A. V.

Having encountered our exalted guest backstage, I was taken back by our sovereign's gentle words: 'I am at fault; I did not congratulate you . . . I heard that you were married. I congratulate you on your marriage' . . .

The emperor knew me personally as a comic, and to the same degree he approved my comic talent it dissatisfied the balletmaster Perrot, who, as I already mentioned, did not sympathize with talented Russian artists at all and for that reason assigned me only insignificant roles. Perrot was very displeased once, after *La Fille mal gardée*, when we were told that the emperor had just said: 'Today Stukolkin made me laugh to distraction!'

We artists should remember Emperor Nikolai with special reverence; he did us all so much kindness, and treated us as his children. For that very reason I keep as a holy reliquary his helmet, glove, and a plaster cast of his hand, given to me by my good friend, retired officer of the Life-Guards Mounted Regiment, the late A. A. Galakhov, to whom these objects passed from his father, the adjutant general, formerly the Chief of Police in Petersburg, who died, I think, in the 1860s.

This year (1852) was also memorable because it was, so to speak, the first year that I taught in various educational establishments. Thus, for example, from this time I gave lessons in the Pension Thibaut; then the number of such institutions quickly began to grow, as I was invited to the Regiment of the Nobility (now the Second Military Konstantinovsky School), the Corps of Pages and the Second Cadet Corps, the Imperial Alexandrovsky Lyceum (from 1859), and to the Patriotic Institute, where for my lessons I was awarded, by the Empress Alexandra Feodorovna in 1854, my first imperial gift—a gold snuffbox, and in 1892 a still more precious award—the Order of St Stanislav, Third Degree; to the Smolnyi (from 1880), the Elizavetinsky and Ekaterininsky Institutes; to the pensions and gymnasia Meyer, Aimée, Lyalin (which prepared for entrance into the lyceum), Kuznetsov, Werther, Zalivkina, where I taught until 1891 when this establishment, after a seventy-five-year existence, closed; to the Stoyunina and many others, which I have forgotten because so much time has passed. I now very much regret that I never kept diaries, and never wrote anything down. Beginning my artistic and teaching activities, I never thought that much of what I saw, heard, and lived through could be interesting to the public . . .

To an outside observer, this list of educational institutions may not reveal much, but I am reminded of how frightening it must have been, before these institutions were established, for a young dancing master and his family to make ends meet on 400 roubles per year.

Leaving that part of my activity, I shall say only that I have continued to give many of these lessons to the present day, as I still teach dancing in the institutes named (except the Ekaterininsky), in the Lyceum, with the Pages, at M. N. Stoyunina, etc. etc.

In 1853 I performed in *Catarina* when the title-role was taken by Carlotta Grisi, in Perrot's ballet (with Pugni's music) *The Battle of the Women, or The Amazons of the Nineteenth* [that is, Ninth] *Century*, as one of Mitsislas's friends, in *The Peasant's Wedding* (the role of the organist's son, Stanislav), and in the minor role of the sergeant in Didelot's *The Hungarian Hut, or The Famous Exiles* (music of Benoist), an elegant, venerable ballet of sentiment revived for Andreanova's benefit performance. I had previously been assigned another fine comic role, but Perrot, seeing its merits in my performance, took it for himself. Of Elena Ivanovna Andreanova I shall say that she, a very fine mime, had such an unbeautiful exterior that some punster at the time of her début said, that our 'cholera is past, but what has appeared is ... a mug!' [a play on the word *rozha*, which means both 'ugly face' and 'erysipelas'].

Her looks, however, did not keep her from having an admirer and protector with great influence in the person of Gedeonov. As *The Hungarian Hut* was very old and God knows how long it had been since it was given on our stage, the management was sensible enough to consult old-timers on the matter of its production—inviting back Dyurova, Shemaeva, and Artemiev, already long in retirement. This circumstance unwittingly reminds me of a revival of *Esmeralda* for a visiting ballerina in the 1850s. Indeed, because the management did not turn to old artists for help with this revival, the new production was distorted almost beyond recognition to people who had seen it when our ballet was flourishing ...

After that, among the works in which I played I shall name: the excellent two-act ballet of Mazilier (music of Adam* and Pugni) *The Wilful Wife* [*Le Diable à quatre*], my role being the good genie in the guise of a blind musician, and then, with Perrot's departure, the dancing master and the basket-maker; thus I danced in this beautiful and original ballet continuously to the end of the 1880s, when it was removed from the repertoire; *Paquita* (music of Deldevez), the old *La Fille mal gardée*, and, finally, the opera *A Life for the Tsar* (in the mazurka). In addition, I usually had to participate in several divertissements, for example, dancing the 'Hungarian Dance' with Shamburgsky, and the 'English' dance with Picheau, etc.

---

* The French composer Adolphe Charles Adam was born in 1803 and died in 1856: he was a student at the Paris Conservatoire and wrote the operas *Le Postillon de Longjumeau*, *Giralda ou la nouvelle Psyché*, *Pierre et Catherine*, *Le Brasseeur de Preston*, *Le Toréador*, and others, and also of several ballets, for example *Giselle*, *The Wilful Wife*, etc.—footnote of A. V.

VI

Then came the sad year of 1854. The war had little effect on our theatre world. Performances went on in the normal way, and on 2 February, for Perrot's benefit performance, in his own beautiful ballet *Faust* (music of Panizza and Pugni), I performed the role of a young peasant, Martha's suitor Peters. This beautiful and poetic, very well-staged ballet pleased the public, assisted by talented execution, of course, but also by the art of its author, who did not depart from Goethe's immortal work, differing from it only in details; it stayed in our repertoire for a quarter of a century. The next benefit performance of this year was Perrot's *Marcobomba*, given for Marius Petipa; in this ballet I played the peasant Lopez.

Fearing to bore the reader, however, I shall not list all the performances in which I took part, but recall only new productions, performances for some reason prominent, or works in which I took a new role.

In the spring, as was our custom, my family and I left for our dacha at Strelna. The Anglo-French naval squadron, which plied the Baltic Sea and the Gulf of Finland at that time, reminded us of its presence by muffled cannon-fire from a distance, apart from which we lived our summer life as usual. Returning to the city in the autumn, I began the season on 27 October at the request of the dramatic actress Linskaya,* by dancing the 'English' dance with Picheau in the divertissement of her benefit performance. V. V. Samoilov sang his 'Finnish Comic Ballad' (or 'The Finn's Lament', as the affiche had it) at the same performance with great success. These couplets, comically describing the sorrow and mishaps of a Finnish fisherman pillaged by a vessel of the Anglo-French squadron, cleverly referred to the concerns of the day and were very apropos. Samoilov enjoyed great success with it in various theatres.

In the same year I figured again in a performance at the Alexandrinka, dancing in the drama by V. Ducange and Dineau, *Thirty Years, or The Life of a Gambler*, and finally, I performed in yet more divertissements and tableaux vivants.

In 1855 we lost our benefactor, the Emperor Nikolai. Because of the general mourning we worked less than normal, and took holiday until August. Therefore my only new piece at that time was the ball from Auber's opera *Gustave*, in which I danced in a galop made up of forty-eight pairs.

1856 also brought me something new: the dances in the drama, *Salvatore Rosa*, translated from the French and produced on 27 January in the Alexandrinsky Theatre (I performed a dance of a polichinelle in the

---

* A talented artist in comic old women's roles in plays about mores, Julia Nikolaevna Linskaya died in poverty in 1871.

carnival scene). A significant event in the history of Russian ballet occurred at the beginning of the same year: N. K. Bogdanova,* a Russian ballerina, first dancer of the Paris Grand Opéra, made her [Petersburg] début with great success in the role of *Giselle*. The season being described was marked by the production of two pieces still performed: Dargomyzhsky's opera *Rusalka* and the comedy of Sukhovo–Kobylin, *Krechinsky's Wedding*.† As I was not present at the first performances of either piece, I can say nothing about how they were performed. In divertissements this season I danced all the same things: 'Sabotière', polichinelle, and a Jewish Dance, in most cases with the ever-present Alexandre Picheau.

In August we left for Moscow for the coronation of the late Emperor Alexandre Nikolaevich. The ballet troupe was supplied with a special house, in which we were located downstairs, and our ladies upstairs. Thanks to the kindness of the Moscow theatre management we were offered, on the day of the coronation, tickets for the numbered seats put up in the Kremlin. Not wishing to miss a rare opportunity to see the tsar's entrance, several colleagues and I, men and women, headed for the place of celebration. When we reached it, however, we saw that we were separated from our gallery by a continuous, tightly-packed crowd, through which we could barely make our way. As it turned out, one of us had a friend there, a uniformed gendarme on whose advice and with whose co-operation we, holding on to the tail of his horse, managed to reach our places, from which we had an excellent view of the whole ceremony. But as the magnificent is but one step from the absurd, we laughed heartily at the sight of a company of cavaliers and ladies, passing through the Kremlin walls at just the time of a cannon salute they were not expecting. The fine company ran frightened in all directions, although some of them could think of no better way of escaping the cannon-fire than to lie down on the ground . . .

While nearly all the best artists were taken to Moscow, they had but one piece to dance, as I recall—*La Fille mal gardée*—and then in a benefit for the dramatic actor Nemchinov, a relative of my comrade in the ballet, A. Dm. Chistyakov. In this benefit Shamburin, Parkacheva, and I danced the *pas de trois* from *The Peasant Wedding*, of which I remember, as a curiosity of backstage life at that time, that in the wardrobe there were no

---

* Nadezhda Konstantinovna Bogdanova was born 1836. Her father was a dancer of secondary importance, a coryphée in the Moscow ballet company. She studied with Didelot [sic] and St-Léon. From 1851 to 1855 she danced in Paris, producing an impact with her dancing. She retired in 1863. She is well to this day, living in Petersburg.

† This opera was produced for the first time on 4 May with the following personnel: the prince—Bulakhov, the princess—Leonova, Olga—Lileyeva, the miller—Petrov, and Natasha—Bulakhova. *Krechinsky's Wedding* was given for the first time on 7 May for the benefit performance of the now deceased Budrin.

boots for me except those worn by one of the other dancers, so he had to take them off, and was then left barefoot for a time.

Parkacheva later gained a certain celebrity not only because she played one of the seven deadly sins—Avarice—in the ballet *Faust*, but also for her romantic participation in the famous Moscow trial of Myasnikov, after which she married him for love, in the prison church, and may even have followed him into exile.

The following year, 1857, was as poor in new roles for me as 1858 was in bringing much that was new. In the earlier year I only danced the 'Sabotière' in divertissements with Bauer and Alexandrov, students at the school, and then the polichinelle again and the comic 'English' dance; the next year I was at work in capital ballets like *Le Corsaire* of St-Georges and Mazilier (in Perrot's production). Thanks to the charming, melodious music of Adam and Pugni, to the story, known to all, based on Lord Byron, and also to the decorations of our best artists Roller, Shishkov, Bocharov and Wagner*—'*Le Corsaire*' much pleased our public and was performed until the 1880s. The mechanical boat, Roller's work, which is wrecked and sinks in a complete illusion, produced a grand effect in this ballet until recent times. *Le Corsaire* was very beautifully and richly staged. We have always been famous for our production, and in this ballet the pasha's costume, which has its own little history, stood out for its special magnificence. It was originally made not for the ballet but by command of and for the Emperor Nikolai, for one of his masquerade balls at court. Subsequently, during the production of *Le Corsaire*, the emperor ordered this costume to be transferred to the theatre wardrobe. There, it seems, this magnificently embroidered robe, with braiding and gold, survives to the present day. In *Le Corsaire* I was assigned the small comic role of the overseer of the harem.

Besides *Le Corsaire*, the following ballets were produced the same year: *Armida* of Perrot to Pugni's music, in which my role was the knight Gaston; the same authors' rather pompous *Eoline, or The Dryad*, in which my little role was the miner Franz; Petipa's *A Regency Marriage*, to music by Pugni, in which I played Sylph, the dancing master; and finally, a work I well remember (as I shall explain presently), *Robert and Bertram* (*Dwaj złodzieje*), by the Berlin balletmaster Augué to music by Schmidt and Pugni. *The Two Thieves*, as it was also called, was given here for the first time on 29 March for the benefit performance of Felix Ivanovich

---

* Anton Yakovlevich Wagner was born in 1810 and died in 1885. From 1855 he was a decorator in the Imperial Petersburg Theatres. Selected by the Academy of Arts in 1873 as an honorary, free member, he worked on decorations for the opera *L'Africaine* and on the ballets *The Daughter of the Pharaoh*, *Le Corsaire*, *The Golden Fish*, *Catarina*, and others.

Kshesinsky, who played Robert. I was Bertram. This merry piece always drew crowds, and Kshesinsky and I reigned in it to the 1880s.

Kshesinsky staged *The Two Thieves*, and we learned it very quickly—in hardly a week. *What* this ballet was like and *how* we played it I will not say, but only that it had no role for a ballerina (!) and was always successful. Often the public burst out in Homeric laughter for several minutes on end, so well did we manage to capture its attention with our sometimes improvised comical escapades. I myself was astonished occasionally at the pranks I got away with, and told myself it was due exclusively to the sympathies of the public. Sometimes it went badly for the poor gendarmes who were chasing us (in later times Chistyakov and Svishchev), as we 'thieves' not only made fun of them on stage, but also clambered through the orchestra to escape them, where they could not bring themselves to continue the chase . . .

I must note that the story of *The Two Thieves* was also used by composers, such as Meyerbeer, and by dramatists. Vaudevilles given in the 1860s at the Alexandrinsky Theatre, and by the Germans from the 1850s to the 1870s, were written on the same theme.* I vaguely remember it also from having seen it as a child at the Mikhailovsky Theatre, in an fine performance by French actors, and finally, several years ago, I was at the Cirque Cineselli at a not-at-all-badly produced and performed pantomime on the same subject, adapted of course for presentation at a circus.

VII

For me the novelties of the next year (1859) were: the role of the Marquis Megrèle in Petipa's little ballet *The Parisian Market*; performing the kazachok in Baron Fitinhof's opera *Mazepa*; and a Jewish comic dance, which I danced with Picheau at one of the divertissements.

The most important event of that year for artists was the arrival of the newly engaged first dancer and balletmaster of the Parisian theatres, St-Léon, whose ten-year stay with us was very beneficial for the Petersburg ballet. Thanks to St-Léon, talented dancers who before his arrival had been considered second rate moved up to their proper places: Lyubov Petrovna Radina, Muravieva, and Maria Petrovna Sokolova I. He staged several ballets here, in a majority of which I participated. Of those I shall name (in the year 1860) *Saltarello*, in which I played the tutor of the pages

---

* In the Alexandrinka it was staged for the first time on 27 September 1861 for the benefit performance of E. M. Levkeyeva under the title, *The Vagrants*, a vaudeville in 4 acts, translated from German by O. Boikov. The roles of the two thieves were played by Markovetsky and Alexeyev. At the German Theatre it was entitled: *Robert und Bertram, oder die lustige Wagabunden*, Posse mit Gesang und Tanz in 4 Abtheilungen von G. Rieder, die Musik ist von Tischer. The vagabonds were, in the 1850s, Hornich and Bruning, in the 1860s, Zimmermann and Lobe, and, finally, in the 1870s, the same Zimmermann and Fielitz—foonote of A. V.

of Louis XIII, *Pâquerette* (music of Benoist and Pugni)—the role of Ignas—and the small but graceful ballet taken from the life of Neapolitan fishermen, *Graziella* (music of Pugni)—the comic role of Don Sicilio. Besides St-Léon's ballets, *Jovita* of Mazilier and after that, *The Blue Dahlia* by Marius Petipa (music of Pugni) were produced that year. In *Dahlia* I had the excellent, responsible comic role of the barber, Beausoleil; otherwise the ballet did not have any remarkable characteristics. Among new operas I played only in Auber's *Gonzago, or the Masquerade*. In *Jovita*, performing the role of the robber Cardovale, I had occasion to see the newly arrived Italian ballerina Rosati for the first time.* This artist was already past the first flower of her youth, but she was an experienced dancer and an excellent mime, who benefited the management not only by her spontaneous dancing, but also with her advice and instructions during the production of a ballet. She performed in Petersburg several seasons running, enjoying the public's sympathies.

Of the works just listed, *Pâquerette* stands out in my memory. I see the following scene as if it were now: having finished my number and bowing towards the small box where the emperor was sitting, I suddenly see him pointing out something to me. Glancing quickly in the direction he indicated, I saw what had happened, and was deeply touched by the emperor's goodness and concern: he was indicating that I was being given a bouquet, which I was not all expecting. Incidentally, in this ballet I also often enjoyed the public's approval in the *pas de trois*; on one occasion I was called upon to repeat it three times; Mr St-Léon, meeting me afterwards backstage, always kissed me and thanked me for my performance.

The Pearl of Seville (music of Pinto and Pugni) and *Météora* were new ballets which St-Léon, their creator, produced on our stage the following year (1861); in the first the role of the burgomaster in the city of Haarlem was entrusted to me, and in the second, that of Godli, a worker in the flour mill. In divertissements that year, meanwhile, I danced a Hungarian dance with Chernoyarova and a comic *pas de quatre* with Troitsky, Bogdanov, and my cousin, Lev Petrovich Stukolkin.

On 18 January 1862 they gave St-Georges's ballet *The Pharaoh's Daughter* (music of Pugni) for Rosati's benefit performance. My roles were John Bull and Pasifont. During the production of the ballet Rosati greatly helped the balletmaster Petipa, and indeed all of us, with her advice. Let me say in this connection that when she left us, even her successors were touched by her influence. Thus Maria Sergeyevna Petipa (the balletmaster's first wife), and Vergina, and Kemmerer, and Vazem, and even Zucchi imitated her in the role of Aspicia in *The Pharaoh's*

---

* Carolina Rosati was born in 1827; before she came to Petersburg she danced in Paris and London—footnote of A. V.

Daughter. To this day the ballet is a great public favourite. On 4 September I received my first benefit performance, for which I chose the woman's role, completely new to me, of Mme Caccemardini in St-Léon's choreographic buffonade, *Misfortune at the Dress Rehearsal*; also new for me was the role of Farmer Storph in the same author's ballet, *The Orphan Théolinda, or the Mountain Spirit*, produced with Muravieva at the end of the year.

The fires from which Petersburg suffered in the summer of this year were the reason for various charity performances benefiting people whose homes and possessions had been burned. In one of these performances, in the theatre on Kamennyi Ostrov, which did not then have its present sad and half-ruined appearance, I took part in a divertissement, performing the same 'English Dance', by now respectable but boring for me, with the same, unchanging Picheau.

1863 little stood out in the course of my artistic activity, as it was remarkable only for the role of Beshir, the leader of the Druse, in Petipa's new but quite weak ballet, *The Beauty of Lebanon, or The Mountain Spirit*.

In the circle of close friends who gathered at my place on Saturdays, St-Léon was often a guest. Discussing the art we had in common, someone asked him on one of 'my Saturdays' why he did not use the world of Russian fairy tale in his ballets, as it contains so much poetry and material eminently suited for the ballet. To this the balletmaster frankly responded that he was completely unfamiliar with the Russian epos. Just then someone present observed that we had, incidentally, a tale most suitable for adaptation into ballet—Ershov's *The Little Humpbacked Horse*, since everybody knew it, even children. Right at the tea table the story was related and translated for Mr St-Léon. After that we obtained a copy of the story and read it together in the presence of musicians who came to my gatherings at that time—Wurm (the trumpeter), Ciardi (the flautist), and Seifert (cellist). Besides them, the following people often visited, to make music: Eduard Frantsevich Napravnik, at the time an unknown piano accompanist just arrived in Petersburg who was, it seems, in the service of Prince Yusupov; Czerny, another pianist and teacher of music; and finally, Life-Hussar Albrecht, an excellent player of the cornet-à-piston. Despite the small dimensions of my lodging, these musical gatherings went very well and I retain the most pleasing recollection of them . . .

Then and there, joking, we set to work on the libretto, which we finished in the course of several evenings. This feat was soon reported to the management, which accepted *Humpbacked*, but only with several abbreviations and alterations.

In this connection let me say that St-Léon grew interested in the Russian language after this time, began to study it, and soon learned it so well that he came to speak it fluently, which is more than many foreigners can do who have lived in Russia, in her service, continuously for many years.

Returning to *Humpbacked*: I was assigned the role of Ivanushka, the chief male comic role in the ballet. But man proposes, God disposes, as they say, and on 3 December 1864, for the benefit performance of Marfa Muravieva,* the first performance of *Humpbacked* was given, but without me. I made my début in that appointed role only after . . . twenty years! I shall explain the reason for such an interval; for the moment let me say that the following personnel took part in the first performance of *Humpbacked Horse*: the father—Golts, the sons: Danilo—Alexei Bogdanov (subsequently balletmaster of the Moscow theatres, Gavrilo—Volkov† (who died in 1891), Ivanushka the simpleton—Troitsky (now retired), the horse—Picheau, and the Tsar-Maiden—the benefit artist. Our inveterate composer Pugni wrote the music. The decorations, less magnificent than proposed, were nevertheless beautiful, and as a result of all this the ballet went off with great success.

VIII

Having several times performed the excellent comic part of Martini the Servant in St-Léon's new ballet *Fiametta* (music by Minkus), I was already beginning to think about summer holiday, as spring had come. On 10 May 1864 a mixed bill was performed at the Bolshoy Theatre, consisting of *The Two Thieves*, *The Peasant Wedding*, and Kotlyarevsky's operetta *The Muscovite Trickster*. Robert and Bertram were, as usual, Kshesinsky and I. In the second act, we thieves were running to escape the gendarmes. Running every which way I, having jumped across a card table, struck my leg on the *reika*, that is, on the slit cut through the stage to slide decorations. Immediately experiencing some discomfort in my foot and instinctively grabbing it with my hand to straighten it out, I felt an unbearable pain and was carried to the dressing-room. The theatre doctor was not in the hall, but by a happy coincidence the artist (now the second balletmaster) Lev Ivanovich Ivanov found Dr Tomashevsky in the audience, who, after examining me, diagnosed a dislocation and break of the fibula. The late Grand Duke Nikolai Nikolaevich the elder and Prince Pyotr Georgievich Oldenburgsky, watching from the tsar's box on the side of the theatre as I was carried out, paid a call on me in the dressing-room. Speaking softly to me and to the doctor, who with the singer Bulakhov's

---

* Mariya [Marfa] Nikolaevna Muravieva studied in the Petersburg theatre school. She danced with great success in Petersburg, Moscow, and Paris—footnote of A. V.

† Nikolai Ivanovich Volkov was born in 1836. He studied in the local theatre school, in which he taught dancing from 1863. He retired in 1884—footnote of A. V.

help straightened out my fibula, Their Highnesses, as matters turned out, sent for their own doctors, who in a few minutes made their appearance. These were Ivan Martynovich Barch (at the order of the Prince), who subsequently gained great popularity in St Petersburg and who died some three years ago in the post of senior physician at the Maximilianovsky Hospital—and Alexandre Leontievich Obermiller* (at the behest of the Grand Duke). All three of them quickly converted our shirts into aprons, mixed up some plaster in the washstand, poured it quickly, and set my leg; after it was bandaged I was taken by sailors who worked at the theatre to my apartment in Galernaya Street. The litters on which I was fetched home were of no small historical significance, as they were used to lift and carry the late Empress Alexandra Feodorovna to her box in the last years of her life.

Lying in bed at home, still dressed and made up, I felt something uncomfortable, something getting in my way. It turned out that I had in the back of my waistcoat a collection of watches and watch chains, stars and jewellery which I, as Bertram, steal during the ball from the guests, hide in back of my waistcoat, and which reveals the theft when I tear myself away from the gendarmes, leaving my frock-coat in their hands ... How close is laughter and grief!

The next day the doorbell rang almost incessantly. Among the visitors who inquired about my health were persons sent by the Grand Duke and the Prince. Coming once again to examine me, though having transferred me to Dr Barch's care, was Tomashevsky—to whom I was indebted even before this visit—he embarrassed me, flatly refusing to accept any fee, and saying that he was only doing his duty. These attentions, together with the sympathy the public had shown me, even people I did not know, partially lightened my sorrow. My situation, especially at the beginning of my indisposition, was grave in the highest degree. Discounting my physical pain, I was constantly tortured and agonized by the thought that, for what it was worth, my artistic career was finished, as it was inextricably linked to my good health, as was the future of myself and of my family.

And how strange ... One laughs at premonitions, does not believe signs, calling them idle superstitions, and yet I recall two events which predicted my misfortune almost directly. The ballet *Lise and Colin* was being given; finished for the evening, cleaned of make-up and already in street clothes—I met Emperor Alexandre Nikolaevich on stage during the interval with Grand Duke Nikolai Nikolaevich. The emperor, in his

---

* A. L. Obermiller. Life-Surgeon and privy counsellor, died 10 August 1892 in Petersburg. He received his education in the local military-medical academy (finishing the course in 1853), under the guidapce of N. I. Pirogov, with whom he worked during the siege of Sebastopol in 1854–5, and at Plevna in 1877. He was a brilliant surgeon and doctor, through whose hands passed tens of thousands of the wounded and the ill—footnote of A. V.

disarming gentleness, turned to me, noting in jest that with the moustache I was growing at the time, without make-up and in a frock-coat, I was 'tout à fait joli garçon', and then praised my performance. He asked if my legs didn't hurt, and said, quite seriously, 'Look after them—you need them!'

And I—I immediately disobeyed my sovereign's testament, and lying in bed sick, recalled the tsar's gentle words. The second conversation which had some connection with the incident was with the chief theatre chemist, Shishko. For some reason relating to his specialty, he intended to go to London and tried hard to persuade me to go with him, on the pretext that the English love comic dancers and sense a deficiency precisely in that area.

'Come on—let's go, we'll make some money,' he would say; I went back and forth on the idea, could not make up my mind, and then the catastrophe hit. Shishko came to reproach me: 'There, you see? You didn't go and you ended up breaking your leg. Had we left everything would have been all right! . . .'

In this connection I must add, however, that St-Léon made a similar proposal, saying that if we formed a small but select company—for example, of Muravieva, Radina, Lyadova, Lev Ivanovich Ivanov, and me, then we would have made a great impression abroad. This project, of course, was left unrealized.

Meanwhile time passed. 1865 came and my health was so much improved that on 26 January I could appear on stage again, in the same ballet in which I was hurt—*The Two Thieves*—and then in *Graziella*, in which I was so sympathetically received by the public that I was convinced of their good will towards me, as I was given a valuable gift, a gold watch. My colleagues also touched me deeply, having gathered up flowers, wreaths, and lanterns in my dressing-room, decorating it beyond recognition. It was drawn and painted in this gala state by our decorator Yu. Bach in water colours, and I have saved the picture to this day.

On 11 April I received a benefit performance at which an act each of the following ballets was given: *The Blue Dahlia*, *Graziella*, *The Pharaoh's Daughter*, and *The Two Thieves*. Then, after the mourning imposed on us by the death of the Tsarevich Nikolai, performances ceased.

On the 29th of that month, on the occasion of my twentieth year of service in the ballet, I was given a yearly pension of 1,142 roubles, 88 copecks, and by order of the Director of Theatres, from 3 May I was retained for further service.

Although I felt almost completely well at this time, and my leg was healed, my doctor advised me to take a trip abroad to rest, and to take the waters at some health resort, among which he preferred, at the recommendation of Prince Oldenburgsky, the one at Wildebad. I could not even think of such a trip were it not for the goodness of the late Prince. The treatment of my illness and the expenses connected with it in general had come to about 800 roubles, a very imposing figure for my modest budget, and when I turned to the management of the theatres with a request for assistance, I was alloted one in the amount of . . . 150 roubles. Meanwhile the Prince, at that time in charge of Section IV of His Majesty's Chancellery, gave me, as a teacher in the lyceum, a reliable subsidy of 600 roubles. Thanks to this help, I quickly made ready, and, leaving Petersburg on 25 May, headed through Berlin for Paris.

6
From the 'Recollections of the Artiste A. P. Natarova'

In her memoirs Anna Petrovna Natarova, who was born in 1835, recalls her life between the ages of seven and sixteen. She gives us an even closer look than Stukolkin did at studies, food, classmates, and the rituals of a highly regimented life. In this heartily good-natured account, important people step outside their official stereotypes, such as Alexandre Mikhailovich Gedeonov, the Director of Theatres who hired Petipa, Leonty Dubelt, head of the Third Section of the Imperial Chancery, and Jules Perrot, whose inspiration once lapsed at rehearsal. Natarova writes with equal sympathy of the classroom ladies, the priests, the teachers, and other less exalted persons who tend to remain anonymous or overlooked in academic histories.

'From the Recollections of the Artiste A. P. Natarova' was published in *Istoricheskii vestnik* [Historical Messenger], vol. 94 (1903), pp. [25]–44, [420]–442, [778]–803. At the head of the first instalment is the note, 'Taken down by V[ladimir] P[etrovich] Pogozhev directly from the narration of Anna Petrovna Chistyakova (whose stage name is Natarova).' Natarova's account is lengthy, and parts of it have been omitted in the following translation: Chapter VI (on the infirmary of the theatre school), Chapter VIII (on visits to the school by the Minister of Court and Nikolai I), Chapter IX (on graduation from the theatre school), Chapter X (on training in drama), and Chapters XI and XII (on the imperial circus).

I

My entrance into the theatre school. External students. School performances. The general routine of the school. Instructors of academic courses. The school priest.

My father was a serf of Colonel Sheremetev. My parents were given their freedom, but they continued to serve the Sheremetevs, preparing myself and my two brothers for domestic service. When I was quite young I was, they said, a very pretty little girl; I loved to make faces and engage in silly play-acting for Madame Sheremeteva. These antics, as far as I can remember, caught the attention of Doctor Bers, who had a large practice in the theatre world. I turned seven. They began to think about what to do with me. A friend of the Sheremetevs came once; her name was Maria Ivanovna—'eternal memory' to her [words of the Orthodox memorial service] for shaping my destiny—though I have forgotten her surname. After watching me for a while, she said:

'Why not put her in the theatre school? Dress her up in a pretty frock. I will take her.'

No sooner said than done. It was, I recall, the summer of 1843. Maria Ivanovna took me to the director's chancery. Next to it was the office of the director, Alexandre Mikhailovich Gedeonov. In the chancery they asked: 'What do you require?' 'I need to see the director.'

Gedeonov came out to have a look at me, asked how old I was, called an assistant and ordered me taken over to the school. In the school I was taken immediately to balletmaster Titus's dance class. Titus examined me and made some sort of note. Then they took me back to Gedeonov and sent for Marochetti, the theatre doctor. He also examined me. After this the director said: 'Enrolled; normal administrative requirements waived.'

My fate was thus decided. How simple it all was then!

External students were taught only the arts, not the sciences, and therefore parents, if at all well off, tried to assign their girls to 'classroom ladies' of the school for upbringing, board and training. But the main reason for turning girls over to classroom ladies was that these women made efforts on behalf of the children so assigned in connection with dancing teachers, requested that they be cast in school performances and divertissements, provided opportunities for them to excel and thus to be readmitted at state expense.

I was assigned to the classroom lady Ekaterina Vasilievna Rumyantseva [who was rumoured to be the illegitimate daughter of the famous Zadunaisky; her mother was a Kalmuck woman]. Besides me, she was governess to seven girls, including Maria Mikhailovna Alexandrova (Schnell), Maria Sergeyevna Surovshchikova, later Petipa, Rumyantseva's own daughter, and others. Her apartment in Theatre Street, with windows opening on the courtyard, on the fourth floor, was very small, two rooms in all. The crush was frightful; girls slept side by side on the floor. Indeed, payment for this was not much, even for that time—6 roubles a month for everything: food and tuition. An instructress in drawing and piano even came; the clavichords were awful. We had barely started to study music before we began to play 'Mein lieber Augustin'. The food was modest: we were given breakfast, lunch, and evening tea with bread.

We went to the school for dance class. Becker taught the beginning students, and then we transferred to Fleury, Stukolkin's predecessor as the comic in the ballet company. After Fleury's departure we went to Frédéric. They made Becker a teacher of dance just to let him serve until his pension. He taught, rewarding us liberally with names like 'numskull', 'strawhead', and the like. My name was 'strawhead'.

In Lent of 1844 I took part in the school performance. I was a very pretty, very blonde nine-year-old; in school they called me 'the white little girl'. I had to dance a Russian Dance in that performance, for which my aunt made me a very fine costume. The director and his friends, the inspector of the school Auber, and the parents of the students all attended these performances. Of the outsiders Gedeonov invited, I recall Leonty Vasilievich Dubelt, who always gave the children the fruit-drops he brought in the back pocket of his dress uniform. I also remember Zherebtsov, a wealthy landowner and theatre-goer, and the landlord Tvorogov. At that time the school theatre was located where the dance class is now. After performances the director would come on stage and all the participants, as appropriate, would receive approval or reproof. The performance would consist chiefly of dramatic pieces, the balletic part being a divertissement. In my first school performance I danced successfully. In my second I danced a cachucha. The director approved, and I was transferred to state-supported status. Such transfers came in the spring, after graduation, which normally took place on Palm Sunday or Annunciation, depending on the management's finances.

There were three dance classes in the women's division of the school: Becker taught the youngest, Fleury the intermediates, and Titus the seniors. All the teachers were strict, Titus the strictest. If ever you went across the hall against regulation, not with an open neck but with a scarf thrown over your shoulders, Titus, saying nothing, would catch you and snap you across the shoulders with his violin bow.

The dance teachers were excellent; they taught gradually and gave the director every chance to notice students who excelled in lightness or grace. External appearance counted for a lot. Alexandre Mikhailovich Gedeonov used to say when children were received into the school: 'If she doesn't fulfil her promise, then she will be lovely furniture on stage.'

And he was right. He had an astonishingly accurate eye for students. Gedeonov dropped in on the school once or twice a day. In the morning he made the rounds of dance classes, given from 10.00 a.m. to noon and even until 1 o'clock. During these rituals the teachers would invite the director's attention to students who were doing especially well.

It was interesting that girls were not provided with dancing clothes even though dances were a speciality of the female students. We washed our own hosiery, and those who were a bit less well off would tuck up their fustian skirts during dancing class. In general they issued linens to us twice a week, and bed-linens once a week. On Saturdays chamber maids brought in bundles of linens and put them on the bed, and on Thursdays they gave out a pair of stockings and a pocket handkerchief. One dress was given to everyone for the week (a very nice one, with light blue ribbing) and a black cotton pinafore. We could wear our own wool pinafore. A fustian slip skirt was given to us during the winter, and in the summer one of dimity, two widths and two gores. The students sewed with their own thread—none was supplied. We had to sew during the summer. The children, to be sure, did not have a dacha, but there was a little garden in the courtyard of the director's home where we worked during the summers. We had to hem six white cotton pelerines, two white pinafores, six kerchiefs, sleeping bonnets, and sheets. Blouses and pillow-cases were given to us ready made; they had only to be marked. All linens were marked—embroidered with a red numeral. But we were not given any red cotton thread. We also had to mark and hem kerchiefs for the boys' division.

The sewing material was given out by the senior supervisor, Alexandra Vasilievna Figareda; she did all the cutting. Figareda was a good woman but strict, and she beat us sore. If something were marked or hemmed badly, she ordered it unstitched and redone, and boxed our ear for our carelessness.

Every month Alexandra Vasilievna Figareda gave us a handful of pins, hairpins, one needle, and thick paper for the darning of dancing slippers. The little children danced in leather slippers, the oldest in proper dance footwear. Every Sunday the chamber maids distributed pomade to us, which was done as follows. Our beds were situated head to head in the bedroom, and between them there was a little chest of drawers, one each for every four girls. The chamber maid would come in mornings with a blue jar containing brown pomade that smelled not quite of mignonette, not quite of vanilla. In she walks, and without the slightest blush she scoops the pomade out of the jar with her fingers and twice plops it down on the edge of the little chest of drawers, each student receiving a plop. The blue jars were occasionally pressed into service instead of a little cup when the girls clubbed together to buy tea.

They woke us up at 7.00 a.m. by means of a loud bell. We would no sooner leap up than be in the washroom—in one jump. And the washroom itself was interesting: there were eight or ten tin wash basins around the edge of a large round table, and above them wash-stands with a raised peg. A large wooden tank was located above the table, from the middle of which a tube supplied water to the wash basins. The supervisor ensured that we washed our hands and neck clean and brushed our teeth. We were not officially provided with soap, only a toothbrush and combs. Nor were there large mirrors; each of us combed her hair in front of her own little mirror. In general there was not much other furniture in the bedroom—two lamps in all. We made our own beds and tidied up.

At 8.00 in the morning the students went to the large dining-room for prayers. Placed on the table for each student by that time were a 'quarter white loaf', a small white bun, and two pieces of sugar. 'White loaf in four' is what we called a fourth of a double French loaf—in other words, it was one and a half copeck buns. Hot concentrated tea was brought in in large copper teapots, milk and boiling water for students who wanted to brew their own tea. Dishware consisted of round pottery mugs with handles. There were no spoons—we mixed our tea with hairpins.

Our academic classes began at 9.00 a.m. We had various teachers.

Vasily Yakovlevich Tyazhelov, a Ukrainian, taught geography and history. When we did not know our lesson, he said: 'Lazeebones, go kneeyahl in the cohrnah,' and right there in class children would be kneeling in the corner. Tyazhelov lived on Vasilievsky Ostrov, and we always waited for ice floes which might detain him on that side. But it never happened—bad weather never held him back, and he appeared in class punctually. In geography class he called us to the board and made us point out cities, lakes, rivers and mountains on the map. We were to do this with a jab of the finger. Naughty children once devised a prank: they took down the board with the map on it, put it on the floor and poured ink all over it. But this was to no avail—the administration bought a new map, and everything proceeded as before. In history Tyazhelov called on six girls at a time, a whole bench's worth, and questioned them in sequence: the first finished to a given point, the second continued, and so forth. The students got the idea and began to study accordingly, distributing lessons in parts among each row of seats. They crammed to learn the lesson by heart and thus were able to respond excellently in class. But Tyazhelov noticed what was happening. Once the first girl responding stopped when she knew no more, and Vasily Yakovlevich asked, 'But why are you stopping? Continue!'

Of course, our whole plan was turned upside-down, and we received a bad mark in the journal. At that we generally studied hard, and at examinations always responded excellently to Tyazhelov's questions. To do that we lost many a night's sleep in preparation.

Klimov taught calligraphy. Sorokin was a wonderful literature teacher in the senior class. In the youngest class the tutor Panov taught gramma' he had a mannerism of speech: 'I, you know . . . '

Olimpiev taught Russian grammar to the seniors. He was very old and deaf. In responding, students took advantage of his deafness: at the beginning of a lesson they answered from the assignment, but then simply recited the Lord's Prayer, and if much was assigned, continued with an 'Our Lady'. They ended by returning to the assignment. Olimpiev listened, apparently with pleasure, smiled, and put a '5' in his book. The senior supervisor Figareda learned of this and told Gedeonov that it was time to replace the old man. Gedeonov called in Olimpiev and said:

'It is time to call it quits, because the girls are ridiculing you.'

Olimpiev answered, 'My deafness can be naught but benefit to the girls.'

'How so?'

'My deafness forces them to speak loudly and clearly, which will be required of them on stage.'

Nevertheless, Olimpiev was made to give up teaching.

The instructor Malinovsky, a good teacher and an amazingly good-natured person, taught arithmetic. He was so agreeable that he allowed the little children to write all kinds of nonsense in chalk on his civil-service uniform. He wore a cap and brought it into class. By turns the girls took the cap, put it on, and paced around the classroom, while Malinovsky laughed.

Sorokin, the literature teacher, taught children to use their heads, required them to practice writing, mainly by drafting letters. He did not like the term 'because', called it trivial, and looked for it when reading student compositions. Later, in 1850, the excellent teacher Matveyevsky was hired as a literature instructor. He was a handsome young man, spoke and read simply, clearly, and intelligently. Something new and good began to be sensed when he was appointed.

Father Pyotr Uspensky was our priest. Quiet and full attention marked his lessons. Still there were naughty children in his class, who made and threw paper darts while he talked. He noticed this prank immediately and said, 'And who threw the dart? Raise your hand!'

The guilty party [Zhuleva] raised her hand.

'Ah-hah! for shame. You answer well but cannot behave yourself in class.'

Father Pyotr, after a holiday or a Sunday, invariably asked Zhuleva which Gospel had been read in church. Ekaterina Nikolaevna's answer was always perfect. Father Pyotr altered the pronunciation of her surname and called her Zhúleva instead of of Zhuléva.

More than anything else we feared confession with Father Pyotr during Lent. He did not even ask us about sins in confession, but required us to answer questions from the short catechism. We all prepared for Communion in the first week of Lent, the seniors in the last. Confession was scheduled before the All-Night Vigil. At five o'clock we were already running to church with the short catechism book in our pockets. We arrived, stood in rows and waited our turn. The good father moved back the screen: 'So then, you, come on.'

And in sixes, the way we sat in class, we went behind the screen and bowed down to the ground in prayer. The screens moved back, and the father began to query us on the short catechism, mostly from the Credo, according to its parts. Father Pyotr's reward for those who answered poorly was this: 'So then, during the All-Night Vigil you bow fifty times before the Saviour, and you—before the icon of the Mother of God.'

Unquestioning, in view of everybody, we did our penitential bows. The punishment was great, so great that the director requested the priest not to assign us such 'rewards' again.

A bishop normally came to examinations in religion, and the students answered excellently. Once the bishop was amazed and said to the priest Uspensky: 'Remarkable. I thank you. They didn't answer me that well even at Smolnyi [Institute for Young Women of the Nobility].'

Father Pyotr Uspensky, giving away his daughter in marriage, yielded his place in the school to his son-in-law, Father Mikhail Bogolyubov. On the day of Father Pyotr's daughter's wedding, after dinner in the school, all students had a Crimean apple at their places. When Uspensky left, the girl students cried and sorrowed, recalling his goodness and tolerance.

The new catechism teacher, Father Mikhail, taught well but was a bit shy at first, for he had to get used to theatre customs. The children did not study too well for him: 'To what point do you know?' We answered, 'To the Flood.' Then he said, 'And when did the Jews cross the Red Sea, do you know that?'

'No.'

'Well then, let us proceed.'

Early in his teaching he made a grave mistake. Wishing to give his students a taste for learning at any cost, he declared that he would excuse from catechism class those who answered questions well from the short catechism.

They believed Father Mikhail. Diligent students, wishing to be exempted from class, learned the entire catechism and answered his questions beautifully. The lazy ones, which included me, answered with tears in their eyes, 'Forgive me, I am not prepared!', whereupon a generous deadline for preparation was given—four months. Father Mikhail told the lazy ones: 'I deplore it and I am waiting,' and he did not excuse the diligent students from lessons, but as an encouragement only reseated them on the first bench. The students saw the deception, and of course got angry.

I recall another episode with Father Mikhail. There was a French student in the school, Victorina Bassin, sister of Laura Bassin, the celebrated equestrian in the circus. She was a friend of mine. Once she asked me to take her to catechism class; I allowed her to sit next to me though of course I had no right to do this. We sit down. The priest questions the class in the normal order, according to which one girl begins, another continues. The priest goes from bench to bench and simply says 'Continue.' My turn came. Victorina Bassin nudged me and said, 'Ay!' I began to laugh.

'Why are you laughing?' Father Mikhail asked.

'She made me,' I said, pointing at Victorina.

'Sit down, and you "continue",' the priest said, turning to Victorina.

'Good Father, you see she is French,' I hastened to say.

'What? A French girl? In my class? . . . *Allèz-vous èn* and *ne venèz jamàis*,' Father Mikhail said.

The embarrassed Bassin left the room, and the entire class laughed at Father Mikhail's French pronunciation.

When it was quiet, Father Mikhail turned to me and told me to continue, but I did not know the lesson. He told me to sit down, adding, 'And I offended the French girl for nothing!'

How ashamed I got after this. The lesson was well learned.

Our French teachers changed. First there was the Frenchman Thibaut, famous for having us say 'un' or 'en' nasally while he held on tightly to our noses. Then there was the Frenchman Descourt; he was bow-legged and we called him a spider. He would come to class and we would cry, 'Spider, spider, cut the hay.' He would ask, 'Where is the spider?', and we would answer, 'It's walked by!'

After Descourt there was Cannelle. Later on they assigned to us the excellent teacher Mr Fèbre. At first they alloted him six senior students and gave him, for that time, a good salary. He was very successful. Then for the same salary they assigned him the entire senior class. This, of course, forced him to devote less attention to the students, as there was not time for so many, and the result was not so good.

There was one other French teacher, Madame Fillet.

German was taught, but only to those who wanted it. Daumier was the instructor.

Giustiniani was the teacher of Italian, but not for long. He was fond of love verses, and we pestered him with them. Mostly we had conversations with him about Mario and Grisi [singers of the Imperial Italian Opera], whom we raved over at that time, and who delighted us.

There were several singing teachers.

Bystrov instructed the girls and all external students, boys and girls, including the famous Daria Mikhailovna Leonova. He also taught the church chorus of little boys and girls, who made up one choir. Fedosya Grigorievna Kovaleva was later the chief inspectress. She taught the drama class for girls. I remember that everyone, including myself, who had no voice whatever, sang 'Through the midnight skies an angel flew' [a song to a text by Lermontov, probably in a setting by Alexandre Varlamov]. Vittoliaro was an excellent teacher in the senior class. Pavel Nikolaevich

Sokolov taught us couplets for school performances. Baveri was a fine teacher—a conductor of the Italian Opera. He taught Lileyeva, later Latysheva, Onechka Lavrova (Anisia), later Bulakhova, Katenka (Ekaterina Nikolaevna) Lavrova, later the famous Moscow artiste Vasilieva.

And Madame Uttermarck came to teach piano, but only to those who wanted lessons.

II

Classroom ladies. An audacious prank. The relationship of senior students to young students. The antagonism of drama and ballet.

Now let me tell you about our classroom ladies; there were several of them.

Ekaterina Vasilievna Rumyantseva, with whom I lived when I was an external student, was a good woman, and showed us how to embroider on canvas—simple and fine embroidery. While on duty she was always occupied with her work and embroidered the most wonderful dogs, which delighted us.

The class lady Maria Osipovna Larionova was formerly a student of the theatre school. She had no artistic abilities but taught very well, so they appointed her as a classroom lady. Larionova lived with her sister, who had no connection with the theatre but was comparatively well educated—she finished a course of study at the Nikolaevsky Orphans' Institute. As a result of this, parents tried to assign external students to Larionova, because they would be well trained with her. In the evenings Maria Osipovna came to the school and helped the girls prepare their lessons. For this she received payment from the parents. In those days it was difficult for the children—there were ninety-five state students, and the regular teachers lacked the time to take with each of them. Larionova was not patient; if someone had a poorly written lesson, she slapped her hands with a ruler. At the same time she was unbelievably bashful. When we went to bathe (the bath house was on the Fontanka; we got there through an adjacent courtyard) and Larionova was with us, she would cry out, as we undressed or came out of the water: 'Girls, girls, don't look!'

Sofia Petrovna Garkush was a very beautiful young woman, very meek and delicate.

Malysheva was very sharp and demanding. For some reason they called her Grimardeau.

The two sisters Hoffmann were Mina Petrovna and Nanetta Petrovna. Nanetta stuttered. After Figareda's death, Mina was our senior supervisor.

Nadezhda Ivanovna Charukhina was hunchbacked. She told us that in her youth she was very good looking and a fine horsewoman, so expert that Emperor Nikolai I was taken with her. Once she was riding along and saw the emperor, also on a mount. To avoid meeting him she quickly jumped, the horse fell, and she injured her back, from which she developed a hump. She was obviously talking nonsense; even though we were little we did not believe her story. It was said that Charukhina loved presents, and through gifts to her much could be arranged. But Charukhina was very useful in her own right: she showed us various elegant kinds of needlework even though doing so was not among her responsibilities. These included satin stitching and mostly the kinds of knitting then fashionable. Students learned them first of all. Also modish at that time were beadwork and bugles [glass beads], from which doilies, purses, and other things were made, and also special purses of fine silk ribbons of matching or near-matching colours. As Christmas approached, the students would begin work on these purses as gifts for their parents. Charukhina also taught us to knit hoods and scarves: she had a practised hand at this.

We had one other very young and very delicate classroom lady, Miss Homan, assigned to the school at the request of the dancer Andreanova, a friend of Alexandre Mikhailovich Gedeonov.

The senior supervisor was, as I have already indicated, Alexandra Vasilievna Figareda. She was already middle-aged, a good woman but demanding and strict. Director Gedeonov held her in high esteem. She was very short, quite stout, and always beautifully dressed: she wore especially nice caps. She had a red face and a very red nose, for which reason we gave her the nickname 'corned beef'. At 9.00 a.m. she arrived at the school from her apartment; meeting her, children walked up to kiss her hand. Figareda went first to the infirmary and then to class, where she made notes, querying the teachers about good points, and she was also present during lunch. Alexandra Vasilievna Figareda looked after everything conscientiously, and nothing was done without her permission.

Figareda's strictness was not to the students' liking, and once a very impertinent trick was played on her which could have ended sadly. We remembered this story for a long time. I did not take part in it as at the time I was in the most junior class. For some reason Figareda delayed the giving out of pins and hairpins, which, as I pointed out, was her exclusive province. It occurred to the senior students that Alexandra Vasilievna might be pocketing them, so they decided to frighten her. Every evening after the students went to bed, Alexandra Vasilievna made the rounds of the sleeping quarters of the school. She normally came out of her apartment with a candle, across the stage to see if the lights had been put out and if the students were in their places. Until she did this she could not go to bed herself.

On an evening the students planned, Alexandra Vasilievna left her apartment in the normal manner, and no sooner had stepped on to the stage than students hidden in the wings blew out the candle. Nevertheless Figareda, apparently paying this no heed, crossed the stage to the sleeping quarters. There were screens at the entrance, and behind them a green cloth. Drawing back this cloth, Alexandra Vasilievna suddenly fell, stumbling on a shaft the students had intentionally rolled up in bedcovers and placed on the floor next to the entrance. When the students heard the fall, the seniors began a frightful noise of screaming, miaowing and cries: 'Corned beef! Figarka! Give us our pins and hairpins!' To prevent being identified, the students had put out the lamps in the sleeping quarters. The duty classroom lady slept in a room next to the students. Having picked herself up, Figareda knocked at the classroom lady's door and described what had happened. The duty lady put the light on and took Figareda back to her apartment.

The next day was a fright for us. Knowing nothing of what happened, we, the littlest students, were astonished at prayers that Alexandra Vasilievna passed through the dining-room, saying nothing, and rebuffed the girls who went to greet her. Figareda passed directly across the buffet, and where she went from there we didn't know. Ekaterina Vasilievna Rumyantseva took her duties that day; we little students asked her:

'What does it mean?'

'God grant that it will come out all right!' she answered. But this told us nothing.

Prayers were finished, tea was over, Figareda returned and passed by again without saying a word. We went to our academic class, and then to dance class. During dance class Alexandre Mikhailovich Gedeonov came in, ordered the class stopped, and turned to the lady on duty, requesting her to invite Alexandra Vasilievna into the class. 'If she can manage it,' he added.

Then Gedeonov ordered that the inspector, Fyodor Nikolaevich Auber, be summoned and told him: 'I want all the students in here!'

Gedeonov's face was frightening when he was angry.

All the students gathered; they saw that Gedeonov was in a frightful mood. Figareda also came in. Turning to her, Gedeonov said:

'I ask you, Alexandra Vasilievna, to point out to me whom you noticed yesterday. Which person stands in your memory?'

'Probably the students will be so honourable as to confess themselves,' Alexandra Vasilievna answered.

But in response to this naught followed but a deathly silence. Gedeonov said, after waiting a moment: 'So then, if you do not wish to come forward, then perhaps Alexandra Vasilievna will be good enough to speak your vile names.'

Figareda named one of the students whom she managed to notice during the confusion the night before. After this the director asked: 'Who else?'

Everyone was silent. Then, turning to the girl Figareda had named, a senior near graduation, Gedeonov asked, 'You took part?'

'Yes.'

'And who else was with you?'

Silence.

'Since you will not speak, then bear in mind: you will be expelled and will find yourself without so much as a crust of bread, let alone a job.'

Again silence.

'I should hope that such a punishment would compel you to tell who else was in this with you.'

Sobbing, the student identified by Figareda turned to the others and tearfully begged them to come forward; she even got down on her knees before her fellow students.

'Take pity!' she said: 'You see the situation I'm in. I alone shall suffer for all!'

She implored so, and cried so, that the others began to weep with her. Yet no one identified either herself or anyone else.

Seeing this, Gedeonov again asked Alexandra Vasilievna to remember who were involved in this impertinence.

'But then, not just on this occasion but before now someone has perhaps been impertinent to you, or inattentive, or wilful? Probably some of that pleiades had a part in this most recent adventure. Give me the names of such people.'

Figareda pointed out several students. Gedeonov said to them: 'Now, you will all be expelled.'

Then Alexandra Vasilievna intervened. 'Punish them as you will, Your Excellency, for this deserves punishment. But don't deprive them of a livelihood.'

They were punished as follows: students found guilty were segregated and placed in a room next to the infirmary. They wore dresses made of ticking material and their names were written on a blackboard located in the large dance auditorium. A red board, listing students who had excelled in their conduct or studies, was placed there as well. Girls being punished were prohibited from visiting the others. The first day they were given bread and water. The director gave all of us a stern lecture, which we remembered for a very long time. He really read us the riot act: when Gedeonov got angry, he yelled loudly and simply lost control. This incident, moreover, especially chagrined him: he did not come to the school for some three days. Then he lightened the punishment, and reduced its term to a week.

After this when Alexandra Vasilievna gave us pins and hairpins she would recall this sad occasion and say: 'Ruffians!' It must be said that Figareda was an honest woman. Once my parents wanted to give her a present, but she brushed aside the idea so contemptuously that they didn't know how to withdraw the offer. She was not mean, but sometimes coolly ordered children to be spanked. In general they spanked us fairly; in those days they instilled more than persuaded, thinking that children would grasp things sooner that way!

They mingled young students with the seniors, to learn lessons from the older students and to follow their example of neatness. The older students, thinking themselves in charge, had their own way of disposing of matters, and every Saturday brought an entire week's punishment for the little kids. Seniors thrashed juniors. This was ordinarily carried out with busks from a corset, steel wrapped in suede. They thrashed us, as is proper, directly on the body; the amateurs, especially when they got angry, thrashed us with the same busk, but by the edge, or rolled up a towel and braided it, soaked it in water and beat us with that. Or this would happen.

'Oh well,' a senior would say, 'I'm not in the mood to thrash you today, so go to Sashenka and ask her to do it for me.'

Normally this Sashenka was an amateur when it came to thrashing. The little girl would go to Sashenka.

'Sashenka! Mashenka asks you to thrash me!'

'Lie down!'

The blows followed, quite powerful ones. When the execution was over, one had to say, 'I respectfully thank you for beating me.' This was obligatory; gratitude was always expressed.

Thrashings always took place after dinner, before going to bed. My how they beat us, until we were sore! Not one classroom lady knew of it . . . which in truth is to say that they all knew of it but made it look as if they didn't. In their time the seniors, like the little students, had also been thrashed, perhaps even worse. This whole ritual originated not in meanness, but in a bad precedent.

There was yet another ritual carried out between the seniors and the small students. We were poorly fed, and to sate ourselves we had to use our own resources. Relatives were allowed to visit on Thursdays and holidays. To the extent they could afford it, relatives brought tea, sugar, sausage, other snacks, and sometimes a little money. The authorities did not prevent it. Thus, after relatives' visits the seniors would shout: 'Little girls, over here!' and we would dash headlong to them.

'Now then, here's the situation: we want some good, tasty tea. You... so-and-so, bring the sugar, you, the tea, you, the rolls.'

From whomever they favoured they requested only boiling water. But even boiling water did not come free—the chamber maid had to be paid 10 copecks or be given a little ribbon, or, if one hadn't either of these, then one had to give them one's breakfast the next morning.

'Now then, step lively!' and the girls ran to fetch everything needed for a good tea.

'To drink tasty tea', in school parlance, was to settle your company down to tea with fine rye bread, Finnish butter, and sausage. The seniors always took part. What was left over they gave to the little kids, but this happened only rarely.

Later on they began to club together for such a feast. We would get up 30–40 copecks and consider ourselves rich. Through the chamber maids we would buy herring at 3 copecks—we especially requested herring with roe, for that would be a free additional treat; then we bought vinegar, linseed oil, for 2 copecks Finnish butter, sausage for 5 copecks and a 5-copeck piece for a boiled potato. The potatoes were bought from an old woman at Chernyshev Bridge, who was always sitting there with hot potatoes in a cauldron covered with a cloth. We had to gave the chamber maid an additional 10 copecks for her commission.

Teas were usually planned for after dinner. When all the supplies were at hand, we gathered in the washroom and first of all washed the herring in the wash basin. Then we moved into the bedroom, where we sat down. We cut the herring into pieces according to the number of participants, poured vinegar and oil over it, cut the rye bread lengthwise, smeared it with butter, placed the sausage on top and then cut it into parts. Then we divided it up and took our pleasure. The herring we took from a plate, not with forks but hairpins.

It sometimes happened that a girl not participating walked around the feasting company.

'Here, girls: give me the fishhead to suck on!'

'It's already been given away.'

'Well then, at least give me the tail!'

'The tail's also been given away.'

We were happy if we got to dip the bread into the vinegar and oil... even though the oil smacked strongly of the lantern.

The snack was accompanied by tea. Two huge teapots were brought in—one with tea, the other with boiling water. Pillows were placed on stools, the teapots put on top of them and then covered with more pillows. The tea was very hot. We drank it biting off bits of sugar on the side—there was never much sugar!

But then, as we were all friends, we tried to support each other, one person helping the next. There was dissension only between the drama and the ballet students of the senior classes, and that was like the Guelphs and the Ghibellines. A fair amount of arguing and bickering went on. The drama students called the ballet students 'dummies', and the latter in their turn called the drama students 'ugly mugs'. The separation into dramatic and balletic specialities took place at the age of sixteen, whereas before then everyone studied dancing, for the development of grace and the ability to wear a costume well.

III

Instruction in dance. Beginning instruction. The intermediate class. Mr Frédéric. Episode with the young Moscow artists. The senior class. Mr Titus.

We were taught dance gradually, not hurriedly but correctly. They started the clumsy children as follows: standing them up at the barre by the wall, a teacher showed them the positions, and then battements on the floor, without lifting the leg. When they could extend their toes and stand 'turned out', as they say in ballet, with feet opened out to the sides, then followed grand battements, with the leg lifted. At first all the external students went through their paces aided by the barre, but then they put you in the centre in pairs: the lead pair were best, the beginners at the back, who were to watch what the lead pair did.

After exercises we moved to 'calm *pas*'—*adagio*. We studied basics: turns of the body, extensions of the legs and curtseys. It was very difficult. More than once we fell on our noses with a thud until we learned how to hold our backs. But understanding came easily—they hit you on the back, and you understood! Thus we proceeded, slowly, without haste. Sometimes within the first year of study the more capable students were given a part in school performances. Those who progressed poorly remained in the junior class a second year, and after that were transferred to the intermediate dance class. They kept you in the intermediate class for a long time—some six years; it was divided into senior and junior sections. You were transferred to the senior section when you reached sixteen.

Frédéric taught the intermediate class. He was an amazing lover of his art. To give more time to state students, he began work with the external students at 8.30 in the morning, but allowed state students who wished and did not have an academic class at this hour to study with the externals. Frédéric was a marvellous teacher. So many fully-prepared students came out of his class that a balletmaster had no difficulty explaining what he wanted—all quickly grasped what was required. Those who excelled in talent and were placed in the ballet were assigned according to their gifts: whoever possessed lightness—to small solos which required elevation, and whoever, lacking lightness but blessed with liveliness and grace—to character dances.

Frédéric started with the preparation of classical dancers, normally in pairs, and led them gradually to their transfer to the senior class. After them, students who did not especially excel were grouped in fours; he made them to go through the same *pas*, took note of their ability and mental quickness, and little-by-little transferred the best among them to the first group. Less able students were grouped in eights for preparation as coryphées. Those who remained, being prepared for the corps de ballet, were not especially pressed in *adagio* but were compelled to do fast *pas*, that is, to hop, leap, jump, and jump some more! But this was reasonable: requiring them to jump, Frédéric increased their endurance for dances of the corps de ballet, where you aren't asked if you are tired. Soloists have breathing spells; dancers of the corps de ballet do not. And in fact, under Frédéric's guidance they were all hardy.

To give us a little boost in our studies, Frédéric presented a list of his students to the director every week, where opposite the surname of each he noted: 'parfaitement', 'trés bien', 'bien', 'passablement', 'mal'.

Frédéric was resourceful in his encouragement of students. He began by sending Viennese pastries to us on Sundays. At that time they were very much in mode: in a simple little basket were placed meringue with fruit. Whoever was designated 'mal' on his list did not receive a pastry.

I recall how this struck us the first time. On Sunday a porter brought us two baskets. 'What's this?' asked the classroom lady. 'From Mr Frédéric to his students. Here's a list.'

The classroom lady, having read through the list, began to cut the pastry into portions and pass it out accordingly. Whoever was listed as 'parfaitement' received two pieces. For a while we enjoyed the pastry and thanked Frédéric very warmly. But we soon got bored with pastries; Frédéric noticed this and began to send glass jars of syrup, raspberry, and cherry. The class ladies distributed it to us with spoons—to some one, to some two, and to some three spoonfuls. The class ladies tired of pouring the syrup: 'Do it yourselves!' they told us. Indeed, we then got bored with the syrup, and began to pour it into inkwells in the classroom. Frédéric noticed this too, and began to send candy. Then he just sent money—from a 10-copeck piece to a rouble every Sunday.

One week—it was in 1851—we all had been especially diligent and awaited a wonderful reward from Frédéric, and he, although we had worked so hard, sent us all of a 10-copeck piece. This angered us. How so that? He could at least have given us a 20-copeck piece for effort!

And we quickly let him know that we were offended. The ballet *Giselle* was being given, and we were taking part. In the second act, when the huntsman, who was being played by Frédéric, falls into the hands of the shades, they torment him, balletically of course—that is, they turn him around when he ought to run up a hill. We, the last on the hillock, were to turn him and throw him into the water. So we really began to turn Frédéric! Throwing him down the hill, we kept repeating: 'There you have your 10-copeck piece!'

Frédéric made a point during class of walking around the lines of students (we normally stood in four lines) and, wiping his hand first, testing each student's forehead for perspiration. He would look, dry his hand and test the next. Having observed his method, one of the students wet her forehead with saliva and began to pinch her cheeks. Frédéric noticed.

'Aha, you are not quite red enough!' and slapped her on one cheek and then the other.

'There, you're red now! Now you're sore and ashamed!'

We were all struck by such an unexpected outburst from Frédéric. He was very angry, but embarrassed after that. When the class ended, Frédéric sent the girl a lot of candy.

'Why so much?' the classroom lady asked him.

'She is still young, and needs to be spoiled,' Frédéric answered. After that he always kept an eye on this girl, an orphan with no relatives. Frédéric had dancing clothes made for her, gave her a so-called 'rose dress' of cotton print, not nankeen, 'for diligence'. For diligence in dancing a rose dress was given first, and the next award was a 'tunic' or ballerina's dress. Normally students wore sand-coloured dresses made of nankeen. Frédéric then gave the same girl satin dancing slippers, at a time when the rest of us were dancing in leather. Frédéric did not, however, lavish such kindness on this one girl alone. At school performances, when separate *pas* were mounted, Frédéric bought flowers for the headwear at his own expense, that everyone might be beautiful and elegant. At Christmas Frédéric also put up a huge Christmas tree, laden with candies from top to bottom.

Pimenov taught the boys; he was not a very good teacher. In 1847 the Petipas arrived, father and son. The father was engaged as a dancer for character roles and a teacher for the boys, and the son as a dancer. Together they produced the ballets *Paquita* and *Satanilla*.

Here is a story in that connection. In the 1840s our male dancers were getting on in years. To freshen up the male staff, Gedeonov in 1849 issued an order to Moscow, where Alexei Nikolaevich Verstovsky [Intendant of the Moscow Theatres] chose seventeen young artists for Petersburg, including Ozerov, Chistyakov, Bubnov, and others.

In May the Moscow youth came to Petersburg by mail coach and stayed temporarily in the theatre school. The day after their arrival they went to Gedeonov and asked: 'Where are we to live?' The inspector of students declared categorically that there were no places for them in the school. Gedeonov thought for a moment.

'See to the following, then,' he told Aubel, 'find them a room with board and a maid, but remember, their salary is only 14.50 roubles per month.'

Aubel found three rooms. Lodging, tea, lunch, and dinner and a servant—all for 7.50 per person. The Moscow youth were assigned to the elder Petipa for training. Once Gedeonov came to class and asked Dudkin, one of the boys from Moscow:

'What class were you in?'

'I never studied dance,' Dudkin answered.

'Never studied it?'

'I studied violin.'

Gedeonov then inquired of the others: it turned out that most of the Moscow visitors had never prepared for ballet. Verstovsky chose them on the basis of height and looks alone.

'My good fellows, what am I to do with you?' Gedeonov said. 'Sorry ... You must each go your own way. Let's wait awhile to see where to place you.'

All of them, however, continued to study with Petipa and in January they even received a rise to 20 roubles per month. After that some went into the drama company, some into the ballet, while many left for the provinces as dramatic actors.

With Petipa a new school of dance was introduced, and attention was paid to mime. Petipa *père* was a very fine actor-mimist.

One should add that when Becker was made Titus's assistant, the elementary dance class was entrusted to Daria Sergeyevna Richard. Daria Sergeyevna had suffered from dropsy for eighteen years and it was difficult for her to move, so she spent most of class seated. She showed us all the *pas* with her hands while sitting in a chair. We could all dance with our hands: show battements, *ronds de jambe*, *assemblés* and glissades. Demonstrating with her hands, Richard simultaneously explained in words. We understood her perfectly, and her explanations stayed in our memories for a long time. Richard taught beautifully. With us in her class was her daughter, Zinaida, subsequently a well-known soloist who even attained the rank of ballerina. She later went to Paris. In the evenings Richard taught us ballroom dances; she did this not with her hands, but explained while standing. She demonstrated with her feet and body how to do things and to sit. A strict teacher, and a fine person.

Sometimes they sent students from Moscow to perfect their talents; first, I remember, they sent Zhuleva with Selezneva, then Lavrova, Lyubskaya, Popova (later Manokhina), Grigorieva, and others. Upon completion of their studies, Lavrova and Grigorieva requested to serve in the Moscow theatres. Some incident occurred in connection with this of which we students knew nothing; I recall that Gedeonov got very angry, and these graduates were sent off to Moscow in barely twenty-four hours.

Titus taught the senior class. We came to him only two years prior to graduation, and only those of us assigned to ballet who distinguished themselves in dance. The others, who had been assigned to the corps de ballet, took instruction in the intermediate class. Becker was appointed Titus's assistant, and taught the class when Titus was away. Titus had a state apartment downstairs in the director's quarters. Every day at noon he went home to drink chocolate, at which time Becker would substitute for him. Later, on Becker's departure, Richard was made Titus's assistant. At Titus's lessons we continued the same *adagios* as before, but now strove for greater perfection. Titus produced a special ballet, *The Two Sorceresses*, at the Alexandrinsky Theatre for students who excelled in dance. Two graduates took the first performance of the two sorceresses: Prikhunova and Nikulina, and Maria Petrovna Sokolova played the role of the little boy. By means of this ballet, so to speak, Titus assigned students their future places in the company. Titus was very old and in my time was ending his career. Soon he left the service and went abroad. The balletmaster Perrot, who replaced him, arrived in Petersburg in 1848, as best I can remeber. On Frédéric's departure the teacher-dancer Volkova replaced him in the intermediate class.

IV

School food. Student participation in performances. Getting to the theatre. The courting of girl students. L. V. Dubelt.

As I mentioned, they fed us poorly in school. At noon we had a second breakfast, that is, they brought baskets with black bread, '4-way buns' and hot water into the dining-room. Whoever wanted tea could brew it. That was our *grand déjeuner*.

At 2 o'clock we had lunch, comprising three dishes, for example: soup with vermicelli, roast meat with potato and sauce, and third, puff pastry with rice. We ate in the following way. Seated ten to a table, one of us gave out the portions on plates and poured two bowls of soup for the two girls sitting next to her. Normally we covered one bowl with an empty plate, but the other we left open while awaiting the second dish—beef with a disagreeable, muddy, thick sauce. The beef was all gristle; we called it *'moire antique'*. We took the beef and the potato from our portion and carefully washed them in the open bowl of soup to get rid of the awful sauce. Then we put what we cleaned off on to the empty plate. In this way we got a quite tasty soup with beef and potato. We happily ate one plateful for two. From the third course of pastries we removed the middle and ate the crust. For supper they gave us the same three dishes again. One cannot say that the portions were huge, but the bread and kvas were excellent, and these we took to our hearts' content.

Occasionally the cooking was unbearably repulsive, at which time the students created an uproar, and the senior supervisor Figareda sent for the steward, who on her order prepared potatoes in their jackets with Finnish butter in little rounds, which we very much liked. This usually happened when they had given us something really awful: 'pea sauce' with our beef. And the potato sauce of which I just spoke, which we left on our plates, got 'caught up in transactions'. The next day we happily gave the maid a '4-way bun' to take it back. In return the maid, adding Finnish butter to the sauce, fried it up in a very tasty way, a delicacy we awaited with great impatience. In the evenings this maid's brother—a Finn—normally came to the buffet. He brought frozen cranberries, frozen apples, sunflowers and mirthfully enough, the '4-way buns' which his sister collected. Since by evening we were hungry, we paid money for all this, and ate with pleasure. How they bamboozled us!

Once—it was summer—a dish of macaroni went flying on to the floor; there was a green worm in it. The steward Komarov was ordered upstairs, where the senior supervisor fairly blew her top, after which, understandably, we were immediately given our favourite jacket potatoes. After Komarov, inspector of students Aubel became steward; at first we ate well, but then things slipped back to where they had been. Several classroom ladies who kept students received official board for money, but their food was cooked better than ours.

At five in the afternoon, after academic classes, hot water was brought in again, and the baker came. Whoever had money drank her tea with rolls.

Ordinarily at about 7.00 p.m. the students were sent to the theatre. They did not make the little students dance very much; we were cast mostly as cupids and nymphs, and were sometimes assigned to an opera. Frequently male tailors dressed the little students at the theatre, and in the ballet *La Sylphide*, for example, it was mostly men who did this, for they put linen corsets on us and had to secure the corset ring skilfully at the back. I remember that the visiting Moscow dancer Sankovskaya danced the Sylphide at that time; she astonished everybody by running around the stage and going through her *pas*, all on pointe. This was new at that time.

The little students' tricot did not quite fit. It was given out in the following way: a huge basket with tricot was stored in the dressing-room beneath the table. A boy was always sleeping on top of the tricot in the basket, a tailor's assistant who passed out tricot and costumes to us.

'Alyoshka, get up!' the tailor Nikander would command, nudging the basket containing Alyoshka with his foot. 'Get the tricot, pass it out! The students have come!' This Alyoshka later became the celebrated costumier Alexei Stolyarov.

The quickest students got the right size tricot; the others, more timid, took what remained. And so if the tricot was too large, it was tucked up and hemmed, if the legs were too long, the slack was taken up in the slipper. The tailor made all these adjustments on us. Beneath the tricot we wore a corset, to which the ring for flights was attached. We put on a lace-trimmed blouse outside the tricot; it was too big and had to be pinned, but since we were given pins but once a month and they were quickly gone, the tailor had to sew it, fastening the lace to the jersey. While sewing our clothes, the tailor would scold us: 'Ah! You sloppy children! Good-for-nothings! Again no pins!'

Students used neither powder nor make-up, but the classroom ladies (two women always travelled with us) kept rouge and cotton wool. When ready in their costumes, the children came down to the foyer, went up to a lady, and she, according to her taste, put on some rouge. Sometimes we didn't come out so very pretty.

Our coiffure was simple. The girls in school curled each other's hair with curling papers and sugar water to make the curls hold better. Combing it was frightfully painful!

Our headwear was fancy gold thread if we were cupids, garlands of roses if we were nymphs. Flower ladies distributed the garlands, draped over their arms. The children smoothed out the garlands and put them on. They looked very nice. We had silk slippers for ballets—with kid soles for dancers, coarse linen soles for students. The first time we wore the slippers undarned, but then darned them down to only a few threads of the original cloth. A pair of slippers lasted five to six performances.

Flights were difficult and unpleasant for us. They were especially numerous during the production of the ballet *La Sylphide*. Valts, assistant to the machinist Roller, tried out the machinery before the children did. He attached sacks full of sand to the apparatus and flew around with them. We were attached to the lines by the ring at our back. It was frightening to fly and difficult to hold the body correctly, let alone to move the legs as required. Once a student whose tricot was too long held it in place during a performance with red suspenders, which were visible from below. Gedeonov ran in and shouted: 'Sylphide! Which sylphide has the red suspenders?'

They found the guilty sylphide, who came to grief over the suspenders.

In the evenings the girl students were taken home from the theatre and given dinner. They took us to and from the theatre in coaches crammed as full as possible. One coach was three-wheeled—one in front and two in the back at the sides. It was wider and lower than the others; it had benches along the sides, and three stools in the middle. The seniors rode in this coach, sometimes with a few juniors. The maids rode in it as well, with ballet dresses for seniors who were to dance a solo. It carried twenty-eight people in all. The three-wheeler, like the four-wheeler, was drawn by four horses. The little students travelled in the four-wheelers. The school porter literally threw them in: 'Now then, go, go!' And when one coach was filled, he cried: 'It won't hold any more!' And they went into the next one.

The classroom ladies rode behind, in carriages. Figareda was present at all departures, counted us and noted where we were.

In the daytime we wore cotton print dresses and leather slippers. Departing for the theatre in winter we wore high slippers lined with flannelette, which laced up the front. On our dresses beneath our overcoats we wore green wool kerchiefs, and the older students, green shawls. We wore coats of green quilted calico, with slits but no sleeves. The coats had pelerines. For headwear they gave us a cap shaped like a bonnet, in red and black stripes, knitted plainly of light wool, and not very warm. The ties were of worsted with tassels. We tied them up in coquettish, rooster-like ways. We were allowed to wear scarves we had knit ourselves, and so we did, and fairly showed off with them.

In the summer-time they took us to the theatre but rarely, only for divertissements. We travelled in the same coaches.

Performances ceased at the end of Butter Week, but on the Monday and Tuesday of the first week of Lent it was German Butter Week; ballet and opera were given on these days.

On Wednesday of the first week of Lent we began to fast, and on Thursday went to the sauna. It was done as follows. The bath in Leshtukov Alley was reserved for the whole day. They took us there in the same coaches, but they made many trips, which began early, 7.00 a.m.; everybody had to be back at school for prayers. Whoever washed fastest was sent home quickest. They sent the maids along to wash us, older students and young. Normally we went to the sauna once every two months, or perhaps more often. There were no classes on days we went to the baths, and we spent the rest of the time as we wished.

During fasts they gave us Lenten bread for breakfast instead of a non-Lenten bun, and a piece of lemon in place of milk. Food during Lent was horrible—mushroom soup, potato and rice cutlets, and watery kissel. On non-fasting days we also had kissel, but cold and cut into pieces—a horrible muck. The only thing we liked in Lent was herring salad.

We usually rode to the theatre along Sadovaya and Sennaya Streets, and we, the little students, were always taken by the fact that officers were riding next to us. These officers usually accompanied the students and rode in their carriages on both sides of the street: they were suitors. Our coach drivers made a lot of money on these trips. Instead of driving along the centre of the street, a driver might veer to the right or the left and crowd out the carriages accompanying us; they were paid well for not doing this. The officers' attentions reached the point that once gendarmes were ordered to accompany the students' coaches, thereby to stay the officers. But this order was revoked because the detachments of gendarmes paid even more attention to the theatre coaches than the officers.

The officers were inventive in devising schemes to see students. Disguised as drivers they once set out for the courtyard of the Leshtukov baths. There they awaited the students and helped them embark. When the students recognized the officers all manner of noise and cries ensued.

'What's wrong?' asked the classroom ladies travelling with them.

'Nothing, we stumbled,' the students answered.

The officers were gone by the time the students left the baths.

Students got acquainted with officers in the following way. In the Bolshoy Theatre officers subscribed to a lettered *loge* in the third row. The officers applauded, and each student somehow knew who was looking at whom. Then the officers, as the students left the theatre, went to the stage door but stood at the side. After that began their daily promenade along Theatre Street at 3 o'clock in the afternoon. Lunch was over at this hour, and the students stood in their windows greeting whomever they wished to favour. In the summer-time officers entertained the students thus: a fruit pedlar with his stand would come by. Wishing to make the students laugh, an officer would buy the whole stand and toss it into the air. The fruit was scattered flying on to the roadway. Out of nowhere hordes of kids, children of watchmen and janitors, would run out of their homes to choose among the treats. A free-for-all would take place while the students laughed and applauded. Gedeonov was told about this, and such events were not repeated. But in general Gedeonov, when told of the officers' escapades, would say: 'True, it isn't good. But to prevent such things is obviously impossible. And anyway, the girls are not being prepared for a convent you know!' And in fact nothing could have been done. The classroom ladies also viewed the officers' courtship of the girls indulgently.

At that time courtship led to marriage, after graduation of course. Dancers who had liaisons with their suitors behaved themselves tactfully, modestly, and would not show off the fact that they became rich. Most of them first of all assisted their poor relatives and helped them out in life as best they could. Sometimes such dancers would be changing clothes for rehearsal in the bedroom with us, and would talk to us and in a cordial way would instruct us to be cautious and not be attracted by glamour. 'Luxury does not bring you peace,' they would say, 'as you silly girls think. If you only knew . . .'

Although they lived together as common law couples, they lived well, like man and wife. Several of these women were remarkable, and did not know how to refuse help to others.

Our religion teacher Bogolyubov for some reason went to the metropolitan (whose name I somehow forget).

'Ah, you, theatre man! So then, how are your unfortunate charges? Perhaps they don't cross themselves?'

This offended Bogolyubov, who answered him: 'Your Grace, when poor people come to me I send them directly to these unfortunates, as you choose to call them. And the truly unfortunate whom I send to them never return empty-handed.'

'Really?'

'Yes, Your Grace. What's more they study scripture, and when they graduate onto that stage they can cross themselves and pray.'

'So then, I am sorry, I did not know . . . And you have taken offence.'

Bogolyubov himself related this to us.

When our dancers took a suitor they did not advertise the fact. They passed modestly, in veils, from the theatre to the carriages where their friends awaited them, and didn't call out to the coachmen. In rehearsal they acted the same as everyone else, and we students did not even know who was taking up with whom.

Suitors sometimes appeared in the building opposite that of the theatre management, the site of some official department. Clerks and officers who visited them often stood in a window and conversed in signs, and the girls laughed. This was especially frequent in the summer, when the windows were open. Once during such a conversation the inspector of the school Auber came into the dormitory unexpectedly; some of the students were changing clothes in the back of the room. Cries went up. Auber went to the window and got fiercely angry at the sight of the clerks opposite: '*Mais comment*! (in conversation he always mixed Russian with French),

Les fenêtres are open! Some clerks *regardent*, and the kiddies are jumping about in their petticoats!'

He ordered the windows closed; with that the practice ended.

In general the mores were patriarchal. For example: mornings or evenings the students would get up, go to bed or change their clothes—and the lamplighter, without the slightest announcement or query, would come in with a ladder to light or trim a lamp. The students were *en déshabillé*. A voice would ring out: 'Girls, girls, a man!'

Not knowing who it was, the girls, alarmed, threw on blouses or whatever they could. Soon a student's quiet voice was heard: 'Ah, how silly, girls, it's only the lamplighter!' And the whole group would calm down again, dressed as they had been before.

Suitors of the girl students were sometimes officials of the school. Once a student missed her ride to the theatre, for the others had already left. One of the school staff said: 'I will take this girl in my sleigh!' She was very pretty. Returning to the school from the theatre, the staffer proposed: 'Come to my place, I'll fix you some tea.'

The girl was flattered, but a short time later came to Figareda all in tears. We never knew what happened to her, only that the director called her in the next day and questioned her at length. To us she never said a word. Then we learned that the person wanting to do her harm was pardoned on the basis of his long service. Before this he had once approached another student in a classroom, and tried to caress her. It was at dusk, after 5 o'clock. Without much ado, that girl slapped his face and knocked off his glasses. She told us about it much later, after the story I've just related.

The courting of students was sometimes encouraged by persons on the management staff. One official in the theatre bureau, quite revered, had state quarters in the same building. They say he set up meetings between officers and students in his apartment. Once two graduating students got kerchiefs from the maids, threw them on, and were making their way out of a back entrance, across the courtyard and through the gates to his apartment. The mother of one of them suspected the affair and caught them both when they tried to crawl through the gates. Nevertheless, other students probably made their way to his apartment.

Meetings—with external students and the children of tutors—were also arranged in the summer, when students took walks in the garden in the second courtyard of the director's home. The son of one of the managing officials met repeatedly in the garden with a student in whom he took an interest. The young man came to the little garden, and one of the senior students conversed with him while we, the little students, walked around to avoid the place where they were standing, thus escaping a swat later from the older girl for excessive curiosity.

Frequently there were serious reasons for the intimacy of graduating artists and their suitors. Before graduation from the school many students never imagined the poverty in which their parents lived. However badly they fed us at school, it was worse at home. On the other hand, the sight of even tawdry luxury developed in us from our earliest years a taste for the magnificent. Our splendid costumes compelled us, in one way or another, to break away from the poverty we suffered in the lower salary grades. In addition, we still had many relatives who looked to us for help. It was difficult. Then suddenly there would be admirers with offers of assistance, and a girl would come together with one of them.

But you also met people who rendered unselfish help. One was Leonty Vasilievich Dubelt. They say that he was a frightening person, but he was wonderful to us. Many of us remember his goodness. At the Third Section he had access to an unofficial account, and at graduation many asked him for help. He rarely refused. One graduating student in dire need wanted 50 roubles for her family. What to do? She went directly to Leonty Vasilievich at the Third Section and explained her need.

'You need money? Sweet young thing, you can always get it.'

'How?' the graduate asked, most naïvely.

'It's easy . . . you'll even have enough for a carriage.'

Suddenly she understood what was being said. Not knowing how to respond, she said: 'But that's a sin!'

'Well, it's a small sin, but what a good life!'

'But that life will not give me one thing—spiritual calm.'

'But the anxiety is not great.'

'But how will I be able to look *others* straight in the eye?'

By this our graduate was hinting at the fact that Leonty Vasilievich was enjoying a close relationship with another ballet artiste at the time. He thought for a moment. 'Well, fine then. I don't have any money at the moment. Come back tomorrow.'

Tomorrow came. The graduate went back to the Third Section. Leonty Vasilievich gave her the 50 roubles and said: 'Take this to your family and don't be angry with me. I was only joking.'

'And so was I,' she answered. She thanked him most sincerely, kissed him, and departed.

Leonty Vasilievich did not like to refuse people. Once I asked him for something personally. He promised to do it. His secretary told me right off: 'That can't be done.'

'Why not, I ask you, if Leonty Vasilievich promised?'

'Ah, my pretty friend. Leonty Vasilievich is so kind that if you asked him for the moon, he would wish with all his heart to give it to you, but he couldn't . . . He just cannot refuse.'

Here is another case. One student's father was involved innocently in some serious and well-publicized [political] incident. They took him to the Third Section. Her mother came to the school in tears, without explaining what was wrong. She told her daughter: 'When Leonty Vasilievich comes, ask him to have mercy on your father.'

Though young, she still asked her mother what was wrong. 'Your father is not guilty, tell that to the general.'

The next Sunday Dubelt came to liturgy at the school church.

On seeing Dubelt, the little girl broke down in tears.

'What is the meaning of this?'

She began to sob still more, so Dubelt led her over to a corner and said: 'Stop crying and tell me what has happened.'

'My mother says that my father was taken in and that you can save him . . . Please save him!'

Dubelt kissed her, consoled her and comforted her and said:

'Don't cry, sweetheart. I will look into the matter and try to help. But you must be silent; not a word to anybody.'

And in fact he looked into it, helped, and the girl's father was set free.

At Easter many theatre people went to pay respects to Dubelt at Zakharievskaya, where he had his home. His valet Zakhar, an old man who knew whom to admit, sat downstairs. Leonty Vasilievich treated us to chocolate, tea, whatever anyone wanted, and gave us small, inexpensive presents: wire cigarette cases, bonbonnières, and the like. Once we asked him: 'And you, Leonty Vasilievich, where do you go on holiday?'

'My dearest,' he answered, 'when the master goes away, he leaves his dog behind on a chain to bark. I'm that dog. And therefore I can never go on holiday.'

Passing out treats, he would ask if one wanted coffee or tea.

Someone would answer him, 'Thank you.'

'And what does "Thank you" mean? Do you want something or not? If you don't want anything, then just say, "No thank you".'

V

A ball in the school. Ballet reheasals at the school. The balletmaster Perrot. The production of Esmeralda.

On the first day of Christmas every year a ball was put on for us. Boy students came to the girls, and later the seventeen artists from Moscow, of whom I have already spoken, also came. We danced. Leonty Filippovich Aubel played the piano; a ballroom pianist was invited besides. The ball began at 10.00 p.m. and was over at midnight. The director, Dubelt, and inspector Auber were invited. The dances began with a polonaise and a quadrille. Some students were too lazy to dance; inspector Auber went up to them and said: 'Children, *mes enfants*, if you don't dance, *demain* you will stay at school and not go home to your parents.' The dances got livelier after that. Almond milk and lemonade were served, passed around by two servants. Then they brought in two large trays with pastilas, gingerbread, nuts, and raisins. Coming in with the trays the servants held them high above their heads. As soon as they let down the trays, boys and girls would immediately empty them. Presents were strewn about on the floor. On occasion Auber would say: '*Mes enfants*, please pick it up or you'll slip and fall on your nose.'

Daria Sergeyevna Richard directed the dances. At this ball the boys would court the girls. The acquaintance of boy with girl students often led to romance and matrimony, as did encounters of the girls at joint rehearsals with both male students of the school and artists. I owe my own marriage to getting acquainted with my future husband at school. A girl expressed favour to a boy by learning the identification number of her favourite's linen, and then trying with special care to mark his kerchiefs and the toes of his socks with that number. And in fact, some such markings were astonishing. The management took a benign view of this type of courtship, foreseeing legal matrimony in the future. In such cases the classroom ladies were also very indulgent. Several men dancers would stay on and talk with us at the end of ballet rehearsal in the large hall while the classroom lady walked back and forth across the room but did not intervene. Other meetings were impossible, since students over sixteen did not go home for holidays. The lady walked around the hall for some time before breaking up the conversation: 'So then, it's time!' And the men left.

Marius Ivanovich Petipa's courtship with Surovshchikova began in this way; she later married him.

Important rehearsals in the dancing rooms of the school were a great and useful entertainment for us. The room was crowded, and the ballet artists who came to rehearsal used the beds in our sleeping room to change; some of them spoiled us—bringing presents and sometimes giving us money. Later two rooms were set aside for ballet artists to change into rehearsal dress. The rehearsals took place during the day, but sometimes also in the evenings. When the balletmaster Perrot mounted a new ballet in 1851, rehearsals often went to midnight. The girls especially loved these evening rehearsals. Usually the ballet artists put up tea and brought snacks. With particular impatience we awaited the smoked whitefish, of which there were tasty bits left for us.

We were allowed to stay in the hall and watch big ballet rehearsals. How good and useful this was! Let it suffice to say that in 1848 we had a chance to see how the balletmaster Perrot, in truth a genius, produced the ballet *Esmeralda*. The composer Pugni arrived with music and showed Perrot what he had written. Perrot had determined in advance the tempi and the number of bars in each piece. And Pugni had prepared it thus: at one end of a piece of paper he had written the motif of a particular number, but if the sheet were turned upside-down, you would find written there another theme for the same piece. He shows Perrot the music. The musicians play it. Perrot listens.

'Well then, will it do?'

'No, it won't,' says Perrot.

And we young girls waited impatiently for Pugni to turn over the page. This very much entertained us.

'And how about this?' asks Pugni.

'That is fine.'

Pugni came with two violinists: Alexandre Nikolaevich Lyadov, and Sokolov. Sometimes other musicians came in their place. I recall Sokolov: he played second fiddle. During rest periods or whenever the music for some reason had stopped, he put down his violin and calmly knitted his stockings.

When all were assembled, Perrot came in. He was a small man, unhandsome, with a broad nose, but he had beautiful eyes and a fine mouth. In general he was very likeable, so much that we called him 'Yuly Ivanovich the Handsome'.

By the beginning of rehearsal the balletmaster had already assigned a number of pairs of coryphées and corps de ballet.

The rehearsal began with a sign from the régisseur Marcel (who was an excellent régisseur). Perrot selected the pairs and then said to the musicians: 'Begin.' Perrot sat curled up in the middle of the hall while the music played; he took his snuffbox, sniffed, and listened. The others stood. Perrot would listen through a page—and the plan of dances was ready in his head! Having arranged the men and women of the corps de ballet in pairs, he said: 'Watch!' and he showed them what they were to do. 'The *pas* is the same for everybody. This side begins with the right foot, and the other with the left. Be so good as to begin!' The corps de ballet would perform the demonstrated *pas*; when everything was going well, Perrot thanked them.

At the first rehearsal of *Esmeralda* Perrot was displeased that the women were dancing in long dresses. 'I can't see their feet. Why are they not in ballet dress?'

'The corps de ballet does not take daily class,' the régisseur tells him. 'They don't have ballet clothes.'

'But I insist that they do.'

'This is quite difficult, as the corps receives so little.'

'True, but have they no tunics? Let them be in tunics if not in short skirts.'

There was no alternative. A new expenditure had to be made in the budget of the corps de ballet for dancing clothes. At the next rehearsal everyone arrived in short dresses.

The corps de ballet was huge, for which reason Perrot began with it at rehearsal. Finishing with the corps, he moved on to the coryphées. And evenings at the school Perrot rehearsed the acting parts and other solos in *Esmeralda*, and later for *Caterina*, which he also produced. There we saw the incomparable acting and mime of such artists as Fanny Elssler (Esmeralda), Perrot, Golts, Didier (Quasimodo), and Marius Ivanovich Petipa. We saw how Perrot demonstrated acting to them. We were struck by what was possible to convey in coherent discourse without words. Of course we all knew the story and the *pas*. We rehearsed the whole thing among ourselves in the evenings, and willy-nilly learned the whole enterprise well.

Perrot did not give outstanding soloists much of their own to do, but held them in reserve for effect. For example, in the first act of *Esmeralda* he replaced coryphées with stars like Prikhunova, Makarova, Amosova, and Snetkova in the first pair of the mazurka. In the second act the same dancers appeared as nymphs and danced their solo incomparably; and in the last act they were part of the crowd at the carnival.

Sometimes music was played at rehearsal, Perrot tried to think, but his creative imagination did not flow. He would say to the musicians, 'Be quiet for a moment!' The music ceased. He would think again, and again nothing. 'I'm sorry,' he would say to the company, 'I can't do it. Rehearsal tomorrow!' And everybody would leave.

VII

The school during summertime. Summer promenades. A performance at Peterhof in 1851.

In 1848, during a cholera epidemic, the students were not allowed home for the summer, and bringing berries and fruit into the school was prohibited. They even stopped giving students kvas, and brought two buckets of water into the buffet instead, pouring into each a bottle of 'Medoc' red wine, which produced an untasty mixture of unpleasant colour. Despite these strict measures, we managed to eat both the berries and the melons which our parents sent us: in those days berries and fruit were very inexpensive. Since the melons had a strong aroma, we ate them not in the dormitory but on the student stage, and threw the rinds beneath the stage, thinking that nobody would notice. But inspector Auber, who entered the school across the theatre, caught the melons' scent.

'*Voyons*! Who has melons when they are *défendus*?'

The guilty party, of course, was never found. They took away the rinds, we continued to eat melons, and nobody got ill.

In general, passing the time during the summer was boring. Our usual promenade was limited to the garden in the courtyard. When it was hot, and the cook came to the ice-house, we asked him for a little piece of ice and dropped it down our backs. This had the effect of a refreshing bath, and nobody caught cold. Of course, we were all interested in food.

'Kind sir,' we asked the cook as he passed by, 'who's eating the salad? The cows?' The cook laughed and said: 'They are.'

'Ah, that means that we'll get some too fairly soon,' we said, assured that our favourite dish was nearly ready.

When they refurbished the dormitory in the summer-time, the beds were taken into the courtyard to be painted. They moved us into classrooms, and we slept in numerical order on mattresses on the floor. This was a very merry time; we considered ourselves free. Dance classes were the only ones we had, and these were given only three times a week.

Walking around the garden in the summer, we always looked into the windows of inspector Auber's apartment. When Auber appeared in his window, normally in his dressing-gown, we curtseyed and said: 'Fyodor Nikolaevich, give us some little books to read!'

'Just a moment, I shall get dressed and I'll call you,' he answered. Auber had a so-called pocket library with translated novels like *The Three Musketeers* and *Countess Monsereau*. He gave these little books to the girls, but insisted that they be returned on time. Sometimes Auber gave us treats, black plums and raisins.

Once a summer they took us for a trip outside the city. At first they took us to Bekleshovka, to Kushelev Garden, but they found this awkward—the students ran around the park and it was nearly impossible to get them together again—the classroom ladies had real problems. Once they decided to take us to the Strelka. They took us in state carriages, and from there we went on foot to the New Village, where the inspector of students and steward, Leonty Filippovich Aubel, lived; his dacha had a large hall and garden. We went in pairs, and the weather was magnificent. We entered Leonty Filippovich's hall as one large gang; tea and buns had been prepared there, to which the hungry students did full honour.

'Do you want to have a little fun?' Aubel asked. We readily agreed.

Aubel himself sat down at the piano, and we danced until our feet gave out. We had to be taken back in carriages.

In 1851 they took us to Peterhof. The Grand Duchess Olga Nikolaevna came this summer, and Emperor Nikolai Pavlovich ordered various entertainments. A huge festivity was set for 1 July, Empress Alexandra Feodorovna's birthday. The balletmaster Perrot was ordered to produce on Olga Island one act of his ballet, *The Naiad and the Fisherman*, which was scheduled for production in the regular season. Roller was assigned to build the stage. Radina (later Galler), Anna Ivanovna Prikhunova, Perrot, Johanson and others, together with the girl students, were to participate. There were evening rehearsals in the school, sometimes to 1.30 a.m. Perrot did a marvellous job with the ballet.

Six days before the performance the ballet and the students were taken to Peterhof, where everyone was lodged in the English Palace. Even our beds and servants were taken there. We students stayed on the upper floor, the artists downstairs. The food came from the court; we had lunch and tea on the *bel-étage*, where there is a terrace and a stairway leading to the park. In the morning everyone took coffee at a personal place setting—with a little tray, cup, cream pitcher, etc. We were astonished at how much china there was in the palace. The classroom ladies came with us to Peterhof, as did the senior supervisor Hoffmann, the successor of Figareda, who had died, and also the dancing teacher Richard. The women were housed with us upstairs.

They fattened us up with breakfasts and lunches, for which Gedeonov joined us. They brought us from Petersburg by steamship, and every day court coaches were dispatched to take us on rides around Peterhof. The tsar visited us every day in the English Palace and always asked, turning to the students: 'Children, are you happy?'

It would be interesting to know whether anybody has the photograph taken in Peterhof at that time! Our sovereign was descending the staircase of the English Palace, and we were all running after him. In general our sovereign Nikolai Pavlovich was very attentive to us and to artists. Despite the fact that smoking was prohibited everywhere, in the theatre and on the streets—the tsar permitted the artists to smoke. 'You needn't be shy—go ahead and smoke,' he said.

The students were allowed to go with the ballet artists to see the marvels of Peterhof, such as the house of mirrors, where the sentries took the artists for high officials, bowing to the waist and saying [hoping for a tip], 'Don't forget us with your kindness.'

We went also to see the pheasants and eagles in their cages. One little girl asked the ballet artist with her: 'What kind of bird is that?'

'An eagle.'

'Ah, you're crazy, monsieur. Even I know that an eagle has two heads.'

Once after dinner the court coaches arrived, and we went for a drive. The driver asked, 'Where to?' 'Just keep going,' Hoffmann told him.

We went to Alexandria. Hoffmann likely either forgot or did not realize that our performance was to be a surprise for the empress, and so theatre personnel were prohibited in Alexandria. Moreover, Hoffmann did not know the area and the driver did not say where he was going. We stopped, got out of the coaches and walked across a park through a small wicket gate to a hedge of acacias. We went along a little road and looked to the left, where garlands and monograms were hanging. We went to admire them—no one stopped us—then looked to the right, and—oh horrors! Before us was a little square, the palace, and our sovereign himself. We cried out and dashed headlong into a run. The emperor yelled, 'Where are you going? Stop!' We stopped.

'Come over here!' And our sovereign went to meet the empress, who had just come out of the palace. He led the empress over to us. She asked our names. The tsar also conversed with us and was especially nice to Radina, who had demonstrated her talent in the preceding season. Polish dancers had been summoned to Petersburg, who danced *The Peasant Wedding*. After them this ballet was performed with our artists, and as a student Sofia Radina danced a solo mazurka in it, beautifully imitating the talented Polish artist Adams. This had caught the imperial eye. In Peterhof she was to dance the naiad. The emperor asked: 'And will you be in male costume?' Radina was very nicely built. 'You are so graceful,' he said, 'we will be admiring you.'

The empress, having spoken to the children, turned to Hoffmann and asked: 'Qui a remplacé Mme Figareda?'

To this poor Hoffmann, distracted with fright, answered, 'C'est moi, madame!'

The empress turned her back to us and ceased chatting, and to end the meeting our sovereign said: 'Now, my children, travel on! Have a good time!'

With this our visit ended, and we left. You had to have seen our unhappy Hoffmann's face; she was nearly in a swoon, and said, in utter despair: 'What have I done? What have I done!? How I answered the empress!'

As I have noted, a magnificent table was set for us in Peterhof; on this occasion Alexandre Mikhailovich Gedeonov kept close watch over the interests of the artists, who dined together with us and with Gedeonov. Once after lunch Gedeonov asked Perrot and Petipa: 'Gentlemen, why are you not drinking red wine?' Perrot said, 'I don't want any,' and Petipa remained silent. 'Perhaps the wine is not good? Let's see, I'll try it!' He tried the wine, flared up and shouted so loud that the walls shook.

'Wine steward!!' The wine steward came.

'To whom are you giving this? Have you forgotten that these are the tsar's guests?' And the bottle flew on to the floor. After this they always brought good wine.

At Peterhof two rehearsals were conducted on Olga Island, where they took us in court coaches.

The performance was set for 8 o'clock in the evening. On the day of the performance the weather was excellent, but towards evening it got damp and raw, and rain began to fall, for which reason the performance was put forward until 9 o'clock. We were chilled to the bone. Everyone dressed in the English Palace and went to Olga Island in dancing dresses, while we threw on a burnous over our costumes. We were helped from the carriages by all the top brass who had come for the performance. After we arrived they told us that a signal would indicate when to begin, and we had to wait for it a long time in the soldiers' tents, pitched along the shore of the lake. We shivered and took refuge in our burnouses.

Finally the word came: take off your over-clothes and find your places on stage! We took our places on the rafts and waited there from 7 o'clock. We were frightfully chilled. It drizzled for a while but then stopped. The signal rocket soared overhead and caused a terrible fright: the tail of the rocket, with its smouldering powder bag, fell on to one of the rafts. But some young sailors quickly put everything right.

The music started. The orchestra was very artfully placed in front of the stage and was invisible, just hidden by vegetation and flowers. The performance began, and everything went splendidly. It was given in natural light, but as it was cloudy, the empress requested that the performance be repeated the following day, but earlier—at 7 o'clock.

The next day there was burning sunshine. We went back to Olga Island, but were not cold now. When the performance ended the emperor commanded us not to leave. We were called to the stage, where the tsar, the empress, and entire imperial family had already gathered. Our sovereign said, 'I beg you, have something to eat.'

In the twinkling of an eye food-laden tables were placed on the stage, and negroes brought in cottage cheese, yoghurt, and fruit on trays. We stood motionless. The tsar turned to us and said: 'Why are you not eating?'

Addressing the sovereign we always suffixed the letter 's' to our verbs as a sign of deference. And the emperor would tease us by returning the compliment. Flustered by such high rank, we answered, 'We are full, Your Majesty.'

But he clearly did not believe us and laughed, whereupon Madame Richard said: 'In your presence, Your Majesty, they are too bashful to eat.' The emperor burst out laughing and said, 'Ah, I don't want to inhibit you,' and turning his back to us, said: 'See to it.' And indeed, we saw to it—the table was picked completely clean.

The tsar was very pleased with this and laughed. The empress asked to see the mechanical apparatus of the performance and the movement of the seashells along the lake, which was very cleverly done. She asked Roller to show her without the actors. And the empress was shown. It was constructed as follows: the orchestra was at the shore's edge, concealed in vegetation, and beyond it the large stage on the water; on the stage were decorations of flowers and greenery. At the sides of the stage there were rafts manned by sailors. We approached in groups, the same groups as in the carriages which brought us; when one pair of rafts went off at one side of the stage, another pair appeared on the other side, and the stage filled with naiads. The rafts were pulled from ashore by ropes on pulleys beneath the stage. After all the naiads were on the stage the central raft approached, carrying the queen of the naiads and the little students. Shaped like a seashell, this raft moved right up to the audience. During the performance the groupings illuminated by the sun were astonishingly beautiful. Throughout the ballet, especially during pantomimes, the song of a nightingale was heard. For this some gentleman was invited who for 5 roubles most tastefully sang like a nightingale. He was located on the shore among the trees.

The Peterhof performance remained in our memories like a kind of magic dream.

After inspecting the stage apparatus the empress again spoke with the students, and then the emperor expressed the wish that the so-called 'blue mazurka' be danced, which had been performed by the Polish dancers. Gedeonov explained to the sovereign that they hadn't brought the proper costumes. To which he responded: 'That is nothing. Do it in those you have. And will the orchestra be able to play it?' he asked. 'The orchestra will play it, Your Majesty,' said the Kapellmeister Lyadov.

At that moment the fishermen from the ballet *The Naiad* straightened out their trousers, and four pairs of dancers performed the blue mazurka. The emperor was very pleased and thanked them. After this everybody dispersed. Gifts were given to the leading artists.

After the performance they gave us dinner, and then all the artists and students were taken back to Petersburg on the steamer. During dinner Gedeonov said, 'Eat heartily, people, but just remember, you don't need too much wine, since everybody is going back on the steamer, students included.'

It was very merry on the steamer. Whoever courted whomever, such-and-such sat with so-and-so. The director, walking along the deck, noted, laughing, 'Yet everybody is in pairs? Oh well, let it go—today's a holiday, a wonderful holiday!'

IV
Fanny Elssler and Jules Perrot in Russia

INTRODUCTION

The six years between the Taglionis' departure and the coming of Fanny Elssler and Jules Perrot were a time of mixed fortunes for the Petersburg ballet.

The Taglionis left after Butter Week of 1842, but before the year was out Petersburgers had seen the first Russian performance of *Giselle*. The dutiful Titus was responsible for this production, having been sent to Paris to study the work, which was hailed as a vehicle for Russian ballerinas and a tonic for the post-Taglioni doldrums.

Throughout the 1840s Raphael Zotov's ballet reviews in *The Northern Bee*—as in the 1830s, our richest source— struck an ambivalent note, wherein the negative effects of official policy were criticized without the management being cited as the cause.[1] Instead, Zotov blamed Didelot's retirement (when mourning the absence of a good choreographer), Marie Taglioni's departure, which was followed by unwarranted public indifference towards Russian ballerinas Tatyana Smirnova and Elena Andreanova, Russian literary figures (chastised for not writing ballet libretti), and a fickle audience (taken to task for switching allegiance from the ballet to the Imperial Italian Opera). According to Zotov these factors, not a want of talent, were responsible for the company's lack of artistic leadership and its inability to spark public interest.

Zotov's critique of the first performance of Titus's *The Talisman* is an example of his ritual lament. In a long preamble to remarks about the ballet, he wrote:

... Travellers who went around to every enlightened country unanimously declared that we had the best ballet troupe in Europe, and to win first place we lacked only Taglioni. Finally she came to us. Ah! At that point balletomania reached its high mark. It was a kind of enchantment, an ecstasy. In a word, it was the apogee of ballet's glory. But as the mountain climber who ascended a high peak with difficulty, and sensed the delights and pleasures of the beautiful view from the summit, must inevitably descend again, even unwillingly, it being impossible to go higher, so our ballet, having reached a similar height, must also come down again. The public got so used to Taglioni that even while she was still performing here the theatre came to be half empty, and when she left, the

---

[1] Oleg Petrov contends that Zotov's attitude was conditioned by his official service before turning to criticism. Petrov cites correspondence of Zotov and Faddei Bulgarin, editor of *The Northern Bee*, which set forth two guide-lines for the conduct of theatrical criticism: not to touch the theatre management, directly or indirectly, 'as you [Zotov] promised to Leonty Vasilievich Dubelt [head of the Third Section]', and in literature and artistic judgements to stay as close as possible to the spirit of *The Bee*, which Bulgarin elsewhere described as 'honourable and noble, Europeanism without vile revolutions and impertinences'. See O[leg Aleksecvich] Petrov, *Russkaya baletnaya kritika kontsa XVIII–pervoi poloviny XIX veka* [Russian Ballet Criticism of the End of the 18th and of the First Half of the 19th Centuries] (Moscow: 'Iskusstvo', 1982), p. 169.

mainsprings set in motion by this passion for ballet suddenly weakened and ran down.

... In vain our good and solicitous management signed Lucile Grahn and Roussel: both had to leave because after Taglioni our public was difficult to please. In vain our excellent dancers, Andreanova and Smirnova, expended all their art to restore the ballet to glory, and they were forced to seek laurels and applause in foreign lands.

Materially the company survives. We have charming women dancers, skilled male dancers, an amazing corps de ballet: a seed plot which daily blossoms with the rarest gifts, the troupe lacks only the pleasure the public derives from looking at it. The sole means of reviving this pleasure would be the creation of new, substantial ballets; but for this a new Didelot is required, a new Prometheus, a new choreographer of genius; but such a person, it seems, does not exist in Europe at the present time.[2]

Zotov's emphasis on the same themes, often at the expense of the work being reviewed, makes it difficult to know the true state of affairs in the Petersburg ballet.

There is no question, however, about the salutary effect of an event which occurred in 1847: the coming to St Petersburg of Marius Petipa, who was followed in a few months by his father Jean. The elder Petipa taught, the younger danced. Marius also had a hand in the first Russian production of Mazilier's *Paquita*, both father and son in the same choreographer's *Le Diable amoureux*, known in Russia as *Satanilla*. If it is fair to claim that Jules Perrot was the 'new Prometheus' for whom Zotov hoped, one must also acknowledge the bracing effect these productions had on the company and its audience in the year before Perrot arrived. Even Zotov was pleased: 'Our beautiful ballet company was reborn with the production of *Paquita*, and the production of *Satanilla* and its superlative performance placed the company again at its former level of glory and widespread affection.'[3] When *Esmeralda* first attracted critical attention, praise was expressed by placing the new ballet in the same class with *Satanilla* and *Paquita*.

A list of the ballets of the winter season prior to the arrival of Perrot and Fanny Elssler offers a statistical perspective on the state of the Petersburg company at that time. Forty-seven performances were given of nine ballets. For a Russian company sensitive to Parisian fashion, it is a respectable selection:

---

[2] R. Z., 'Feuilleton. Theatre Chronicle', *Severnaya pchela*, 1847, No. 21 [27 Jan.], p. [81].
[3] 'Feuilleton. Annual Review of the St Petersburg Theatres. Second Article', *Severnaya pchela*, 1848, No. 60 [16 Mar.], p. [237].

Paquita (Mazilier) — 18
La Péri (Coralli) — 8
The Simpletons (Auguste) — 3
Giselle (Coralli/Perrot) — 4
L'Elève d'Amour (Taglioni) — 2
La Fille du Danube (Taglioni) — 2
The Millers (no choreographer specified, probably Jean-Baptiste Blache) — 1
Revolt in the Seraglio (Taglioni) — 1
Satanilla (Mazilier) — 8[4]

Fanny Elssler came to Russia in the autumn of 1848 and performed there for the last time (in Moscow) in February of 1851; Jules Perrot arrived a few months after Elssler, and served as balletmaster of the Imperial Theatres until 1860.[5]

They both made lasting impressions. Elssler was one of the most fondly remembered ballerinas in Russia, whose art for decades served as the standard by which that of other dancers was measured; some of Perrot's ballets were performed until the October revolution. Perrot was like Didelot in many ways: he danced in *Zéphire et Flore* (in Paris in 1831, as Taglioni's partner), he was celebrated for composing heroic narrative ballets, he suffered as a Frenchman in Russia at a time when Russia was at war with France, and he spent his last years there in vexing disputes with theatre officials. Unlike Didelot, Perrot did not espouse the cause of Russian artists, a point to which Stukolkin bears witness (above, pp. 118 and 123).

After an awkward moment at her first appearance, which was 'for some strange reason cold,' Elssler won over her audience with her 'fire, vivacity, animation, splendour, and attractiveness.'[6] It was precisely Elssler's art of characterization, especially of earthly heroines, which complemented Perrot's reputed gift for creating choreography in which dance was permeated with gestures meaningful to the story. This combination gave *Esmeralda* a particular animation and expressiveness, as Zotov recalled in his review of the season three months after the first performance:

---

[4] 'Feuilleton. Annual Review of the St Petersburg Theatres. Second Article', *Severnaya pchela*, 1848, No. 60 [16 Mar.], p. [237].

[5] The carefully researched studies by Ivor Guest, with extensive bibliographies, make unnecessary any prolonged treatment of Elssler and Perrot in these pages. See his *Fanny Elssler* (London, 1970) and *Jules Perrot. Master of the Romantic Ballet* (London, 1984).

[6] R. Z., 'Feuilleton. Theatre Chronicle', *Severnaya pchela*, 1848, No. 224 [7 oct.], p. 894. 'The audience tried to correct this error, when she had already left the stage, with calls and applause,' Zotov continues, 'but the first impression must have astonished the famous dancer. Nowhere has she met such an indifferent reception.'

Here we saw not just a ravishing dancer but also a sublime artist. Her dances themselves (the *pas d'action*) revealed a new dimension of this art. Up to now we required only graceful poses, supple movements, lightness, speed, strength: here we saw *acting* in the dances. Each movement spoke to mind and heart; every moment expressed some feeling; every look was in keeping with the action. It was a new, charming revelation in the sphere of choreography.[7]

Ekaterina Vazem, the reigning ballerina of the 1870s, remarked on Perrot's ability to integrate dance and mime:

Perrot's ballets were sharply distinguished from the works of other balletmasters contemporary with and subsequent to him by the predominance in them of the dramatic side over the dance. Perrot, who always composed his own scenarios, was a great master of conceiving effective stage situations, which fascinated and at times even stunned the spectators. There were comparatively few dances in his ballets, far fewer than in works of much later origin. At that, composing these dances the balletmaster took care not so much to give the performers a more effective number as to make the danced numbers supplement and develop the dramatic action.[8]

Like Didelot's *The Captive of the Caucasus*, Perrot's *Esmeralda* was adapted from an earlier source quite different in medium and content:

The person of Esmeralda, overflowing with poetry, is a special type, and has served for many stage works. They all depart radically from Hugo's novel, of course, because it is impossible to produce in a theatre all the astonishing scenes that Hugo created. The balletmaster Mr Perrot took several episodes from Hugo, assembled them, and as a choreographer of taste, altered the denouement, which could never have been put on stage. He ends the ballet with the marriage of Esmeralda and Phoebus, and the dispatching of Claude Frollo to prison. Perrot transformed Quasimodo into a completely insignificant character. Sparing the sensitivities of the lady spectators, he even put Claude Frollo's attempt to kill Phoebus in the wings, not on stage.[9]

In the rest of his review of the first performance, Zotov was long on praise but short on description. 'Is it really necessary to talk about the incomparable benefit artiste?' he asked. 'The entire roster of adjectives which serve to express universal delight has long since been exhausted, and we shall therefore not repeat them. Fanny Elssler danced—consequently it was charming, delightful, amazing!!' He complimented Perrot's performance as Gringoire, then pointed out that 'the genuine benefit deriving

---

[7] R. Z., 'Little Bee. Theatre Chronicle. Review of the Activities of the St Petersburg Theatres' Artistic and Literary Components for the Current Theatre Year. (Completion)', *Severnaya pchela*, 1849, No. 46 [1 Mar.], p. [181].

[8] E[katerina] O[ttovna] Vazem, *Zapiski baleriny Sankt-peterburgskogo bol'shogo teatra, 1867-1884* [Memoirs of a Ballerina of the Saint Petersburg Bolshoy Theatre, 1867-1884] (Moscow and Leningrad, 1937), p. 53.

[9] R. Z., 'Feuilleton. Theatre Chronicle', *Severnaya pchela*, 1848, No. 293 [30 Dec.], p. 1170.

Introduction

from him will be in choreography; for a role like Gringoire we already have sufficient personnel.' In all, Zotov found one fault:

Least of all were we satisfied with [Elssler's] *pas de deux* with Mr Johanson, not for its execution, which, of course, was as enchanting as everything else, but for the very genre of the *pas*, which seemed to us quite unsuited to Esmeralda, a gypsy living in the period of Louis XI. As a benefit artist, Fanny Elssler was absolutely right to choose a noble, grandiose *pas* to dance at a gathering of distinguished company; but we doubt that Esmeralda ever saw a dance like that in her time and circle.[10]

Like *Giselle*, *Esmeralda* found a home in Russia, where it was a test piece for ballerinas from Elssler and Carlotta Grisi in the 1850s to Virginia Zucchi and Mathilde Kshesinskaya in the 1880s and 1890s.

[10] *Severnaya pchela*, 1848, No. 293, p. 1170.

7

Libretto of *Esmeralda*

Jules Perrot

ESMERALDA
Grand Ballet
in Three Acts
and
Five Scenes

Saint Petersburg
At the Typographer Ilya Glazunov and Comp. 1848

Printing permitted. St Petersburg, 19 December 1848
Censor A. Mekhelin

Esmeralda
Pierre Gringoire, *an impoverished poet*
Claude Frollo, *a syndic*
Quasimodo, *bell-ringer of the church of Notre Dame*
Phoebus de Chateaupers, *a young officer*
Aloise de Gondelaurier
Fleur-de-Lys, *her daughter and fiancée of Phoebus*
Diane de Christel, *her friend*
Clopin Trouillefou, *head of a band of vagrants*

Soldiers, vagrants, beggars, gypsies, and people

The action takes place in Paris, at the end of the fourteenth century.

ACT I

SCENE 1 The Courtyard of Miracles

(*A crowd of gypsies and beggars fills the stage. Their leader, Clopin, spokesman for the gathering, is sitting in the middle of the stage on a barrel.*)

Pierre Gringoire runs in. Barely catching his breath, he falls at Clopin's feet, hoping to find there protection from the crowd of vagrants and beggars who are chasing the unfortunate poet. Clopin orders him to stand up; the beggars surround Pierre, wanting to rob him. But after a thorough search all they find is a poem, precious only to the poet; he has nothing else. Cheated in his hope of spoils, Clopin condemns Pierre to death, and despite the latter's entreaties, announces that he will be hanged, adding, however, that marriage to one of the women standing around him can save the unfortunate lad from death. Alas! They refuse Pierre's hand. He faces inescapable perdition, and despairing, prepares for death.

At this moment Esmeralda enters. Walking past Gringoire she observes his plight. Touched, Esmeralda declares that she is prepared to marry Pierre. The poet cannot believe his luck. The wedding rite is begun. At Clopin's command, Esmeralda hands Pierre a clay vessel, which, according to the beggars' custom, must be thrown to the ground and broken into four parts by the bridegroom. After this Clopin gives his blessing to a four-year marriage of the gypsy Esmeralda and the poet. The newly-weds take part in a celebration in their honour.

Meanwhile bells sound in the distance, calling the faithful to evening prayer. This disturbs the beggars' unbridled merriment, and they disperse. A patrol led by Phoebus makes its way around the stage, followed by Claude Frollo. The soldiers proceed on, but Frollo and Quasimodo hide nearby, waiting for Esmeralda. The charming gypsy enters. Quasimodo and Frollo rush from their ambush and seize her. The poor girl tries to defend herself in vain; she is at Quasimodo's mercy when suddenly Phoebus returns with the soldiers; he runs to the bell-ringer and wrests the girl from his arms. Frollo flees, leaving Quasimodo in the the soldiers' hands.

Esmeralda doesn't know how to thank her rescuer, who wants to know whom he has assisted. 'I am an orphan,' says Esmeralda, 'shunned by the world, and I never experienced the tender care and love of parents and friends.' Phoebus, captivated by the gypsy's beauty, asks her to accept his scarf as memento. And the magnanimous Esmeralda, seeing Quasimodo with the soldiers, who mock him, asks Phoebus to free the wretched hunchback. The officer agrees. Esmeralda brings Quasimodo a cup of water to quench his parching thirst, and herself gives him his freedom. Phoebus is delighted with the girl and wants to embrace her. But she deftly evades him, and Phoebus, all the more enchanted by her resistance, must withdraw.

ACT II

SCENE 2. The Newly-weds
(*The stage represents a small room, in which there is a bed, a chair and a table.*)

Esmeralda enters. She cannot take her eyes from the scarf Phoebus gave her. Memories of the young warrior will not go away; she takes several tin letters, with which she spells her beloved's name, and abandons herself to sweet reveries. Just then Gringoire enters. He hastens to his young wife, and full of rapture and passion tries to embrace her lithe form. But the frightened Esmeralda rejects his embrace and moves away. Gringoire pursues her and catches her despite her agility. The poet is at the point of embracing her again when the dagger which flashes in Esmeralda's hands stops him.

After her fright has passed, Esmeralda explains to Pierre that she agreed to marry him because she was touched by his situation, and hoped to find in him not a lover or a spouse but a protector to fend off the dangers the young gypsy faces every day as she dances in the streets. Gringoire submits to her will, and they try out the new dance which she plans to perform the next day. Soon, despite the pleasure Pierre experiences in Esmeralda's company, he feels he must rest and sleep. Esmeralda shows him the room where he must spend the night. He retires.

At this moment Frollo enters, and Quasimodo is seen in the doorway. Frollo falls at Esmeralda's feet, declaring his love for her and begging her to return it. Esmeralda orders him to leave and says that her heart already belongs to another. Jealousy and fury bring Frollo to despair, and he resolves to abduct her by force. At his sign Quasimodo runs in. Frollo looks to see if the doors are locked, and leaves Esmeralda with Quasimodo, who has not forgotten the favour Esmeralda did him, and who therefore lets her get away through a secret passage. Frollo pursues Esmeralda.

SCENE 3

(The stage represents a garden. Preparations for the wedding of Fleur-de-Lys and Phoebus are underway.)

Fleur-de-Lys enters. She has left the ballroom and is quickly finding her way to the garden, consoled by the thought of meeting her beloved fiancé. Seeing that he is still not there, she calls on her friends to prepare some bouquets for Phoebus. Just then Aloise de Gondelaurier comes in. The women respectfully greet her, and Fleur-de-Lys goes to her and shows her all the wedding preparations. Pages enter, announcing the arrival of her fiancé. Phoebus comes in next; first he bows to Aloise, then goes to his beloved and kisses her hand. But the discerning Fleur-de-Lys immediately notices that Phoebus is not wearing the scarf she gave to him.

The celebration begins; Esmeralda's tambourines are heard in the distance. Fleur-de-Lys wants to see the gypsy; the company hasten to satisfy her wish and call Esmeralda, who enters with Gringoire behind, holding her guitar and tambourine. Fleur-de-Lys, struck by Esmeralda's beauty and shapeliness, begins to chat with her. Esmeralda tells her that she knows the science of predicting the future. Fleur-de-Lys is quick to take advantage of this rare art, and Esmeralda predicts she will have wealth and joy in marriage. Fleur-de-Lys, pleased with this horoscope, gives the gypsy a ring and asks her to begin the dances.

Meanwhile Phoebus tries in vain to hide from Esmeralda's glances; forgetting all propriety he runs up to her, then remembers that his bride is watching, stops and asks Esmeralda to continue her dance. The glances, full of fire and love, which he casts at Esmeralda clearly indicate his passion. Fleur-de-Lys, angered by her fiancé's actions, reproaches him, reminding him of his vows to her. Thinking only of Esmeralda, Phoebus responds coolly to his bride's rebukes and tries to assure her that his love for her has not yet waned. Esmeralda continues to dance with Gringoire; but to signal to Phoebus that he alone rules in her heart, she points to the scarf which the young officer gave her, explaining that she will keep it forever.

At the sight of the scarf which passed from Phoebus's hands into Esmeralda's, Fleur-de-Lys can no longer contain her jealousy: she seizes the scarf, and faints at Phoebus's feet. They carry her out. Gringoire leads Esmeralda away together with Phoebus, who cannot leave the enchanting gypsy.

ACT III

SCENE 4. Jealousy and Love

(*The stage represents a room at an inn. It is night.*)

Frollo enters, hunting for a place to hide. He is holding a dagger, with which he wants to avenge himself. Hearing something, he quickly takes cover. Esmeralda and Phoebus come in. Phoebus declares his eternal love for the gypsy but she does not believe him; she is jealous of Fleur-de-Lys, and throws down the ring his former fiancée gave to him. Esmeralda tells Phoebus that all his vows and desires will vanish with time, like a feather she throws into the air. But Phoebus's words are so enticing, so eloquent, that the innocent child of nature falls under his spell, and he embraces her.

Frollo rushes in at this moment. Seeing the lovers retiring into the next room, he runs after them. Behind the scenes a groan is heard. Frollo has satisfied his need for vengeance. Phoebus has fallen his victim. But Frollo is still not content: he summons the people and declares Esmeralda guilty of Phoebus's death. They seize her and put her in chains.

SCENE 5 The Feast of Fools

(The stage represents the banks of the Seine. To the right a fortress. In the background the towers of the cathedral of Notre Dame are visible.)

A crowd of beggars on stage are celebrating the so-called Day of Madmen. They elect as king of the celebration the bell-ringer Quasimodo, whom they carry in triumphantly. Frollo enters. He makes his way through the crowd, but suddenly the sight of the hunchback gives him pause: he recognizes Quasimodo. Frollo indignantly rips away the robes with which the bell-ringer had been invested.

At this moment the gates of the fortress open and reveal a contingent of guards leading away Esmeralda, who has been sentenced to death. Gringoire begs the soldiers accompanying Esmeralda to let her be buried with the scarf Phoebus gave to her. Frollo goes to Esmeralda and explains that she can still be saved, that one word can deliver her from death—one word and Frollo would be happy, Esmeralda free. That word—is a word of love. But Esmeralda is uncorrupted; she prefers death to treachery, pushes Frollo back, and her gaze, turned heavenward, clearly expresses the trust she puts in heaven's justice. In despair and malice, Frollo orders Esmeralda to be taken away.

Suddenly a soldier enters and approaches Esmeralda; at the sight of him she falls in a faint—at her lover's feet. The warrior is none other than Phoebus, the same Phoebus whom Frollo thought dead but who was only seriously wounded. He reveals the true culprit and announces that Esmeralda is innocent of any crime. Frollo rushes at them in an effort to kill both Esmeralda and her lover with one blow. But Quasimodo wards off the villain, and wresting the dagger from his hands, stabs him. Esmeralda, coming to her senses, rushes to Phoebus's embrace and offers him her hand and heart.

V
Nadezhda Bogdanova

INTRODUCTION

In 'Two Forgotten Dancers', an article published in the wake of Diaghilev's first season of ballet in Paris, two Russian ballerinas of an earlier time are described, who, like Diaghilev's dancers, performed with great success in western Europe. One was Gedeonov's favourite, Elena Andreanova; the other was Nadezhda Bogdanova.[1]

Of the two, Bogdanova had decidedly the more curious history. Unlike other Russians who danced abroad, she trained abroad, being taken to Paris to study after initial work with her father, a one-time régisseur at the Moscow Bolshoy who claimed to be a student of Didelot. The family's travels and tribulations are the subject of 'The Bogdanov Artistic Family', the richest nineteenth-century source about Nadezhda Bogdanova.[2] In addition, this article provides us with a rare glimpse of balletic life in the provincial cities of Russia at mid-century, and with the reason why the family decided to go abroad: young Nadezhda performed for Fanny Elssler, who encouraged her father to make the trip.

It is difficult to know if Bogdanova's father had planned to return to Russia in triumph, taking advantage both of Nadezhda's nationality and her budding status as a Parisian celebrity.[3] In any event, he did not reckon on the Crimean War, and ended up losing on both counts: Bogdanova's celebrity in Paris was blunted by her nationality, and they returned from France right after that country had defeated Russia in the war.

Bogdanova nevertheless joined the Petersburg company and made her début, to great acclaim, in *Giselle* on 2 February 1856. For the next two years her success continued, but not without difficulty. 'Unpleasantness has pursued me in Petersburg from the very beginning,' she wrote to Director Gedeonov in October of 1857, attempting to resign,

---

[1] M. V. Karneev, 'Two Forgotten Russian Dancers. (Apropos the Success of the Russian Ballet Abroad', *Ezhegodnik Imperatorskikh Teatrov* [Yearbook of the Imperial Theatres], 1909, pt. iv, pp. 120–9.

[2] On Bogdanova see also Ivor Guest, *The Ballet of the Second Empire 1847–1858* (London, 1955), pp. 57–8, 95–6; Guest, *The Ballet of the Second Empire 1858–1870* (London, 1953), pp. 84–5; Vera Krasovskaya has written the most substantial modern account of Bogdanova's career in 'The Dancer Nadezhda Bogdanova', *Uchenye zapiski* [Scholarly Papers], vol. i (Leningrad: State Scientific Research Institute of Theatre and Music, 1958), pp. 295–322. The biographical data given hereinafter are taken from Krasovskaya.

[3] Vazem recalled: 'She [Bogdanova] was in fact a stage personality more famous than outstanding. Her family (a father who was a former dancer, one brother who was a dancer and the other a violinist) very much loved publicity, and the career of Bogdanova the ballerina was made much more by publicity than by all her artistic resources, taken together' (*Zapiski baleriny Sankt-peterburgskogo bol'shogo teatra*, pp. 77–8).

And all the while I had the boldness to hope that through all vissicitudes my zeal and good intentions, the cordial reception of the public, and my recognized success on foreign stages would lay all this to rest.[4]

The issues here, as clarified in later correspondence, are salary disparities between Bogdanova and visiting celebrities, and that she was not scheduled for leading roles in new productions. As Bogdanova returned to Russia in the middle of Perrot's term as balletmaster, his preferences, and later those of Arthur St-Léon, may have played a part in official decisions.[5]

Disputes notwithstanding, Bogdanova remained on the rolls of the Petersburg ballet until 1875, but no longer danced there after 1864, and even before that, only such as to allow herself a peripatetic life of appearances elsewhere: Budapest in 1858, Naples in 1859, Moscow in 1862, Paris in 1865, Warsaw in 1866 and 1867, the year in which she gave up performing.

As a document, 'The Bogdanov Artistic Family' is exceptional in the balletic literature. Its patriarchal tone, unsparing of severity, its stress on virtue and sacrifice, suggest that the author may have been a seminarian. This possibility is enhanced by the unusual construction of the piece, which is cast in a classical *panegirik* form, also used in Byzantine and Russian literary lives of the saints. The formula of the *panegirik* is revealed in the following elements of the article:

A formal, rhetorical introduction.
The comparison of a life or talent with stars in the heavens.
The description of a good, pious family.
The description of the subject's childhood.
The description of the beginning of spiritual trials and labours.
The telling of problems encountered in the world from uncomprehending persons, from malice, envy, and scorn for humility.
The account of constant work towards perfection, an ideal.
The reward for work in triumph over adversity.
The concluding reference to the Lord and to the monarch of a state.

There are other religious overtones in the account. Like a holy fool, Bogdanova remained mute at class in Paris, her presence there at all construed as the result of miraculous coincidence. Even her name has spiritual connotations, 'Nadezhda' meaning 'hope' and 'Bogdanov' meaning literally 'gift of God'.[6]

[4] Leningrad, USSR, Historical Archive, *fond* 497, *opis* 2, *delo* 15910, fo. 22, quoted in Krasovskaya, 'The Dancer Nadezhda Bogdanova', p. 313.

[5] Krasovskaya suggests that St-Léon, who aided Bogdanova's cause in Paris, may have turned against her after they appeared successively in Budapest, unable to forgive her for making a stronger impression than he had. Yet the only new role she received in St Petersburg was in St-Léon's *Météora* in 1861 ('The Dancer Nadezhda Bogdanova', p. 318).

[6] I am indebted to Natalia Challis for pointing out these connections, and for information concerning the *panegirik*.

'The Bodganov Artistic Family' was printed in *Muzykal'nyi i teatral'nyi vestnik* [Musical and Theatrical Messenger], 1856, No. 30, pp. 538–44; No. 31, pp. 555–62, and was subsequently issued separately.

8

'The Bogdanov Artistic Family'

I

Talent is a beautiful gift from heaven—an old thought, but one which will never age. It is a gift from heaven because talent always strives for the good and the beautiful in the service of humankind, which is its true calling and prescription. Of course, every person must serve humanity according to his powers, but the talented person especially: to whom more is given, more will be asked. And if every mortal who has forsaken his true calling for an egotistical life warrants reproach, then the talented person deserves a bitterer reproach for having despised and squandered his powers to no purpose. Is it not his part to awaken the best in humanity, to instil the highest strivings in man, to tear him away from the dust of the earth, to give him pleasure and lighten his earthly journey, often most burdensome, in every possible way? Neglecting this is a sin, and as a sin it will be judged! So then, glory and honour to the talented who can develop themselves, who have made themselves outstanding, guiding stars. Talent is like a star in the heavens: just one flickering, sparkling little star enlivens your sight to the whole celestial expanse of blue; recall how a thousand stars enticed you as they stirred your musings and diverted you from the unbroken, sometimes heavy chain of everyday thoughts. How comforting and bright your soul was, uplifted in this golden expanse, there to sense for a moment being removed from base mankind, superior to it, alas . . . though perhaps only for a moment. Such moments are not forgotten, and one cannot but recognize their beneficial aspects.

True talents, of course, inspired talents, are very rare; that is why one must consider blessed the family into which such a talent is born and which, understanding its calling, is able to nurture it. But to see a whole family of talent is especially rewarding: it makes a beautiful unit, attracts attention unwilled, and inspires the sincere desire that this unit be not only the glory of its native land, but also an adornment of mankind, for which the first condition is: to be *human*.

There are several talented Russian families, on which it is pleasing to pause and rejoice the heart. The Bogdanovs now join them. First place among the Bogdanovs belongs to Nadezhda Konstantinovna, renowned in Paris, Vienna, and Berlin for her enchanting dances, who now has come to delight her countrymen. Having devoted herself to art, she promises still further development, thereby to comfort us more. She does not, it would seem, look lightly upon her art, which exists not just for the eyes, but like every art has a more important side—its pleasant effects on the soul. With her gracefulness she therefore speaks to your spirit, clearly and tenderly, as very few persons can. She lacks the exclusively physical appeal which most women dancers have, even talented ones, for which reason some individuals will not like her; but people who seek true art, not voluptuous distraction for the eyes, will appreciate her. Her younger sister and two brothers are still developing, and show promise for important successes, the first as a dancer, the latter as musicians—the second brother, Nikolai, being a talented pianist and a dancer as well.

Their esteemed father, Konstantin Fyodorovich Bogdanov, was first dancer in the Moscow ballet company. A student of the famous Didelot, he became a teacher of ballet in the Moscow theatre school, the site of his own first training. In later years he was also senior régisseur. In 1836 a daughter, Nadezhda, was born to him. This charming child grew up carefree, as any child in any class and estate: neither her parents' shortcomings nor fears, nor memories of the past, nor repentance, clouded her youthful soul. The first years of childhood normally pass this way, those happy years which, alas, stay but vaguely in the memory. After that comes another period, when one begins to distinguish the flowers from the thorns, and to understand who is called to tread upon the roses, and who upon the thorns. Little Nadezhda Bogdanova had to begin developing her talent on the thorny path, afterwards to appear, in brilliance and triumph, before cultured European society. It is pleasant to recall the difficult road now successfully travelled, but hours spent in bitter tears cannot be effaced from these memories.

Meanwhile Bogdanov's family increased every year: the children were growing up, and with them grew concerns about their education and future. The father began to teach young Nadenka dancing, but in doing this was not thinking about his daughter's promise, even in its budding stages. And at this early stage it did not occur to him to press her talent, or better said, to force it out of the child, as some parents do, supposedly fond of their children but wishing to find in them a plentiful source of income. His goal was more modest: to give his daughter, in time, the means to find work and earn an honourable living. He was preparing her to be a dancing teacher, an unassuming, insignificant fate. But one event changed her father's plans; suddenly flattering hopes inspired him, and he was not deceived. At a merchants' assembly in Moscow a children's ball was given; the gracious Nadezhda was among the children. It was perhaps the first time that she had appeared at such a large gathering, in a crowd of merry children flitting about; perhaps timid and curious, she watched the busy activity around her; here she was, like a child, not thinking to entice or attract anybody. Caught up in the moment, she could not think of a suitable way to show her talent in front of others. But an opportunity arose: they made her dance the cachucha. Shy and inexperienced, the bashful little girl came to the centre of the hall, surrounded by a throng of curious guests. She began her lively dance, but when she saw everyone's gaze turned upon her she got confused, stopped, and was completely at a loss. The charming girl's graceful appearance awakened warm sympathy; everyone began to encourage and to coax her. And so she begins her dance again, and this time performs with special animation. This was the first flash of her budding talent, and having brightened her, the rays of that flash were reflected in all the spectators. Everyone was delighted, astonished at the child's talent, and showered the child, again embarrassed, with heartfelt praise. Nor did it pass unnoticed that the girl was not rich: one hundred roubles was immediately collected for her, her first gift. . . .

Her father was astonished, watching his daughter's lively dance: for the first time it struck him that her talent might be extraordinary, and to nurture and take solace in that thought is pleasant to every father. He began to ponder other, more brilliant plans: he wanted to pursue the serious development of his daughter's talent, but was concerned about how to pay for it. And at this time his two sons, Alexandre and Nikolai, began to display musical abilities, one on the violin, the other on the piano, and both showed an aptitude for dancing, and the younger daughter Tatyana promised to follow in her sister's footsteps. The most important question facing the solicitous parents was: where can we find the means to finance our children's education?

After much deliberation and many plans, they decided to test their children's talents in the provinces. The idea proved to be a good one, and was not long put off. They set about the task immediately: one brother learned ballet music, the other ballet dances with his sisters under their father's supervision, the mother got everything necessary for the road, and soon the family ballet troupe, except the father who was still in official service, left for Yaroslavl [about 100 miles north and slightly to the east of Moscow].

II

It was the winter of 1848. No outsider could have divined that the travelling company, with small children and modest resources—not a spectacle brilliant to the eye—was setting out on its slow journey not for subsistence, lacking other pursuits, but with the highest striving for education, hoping that God would bless their efforts. Perhaps on this snowy road in winter they dreamed of another road, green and blooming, enlivened by the charms of a beautiful spring.

Finally they reached Yaroslavl. Although this city can be considered the cradle of the Russian theatre because of its association with Volkov and Dmitrievsky, the first Russian actors, that coincidence did not influence the further development of the civic theatre there. Like all provincial theatres, it is maintained by a private entrepreneur. This was the first person with whom the newly-arrived family troupe had to deal. Having found the theatre, Madame Bogdanova left her tired children in a room and went to the proprietor.

'What is it you need, my dear?' he asked somewhat rudely, receiving her with a stern look.

Mme Bogdanova explained her situation and the wish to have her daughter make a début at the Yaroslavl theatre.

'Excellent. Come back tomorrow,' he said, again somewhat rudely, to the mother's request, 'Let her dance, but look, I won't pay her anything.'

Saddened by the proprietor's tone, she left, but would not abandon the family's plans. She promised to bring her daughter to rehearsal the next day. Meanwhile news of the young dancer's arrival spread from the proprietor to his company. The envious provincial actresses started malicious chatter about the débutante before they even saw her. Everyone crowded into rehearsal, impatiently awaiting the new arrivals. What whispering, what doubting smiles greeted mother and daughter on stage!*

Finally, looking important and theatrical, the proprietor came out in a dressing-gown and with a pipe. The little débutante was presented to him as 'Nadine'; he pointed to some attic which served as a dressing-room, where the two new arrivals went, humbly, in the unhospitable gaze of the women who were complete strangers to them. Let us not speak of the feelings which must have stirred within mother and daughter at this moment... The actresses followed them, surrounded the timid Nadine, began to dress her, stick on white make-up, and whispered, 'We're going to get her,' clearly among themselves. But their words were the usual, empty self-advertisement. Having prepared dances from the ballet *La Péri*, Nadine fluttered out on the stage, performed them, and compelled everyone to be more thoughtful.

'So then, that will do!' cried the director of the company in his rough voice: 'She reminds one of Sankovskaya. Excellent, I accept her. How much do you want, my dear?'

'Half the take!' Madame Bogdanova responded, artlessly, unaware

---

* All this the author took down in conversation with N. K. Bogdanova and with her esteemed mother—there is no romantic elaboration whatever in the entire biography.

that her demand was clearly impossible in the eyes of a provincial entrepreneur.

'What! Half the take?' he cried out in vexation. 'So clear out, my dear, I don't need you . . . If you want it, here's a contract: ten performances at twenty-five roubles in paper plus a benefit? . . .'

Madame Bogdanova thought for a moment and agreed: in this situation only famous people could argue and haggle, and first steps are never lucrative. Mr Bogdanov himself, who later arrived from Moscow, produced *La Sylphide* on the Yaroslavl stage for the benefit performance and played the leading male role, taking pains that his daughter's talent be shown in its full brilliance. The public received her enthusiastically and filled the theatre for each performance, to the director's great pleasure. His attitude softened, and he looked at the new arrivals with affection. When Nadine's two brothers wanted to play a concert, he offered them 100 roubles in paper for it. But there was an obstacle: one of the brothers had to have a piano, and nowhere could one be found. The concert would not have been played had not Messieurs T and M taken pity on them and lent them their instruments. The benefit performance and the concert brought the Bogdanovs a small sum, which they somewhat increased by giving dancing lessons in the city.

With this modest amount the artistic family could make a further trip: Mr Bogdanov had to return to service in Moscow; the others set out for Kostroma [about 40 miles east of Yaroslavl], having left a pleasant impression in Yaroslavl, where Nadezhda Bogdanova's artistic career began. At this time Kostroma had nearly been levelled by a fire, which destroyed most of the city. At night and in a single carriage, moving rather slowly, the family was suffering severe cold and frightful weariness, and, fatigued, approached the smouldering city. With great difficulty they found a dirty little room in a scorched house, open to the wind on all sides.

Having put her freezing and exhausted children to bed, Madam Bogdanova went to the governor, Count Suvorov, to whom she had a recommendation. Inquiring about him, she learned that he had gone to a concert which women of the nobility were giving for the benefit of people burned out of their homes. She went there and was presented to the governor. The count received her affectionately and advised her to take advantage of the opportunity: to present the children to the public immediately, after which he would do something for them.

'Take my carriage and fetch them!' he further counselled.

Reassured, kindly treated, the mother hastened to her children. They had fallen into a deep sleep after the tiring journey. With difficulty she awakened the poor creatures and got them on their feet; almost asleep, tears in their eyes, they dressed and reluctantly left the little house which gave them shelter, more precious to them at that moment than an illuminated, magnificent hall. At the courtyard, wind and the blizzard, darkness, the barking of dogs from the fire site—all this, unpleasant and depressing, affected the dozing children, and despite this they were being taken to a large public gathering, where they were to demonstrate their talent brilliantly: need does not ask about time. Thus they arrived. Little Nadine was quickly placed on the platform; emboldened by the favourable glances of the public, she was prepared to begin her dance, waiting for the music, but noticed some confusion between her brothers, who were her orchestra: the key to the violin case had been left behind. For now they broke the lock, Nadine looked around at her public, which looked back at Nadine with impatient curiosity, awaiting something special from the graceful girl. Finally the difficulty was resolved. Nadine did not disappoint expectations and delighted everyone. Loud applause and praise accompanied her as she was leaving the platform; the young boys also drew attention to their music, and many people took a genuine interest in the artistic family. Two evening performances were arranged, from which nearly 500 roubles in silver was made.

On this trip the family was limited to two cities, but it was, material advantages apart, extremely important for them: father and mother confirmed their children's talents, which could awaken sincere sympathy in the Russian public.

III

In Moscow Mr Bogdanov put his affairs in order, paid his debts, retired with a pension, as he had already served his obligatory term of service, and prepared for a second journey. Nadezhda's fate was decided: she was to have a dancing career; her future was based on it. Excerpts from the ballets *La Fille du Danube*, *La Gitana*, *La Sylphide* and many others were studied assiduously; her younger sister, little Tatyana, took part in this effort, representing genies and cupids: of the brothers, Alexandre conducted, Nikolai danced.

Having learned everything and made all preparations, the family left on tour again, but in a different direction—to Kaluga [about 100 miles south and to the west of Moscow]. Arriving there they agreed with the theatre director to eight performances for a third of the receipts plus a benefit. The public of Kaluga was pleased with this news and lavished praises on the entire family, most of all, of course, on Nadenka. From Kaluga they went to Tula [about 110 miles due south of Moscow]. This city could hardly boast its own theatre: small, crowded, it was more like a marionette theatre; it was difficult to put two musicians side by side in the orchestra, and the double-bass had to play in the boxes; the dressing-room was located beneath the stage, which one entered on a rope ladder. Lacking anything better, the family troupe had to make the best of things, even this infantile theatre. The director of a troupe of actors had already fulfilled his contract, for which reason the stage was available; but from ill will he left with the curtain and wings two hours before the performance, making it impossible for Bogdanov to give a ballet. After many requests and much trouble, the situation was somehow resolved, and Nadezhda triumphed on the miniature stage in *La Péri*. The production was barren and insignificant, the lighting was gloomy, the atmosphere smelly and dismal, but here also the public was drawn to the performance: Nadezhda's name was loudly and enthusiastically repeated in the auditorium.

From Tula the family went to Kursk [about 300 miles south of Moscow], where the marshal of the nobility took an interest in them. In the theatre at Kursk a benefit performance for some actor was being rehearsed; the drama *Don Quixote* was being given. Making use of the opportunity the newly-arrived dancer afforded, they added a Spanish dance to the play. The theatre here was not distinguished by its amenities. On the evening of the performance no dressing-room could be found for the poor girl other than the corner where a horse was standing, adorned with donkey's ears and assigned to Sancho Panza. In this company Nadine had to make ready for her appearance on stage after the pretend donkey, which lost its ears as soon as it went before the public, and for which it was received with bursts of laughter. The performance lasted until 2.00 a.m., and the poor girl had to wait until this hour, in what was practically a stable, to delight the spectators with her lively cachucha. Life's turns are strange, as strange as fate itself!

Having given several very successful performances in Kursk, the family went to Kharkov [about 400 miles south of Moscow]. Here the theatre had more amenities: it would be strange if this university city, rapidly developing and perfecting itself in many ways, had lacked a respectable theatre. It even had its own ballet chorus, organized by an art-loving estate owner. The theatre manager received the family of artists with misgiving: in Russia, provincial news does not spread through newspapers, and he knew nothing of our artists' successes in other cities.

'So, what can you do?' he asked, in doubting tones.

They listed the ballets they were prepared to dance, which could be produced without much fuss, even fuss over the music, which the brother had ready to hand. The distrustful manager, who had perhaps heard many unlikely promises from all manner of itinerant troupes, was not convinced.

'This might be like what happened once with a musician,' he said, 'who promised to play on twenty-four instruments; the public expected a full orchestra, and came to the theatre in crowds just to find twenty-four drums.'

Despite his misgivings, he agreed to let the new arrivals on his stage, and was not sorry for it later. Their success was all the greater because our artists found more facilities for producing a ballet, although the roster of personnel was not as large as on the previous stages. In Kharkov, rich in poets and nightingales (as an Odessa *feuilletonist* of the time expressed it), the Bogdanovs were showered with flowers and verses.

IV

With this the furthest wanderings of the artistic family ended. They returned to Moscow with various objectives and intentions. But Mr Bogdanov miscalculated his accounts, whereupon he deemed it advisable to visit Odessa and other cities in southern Russia, to amass the largest possible sum and put it towards the training of his children abroad. So thought, so accomplished. They appeared in Odessa [about 720 miles south-west of Moscow, on the Black Sea] in the summer of 1849 and stayed there three months. The enthusiasm they inspired is expressed in Odessa's *feuilletons* and verses of the time, Russian and French, published and manuscript. Odessa is an artistic and literary city and so considered itself obliged to honour, as it should, its talented guests and compatriots. Here Mr Bogdanov produced several ballets, in excerpts because here too he lacked the facilities for a full production. And here also Nadezhda Bogdanova was the greatest public attraction. At that time she was only thirteen, but she was received as a sixteen-year-old young woman. Here is how one of the *feuilletonists* described her. As we happen to be admiring her at this moment, this description is of interest: 'N. K. Bogdanova, who is only sixteen years old, is beautifully built and unusually well proportioned; her face is nice-looking and full of expression; in a word, she possesses all the gifts without which the most artful dancer cannot produce a complete effect. As for her art, speaking without overstatement, it fully corresponds to her richly endowed nature. Without the slightest effort, freely and lightly, she performs the most difficult *pas* and movements; there is absolutely nothing mechanical and hurried, nothing awkward—everything is rounded, lively, and proportioned, permeated with grace. This very gracefulness is, in our opinion, N. K. Bogdanova's finest quality and points to the true artist in her . . . We ask you to look closely at the dancer, not only when she is performing something difficult, but especially when her movements are simple, when for example she runs, stops, ponders, expresses agreement, refusal, astonishment, fright, love. Observe her face in these moments, her pose; everything about her expresses the feeling she is projecting, and makes her a charming model for pencil and chisel.'

Other Odessa notices about the young artist are offered in the same spirit. It is difficult for us now to judge if there is any exaggeration in these notices, but one can easily believe them, bearing in mind the qualities reported then which we still see now in Miss Bogdanova, though much perfected. We must add, however, that we can believe Odessa in what cannot always be believed in other non-capital cities.

The Bogdanovs stayed in Odessa until the end of September. Their farewell performance, as evident from the descriptions, was sensational. The theatre could not admit all who wanted to enter; many did not get tickets and were deprived of the pleasure. It was enjoyable to see our theatre at that moment, a *feuilletonist* says, after describing the city's chaos and troubling over tickets: the bright light, the numerous public, double the regular number of spectators in each box . . . The ballet began. Giselle fluttered about—and the theatre nearly burst from the applause; Giselle lifted her little hand—bravo! bravo! Giselle lifted her little foot—shouting and applause again (a Ciceronian turn of phrase!); Giselle rises up on tiptoe, and the audience's enthusiasm reaches its height. Shouting and applause continue throughout the ballet, and the divertissement from *Paquita* which followed it . . . Nadezhda Bogdanova with her brothers, sister, and father took fifteen calls in the course of the ballet. At the end of the performance, when the departing artists came out on stage for the last time at the public's delighted call, the enthusiasm knew no bounds; shouting, noise, applause, stamping in the stalls and the parterre—all this was new to the Odessa theatre, which has not seen such celebrations for a long time.

It is hardly necessary to mention all the bouquets, flowers and verses thrown at the young dancer's feet; delighted Odessa did not grudge her this. Speaking of Miss Bogdanova's art, the same *feuilletonist* notes: 'Judging dispassionately, let us say that she is an artist who promises to go far; but she needs to study still more, not because she has acquired so little, but because what she has acquired clearly points to the level of artistry to which she has a right, being so richly endowed by nature: she has to see first-class dancers, that she might take her rightful place among them.'

This advice did not go unheeded. The *feuilletonists* also had praise for the other members of the family: the nine-year-old Tatyana was called a fluttering humming-bird, an enchanting child in whom astonishing grace and lightness and a good-looking face come together. After such an enthusiastic reception and public notices, our artistic family could hope for the fulfilment of their well-laid plans.

From Odessa they went to Nikolaev [about 75 miles to the north-east] and found the theatre in wretched condition. It was broken down, in a basement; before the performance an architect checked to see that it posed no danger to the public. When it rained the audience had to sit beneath umbrellas, so little protection did the roof afford. Father and mother made every effort to make the stage as attractive as possible. The wife of Admiral Lazarev helped them, sending flowers from her greenhouse to decorate the barren stage. Despite the dilapidated theatre, young Bogdanova here also enjoyed a triumph with her performance, although one not free of dangers, as when she was getting dressed a beam fell and nearly killed her. Her art attracted the attention of Admiral Lazarev, who took an interest in the entire family. He advised them to go to Sebastopol [about 800 miles south of Moscow, on the Black Sea], for which a government steamer was about to leave. The admiral's advice was accepted with thanks. Having obtained a letter of recommendation from him, the family boarded the steamer, bolstered with new hopes in which they were not disappointed. Sebastopol was enchanted by the young dancer: she appeared now as Giselle, now as the Sylphide, both accompanied by unanimous applause and enthusiastic praise. In no other city did Mr Bogdanov receive such receipts as here: 5,000 roubles in silver significantly increased the capital being collected for the education of his children.

V

Leaving Sebastopol with pleasant recollections, they took the road to Kiev [about 475 miles south-west of Moscow). This trip was memorable: the snow-covered winter roads, storms and blizzards—all the rigours of winter came together at once, and often placed our travellers in most difficult situations. After great labour they reached their destination, and the first order of business was to fulfil the Christian vow they had made on the road: to fast and preface for communion if God allowed them to reach the capital of ancient Russia. Kiev at that time had famous trade fairs, to which many artists came, taking advantage of the large crowds of people. In those days various performances, concerts, and all kinds of entertainments were given almost daily. The Bogdanov family managed to attract widespread attention above all other artists. By turns the father with his sons and daughters now gave ballet performances, now concerts in the halls of the fair or the university. Their mazurka produced inexpressible delight.

Having amassed substantial capital, the family, pleased at this, set out for their native Moscow via now familiar Kharkov, where a crowd of poets received them with enthusiastic verses:

> At our hearts' secret call, to our purest desire,
> This talented family came to us again:
> We are filled with joy at seeing them.
> And I, despondent singer, am again inspired,
> Again before me are those children of Terpsichore,
> Of whom I sang with delighted heart:
> Again before me the choruses of sweet visions,
> And a ray of inspiration brightens my eyes.

Thus sang the *despondent* poet, but others, not despondent, were also inspired at seeing the children of Terpsichore again. Kharkov showed its poetical side: it expressed the joy of meeting in various ways, and then an elegiac feeling at the time of farewell . . .

In March 1850 the Bogdanovs returned to Moscow, just to learn that the European celebrity Fanny Elssler was to visit the city. Everyone knows what magic this name held for them: she was passing gloriously through the career for which Nadezhda was only just preparing—talent's celebration and talent's hope. To an outsider's mind, what could they have in common? The art they both serve they have in common, albeit at different levels, the art which inspires reverence for a beginning talent by a talent already developed and glorified. With what feeling the young Bogdanova must have looked upon Fanny Elssler; with what feeling her entire family looked upon her; the famous dancer's verdict was more important to them than all the delighted applause of provincial cities. Our artistic family impatiently awaited the guest. And then she arrived; she heard the Bogdanov brothers at a concert and responded to their talents in a flattering way. This emboldened the solicitous mother to approach the European celebrity and ask her to judge her daughters' talent. This did not happen immediately, but finally she managed to have Nadine dance for Fanny Elssler. The glorious artist's verdict was most favourable: she saw great ability in the young dancer and advised the whole family to go to Paris, there to study, work, predicting celebrity for the sister and brothers in time.

VI

Nothing could have stopped them now. They prepared for a lengthy journey and a long stay. The summer of that year they bid farewell to Moscow and to Russia. On the road to Paris, they gave concerts and performances: local journals wrote enthusiastically of the young artists from *Muscovy*. Especially amusing is a notice from the city of Chambéry, where Bogdanova danced at the royal theatre. It shows by what antediluvian legends many foreigners know us: 'It is above all of Russia that one can say that extremes touch: the minarets of the Kremlin—do they resemble the minarets of the Orient? (!!) Do you believe that the polkas, polonaises, the Livonian mazurkas don't bear some likeness to the dances of Isfahan? (!!) Oh! the beautiful odalisque, the delicious bayadère Nadezhda! . . . Come to see her Friday: the charming ballet in which we saw her dance yesterday evening will be given again. Yes, you beautiful ladies so competent in the practice of graces, come see grace personified. And you, our friends who love beautiful angels, come see the white daughter of the North.'

At the waters at Aix (Aix-les-Bains) on 19 August, Mr Bogdanov gave a musical and choreographic evening at which our dear compatriot delighted the spectators with her cachucha, krakoviak, mazurka (called 'Souvenir d'Odessa') and the *grande scène mimique* from *La Sonnambula*; her younger sister Tatyana danced a Tyrolese dance, and the brothers meanwhile performed variations for violin and piano on the opera familiar to us, *Askold's Tomb*.

The *Journal de la Savoie* was delighted by this performance: 'Many concerts and evenings of entertainment have been given in Aix this summer,' we read in it, 'but not one compares with the evening given by Mr Bogdanov's interesting family. In fact, one must have a brilliant talent to surpass the expectations of the select public which gathered in the large auditorium of the Casino. What a pleasant vision this Sonnambula! What nobility! What decorum! Everything in her poses speaks to the heart. Nadezhda speaks the language of the ideal, she poeticizes dance. How well she demonstrated the antithesis between dance on stage and in the ballroom. In Chambéry she astonished, in Aix she forced us to cry. Where next will we see her? Who will stop her in her rapid flight? But one evening and she is no more, nor the graceful Tatyana, this fluttering little hummingbird who melts every heart. What beautiful laurels these children are preparing for their father.'

Following this the playing of both brothers was praised. In Lausanne, they gave three evening performances. Here the newspaper announced Miss Bogdanova as its countrywoman. What an honour! Thus not only our provincial newspapers but also those of foreign lands praised her prior to her appearance before the judgement of the high Areopagus in Paris.

VII

In December they reached their destination. Bad weather notwithstanding, a joyful feeling ought to have stirred the family: they were in the capital where flattering diplomas were issued to talented people, the capital whose verdict proffered European celebrity. On the other hand, they were not immune to an inexplicable sadness about an unknown future. Far from their homeland, in a noisy city amidst bustling humanity where they did not know a single soul, where everyone was foreign, unrelated to them by any common interest—they could not but ponder their situation. Of course, this capital of art, science and education had always received their countrymen cordially, but their predecessors had come to seek diversion and pleasure, to satiate their idle curiosity, and had brought purses tightly stuffed with silver and gold, which enabled them, merrily and easily, to join in with the leading crowd. Our Russian family came not for that but to serve art, to seek perfection in it, to hear the righteous judgement about them. . . . And this was not easy where so many first-rate talents were already developed and brilliant, stars of the first magnitude, where so many citizens were cultured, where they so love and take pride in their own that they acknowledge excellence in a foreigner reluctantly, even one who has won glory at home. Pondering all this, anyone would grow pensive. What is the first step, how to begin, to whom to turn? An unexpected occurrence helped Bogdanov resolve these perplexities.

Soon after they arrived Tatyana fell ill. They called a doctor. From the conversation he discovered what had brought the Russian family to Paris, expressed an interest in them and offered his co-operation. He turned out to be Mr Mann, the chief physician of Princess Murat. In a few days the Bogdanovs received an invitation to appear at the princess's home. It was their first step in Paris and a very fortunate one, for which they could not have hoped at the beginning. They were very hospitably received by their noble hostess, whose guests on that occasion happened to include several German princes. The test of the Russians' talents began in this setting. Nadine attracted particular attention. At first they could not believe that she was a dancer, which, they said, she did not look like at sight. But her Russian dance won everybody over to the opposite point of view. This dance was new to the princess's guests; the tender, swan-like gait, the graceful unhurried movement, had a pleasant effect on the spectators. They praised the young dancer, and also approved her brothers, who had played several musical selections, and sincerely wished them all the greatest success. But their attention was not limited to words. Princess Murat wanted to give them substantial help. She promised to intervene with Roqueplan, the director of the Opéra.

Warmly received, the family went home with great hopes, which were not disappointed. In a few days they received an invitation to the foyer of the Opéra, where rehearsals were given. Respecting Princess Murat's intercession, Roqueplan also invited the best balletmasters, members of the opera commission and important journalists. The Russian children had to perform before these judges: here they could sooner count on prejudice than sympathy, but they remembered that not long before

'The Bogdanov Artistic Family' 201

Fanny Elssler had been their judge, who in this very theatre astonished and delighted all Paris with her art. Fortified by this thought, they plucked up their courage and went boldly to the trial. Nadine danced a mazurka with her father, then the Spanish dance 'La Rondeja', a *pas de trois* and a scene from *La Sylphide* with her brother and sister. When it was over and those present had conferred, the director turned to them, who modestly awaited impartial judgement.

'Your children are charming,' he said to Madame Bogdanova, 'with talents that promise much, but they are nevertheless children: they have wonderful school, but lack the strength required to command their art completely. If you want to make them genuine artists, then let them study for a year, or two or three, let them work to gain in strength, and at that time I shall answer for their success. Think about this; consider if you have the means to live in Paris that long; and then I shall see. Perhaps in a year's time your daughter will be ready to make a début at the Grand Opéra. I do not dispute that in time she can become a celebrity like Taglioni and Cerrito, but for that there is one condition—to study and to study. . . .'

After this Roqueplan promised to recommend both young Bogdanov musicians to Auber, the director of the conservatoire.

It is hardly necessary to say that this flattering report, and the concern with which the family was received, decided their fate. They could think only about finding the means to stay in Paris. Returning home and considering everything that the future held, the parents resigned themselves to the most frugal possible life, and to use all their money for the benefit of their talented children. . . .

A difficult and monotonous life began, which one must also call industrious in the highest degree, devoted to a single goal, one thought—a life for art. But it was not cheerless: their sole and best consolation was continuous success, which moved them closer and closer to their goal.

VIII

The foyer of the Opéra, where dancers practise, is a huge hall lavishly decorated with mirrors and gilt. The stage is on one side, dressing-rooms on the other. Evenings during ballet performances, the hall is brightly lit, and normally filled with Parisian aristocracy and famous visitors who come here to watch as dancers pass from their dressing-rooms to the stage, and if possible to strike up a conversation with them. At one end of the hall there are compartments for students' ballet exercises. Every morning at ten o'clock they must gather here—whatever the weather, holiday or no—continuous study and exercise being the foundation of training, which proceeds year round without interruption. A student thrice absent without a satisfactory reason is peremptorily expelled. From ten to eleven every girl exercises in her compartment. Then the professor arrives and takes the class to one o'clock.

An unlikely magnificence reigns among Parisian dancers: one tries to outdo the other in foppishness and brilliance. Many of them pay 3,000 francs or more to decorate their dressing-rooms. They appear in the foyer in expensive dresses, decked with lace, diamonds, and other expensive items. They drape themselves with layers of cashmere shawls, trying vaingloriously to display their riches to best effect. They decorate their ballet clothes in a similar way.

From her first appearance our young compatriot was clearly different from the others. Unsullied by pathetic vanity, she was not drawn to brilliance, knowing that she left her homeland not for pleasure but for intense, continuous work, realizing that a brilliant future comes only of a modest present. For lack of money she could not begin to compare in her dress with her extravagant fellow students. Her dressing-room was undecorated, her clothes did not dazzle with satin, lace, or anything else valuable. Always modestly and properly dressed, she rose above all the others in something that no money could buy, namely the fresh and natural colour of her face, on which was never detected the slightest trace of fatigue and weariness from sleepless nights passed with friends and noise and perhaps immodest enjoyment. This was her finest adornment, and for it all the others rumoured to hate her, although she bore no hatred to any of them.

To spare Nadezhda such company, competition with it, and burdensome or unpleasant clashes, it was agreed in advance among the Bogdanovs to pretend that Nadine did not understand a word of French. In this way she exempted herself from conversations with people with whom she ought not have anything to do, and she played this role successfully to the end. From the first she was received with mockery and ill temper. Not finding in her what they were accustomed to glorifying in themselves, the other dancers looked her scornfully up and down, remarking caustically about every detail, even down to her slippers, which did not compare with their own expensive adornments. They always came in noisily and with each other, and so it was strange for them to see this little girl accompanied by her mother, who waited patiently in the back of the hall until the end of the lesson. During rest periods Nadine joined her constant companion, while the others took to their elegant bonbons and filled themselves with sweets, conversing about their worldly escapades; to others chamber maids brought snacks. This took place noisily on divans, on which ill will towards our dear compatriot was often the topic of discussion.

One can understand how difficult, offensive, and sad it was for her in such company: but she persevered, firmly and persistently bore the mockery, manifest scorn, and even coarse jokes, bearing in mind the beautiful goal towards which she always strove, and perhaps without even trying to learn the cause of such malice from women to whom she had never done anything. But she was guilty before them of modesty, of purity of thought, of irreproachable behaviour, of the fact that these virtues subsequently won her over in the eyes of others and warranted the esteem of people who value the union of talent and morals. Having said this, we must point out that our sweet compatriot's schooling was not without labour and grief. But this work and these sorrows place her even higher in our estimation, and we thank her with all our heart that she could maintain her virtue, in a foreign land, on the straight and narrow path, to keep her Russian name above reproach. May she be proud of this, and we be proud of her.

IX

Meanwhile the Bogdanov brothers were presented to Auber, director of the conservatoire. He found that they played well but could not read music fluently, and advised them to study for a year with good private teachers to prepare for the conservatoire entrance examination. This was done.

Here a few words are in order about the make-up of the conservatoire. The day of the acceptance examination is announced in the newspapers. Anyone wishing to apply must put his name on a list with the porter. This right is not limited by age. Parents can even sign up small children if they think the latter can pass the examination, the first condition of which for music is to read and perform satisfactorily. The stage of the small theatre in the conservatoire is the site of these trials, its judges are professors, presided over by the director. On the appointed day and hour the candidates assemble in a special hall, and an assistant to the director calls them on to the stage in the same order they had signed the paper. By means of a little bell the director lets a student know when to begin the piece given him to perform, and when to stop. Each professor then writes down his opinion. Judgement is very severe, and many candidates who apply fail to satisfy the requirements. Those found wanting, however, are not denied the right to another trial the next year.

The conservatoire has four departments: (i) composition, (ii) winds and strings, (iii) singing, and (iv) dramatic art. For each a day is set for auditions and appropriate professors invited. Women students for piano and singing are admitted on an equal basis with male students.

On the day after the examination each person accepted receives a notice to that effect. A month's holiday is granted to each student, following which he must dedicate himself to constant work for several years. Every day, holiday or weather notwithstanding, he must appear in class at the conservatoire, where students are strictly monitored. All students are non-resident except for six students maintained at state expense in the opera department. These receive room, board, and all necessities, but cannot be absent from the conservatoire without permission. At the end of the course a student is not bound by any obligation and is granted complete freedom. State students, of course, are people with indisputable talent.

Every six months all students are given an examination to determine who is qualified for further study. The same procedure is used as at the acceptance examination; the students' ordering is determined by lot—who comes on stage first, and who later. Anyone showing no promise of success or talent is excluded from the conservatoire; thus the number who finish the course is far smaller than the complement of entering students. Every year a competition held among the best students for prizes. By general consent of the conservatoire, the pieces to be played, always strictly classical, are assigned six weeks in advance. Each person is obliged to acquire one of the assigned pieces. But to win this contest it is not enough to perform that piece to perfection: the student must also, immediately and without mistakes, play through music just written by one of the luminaries present, among whom there are always several from outside the conservatoire, either Parisians or visitors. This music, by design, contains various difficulties by which the art and talent of the young artists are judged.

There are four prizes in all. To win first prize is considered a great honour which incites great competition among the students. Sometimes the judges must award first prize to three or four candidates, finding absolutely equal abilities in the contestants. This competition also takes place in the theatre of the conservatoire, with the public in attendance. Anyone who wishes has the right to be in the parterre; the boxes are distributed by the director in advance.

There are two classes in the division of instrumental music—one for winds, the other for strings; from them young musicians assemble in an orchestra (*musique d'ensemble*). Here a special professor works with them. In the voice or opera division there are also two classes: grand opera and *opéra comique*. In the drama division there are three: tragedy, comedy, and vaudeville. The course in all these classes lasts from two to three years, depending on a student's progress. Nobody is prohibited from studying in two or more classes, but those wishing to do so must take an examination in each. The final examination and the last competition proceed exactly like the previous ones. To receive a prize, a contestant must have the votes of five of the nine judges. The balloting comes immediately after the competition, before the assembled public. The person who has surmounted all difficulties and performed the music perfectly receives a gold medal, the goal of all talented students.

Students of composition must spend at least five years in the conservatoire and pass the following classes: (i) solfège, (ii) composition, (iii) general bass, (iv) counterpoint, (v) fugue. First prize in this division opens the way to the Institute of Fine Arts. For this the winner must write a cantata, the subject of which the members of the Institute provide. He is taken to a special room where he stays until his composition is finished; six weeks is the maximum time allowed. While composing he must break off communication with everyone. His cantata finished, he presents it to the members of the Institute. If it warrants general approbation, they send its author to Italy for five years, where he must perfect himself yet further and obtain experience: his yearly stipend is 5,000 francs (1,250 roubles). In Italy he must write an opera and send it to the Institute immediately upon his return to Paris; the quality of the opera determines whether he will be a member of the Institute.

Let us turn now to our young musicians. They prepared for the conservatoire examination for a year; they passed it successfully and inspired confidence in their talents: they were accepted into the music division, and the fees foreign students were required to pay were waived. Like French nationals, they took lessons free of charge, which lightened the family's burden. For this we express sincere gratitude to the Paris Conservatoire and thank it in the name of all Russians.

X

A whole year was spent in study.

Meanwhile, the revolutions which were starting in Paris forced the family to ponder the possibility of any Russian staying in France. They decided to make an effort to hasten Nadezhda Konstantinovna's début. They encountered a mass of intrigues, and only the intercession of the Russian ambassador helped their situation. To his letter Roqueplan responded in a letter of his own handed to the Bogdanovs personally: this again emboldened and consoled them. The director spoke generously of the young Russian dancer, and promising to arrange her début soon, concluded: 'Our French artists receive such a warm welcome in Russia that I do not want to let pass the opportunity to assist one of your compatriots.'

Shortly after this Nadezhda Konstantinovna began to prepare her début in a *pas de deux*. At this time several changes were taking place in the ballet of the Opéra. The well-known balletmaster and professor St-

Léon, recently divorced from the celebrated Cerrito, returned from Spain. He replaced the teacher Mazilier, who for personal reasons had tried to put all possible obstacles in the way of his student's début. From his first look St-Léon valued our compatriot's talent and placed her above the other students. This was her first triumph. Learning that she had been assigned a *pas de deux* for her début, he announced that she must début in a ballet, not some *pas*. 'She will dance with me in *La Vivandière*,' he said, and zealously set about rehearsing her. This was a most difficult and wearying time for our artist. From ten o'clock to one she studied in the foyer of the Opéra, then from two to four in separate lessons with St-Léon to hasten progress. Success came quickly with such labour—under the eye of an experienced and concerned tutor her talent developed beautifully and at last attracted widespread attention. Even her fellow students, who had been unfriendly to her for a year, came to applaud her. They softened their tone and expressed the wish to get to know her better. Thus talent triumphed, and opened before her an honourable path.

On 20 October 1851 the young Russian artist made her first début before a select Parisian public. St-Léon brought her out in the leading role of *La Vivandière*. The French were taken with the Russian name, so rare among artists abroad, and her audacity, as they called it, to appear in a role formerly taken by Cerrito and Plunkett, who had delighted the whole capital not long before. Small wonder that the audience planned to receive our compatriot with severe critical scrutiny. But no sooner did she come out than misgivings immediately vanished; 'No sooner did the spectators notice,' wrote the well-known *feuilletonist* Jules Janin, 'that the débutante was still in her first youth, charming and light, than they immediately forgot her predecessor.' Her youth, petite stature, build, attractive appearance and natural merriment, free of the slightest strain often unpleasantly expressed in the smiles of many dancers—all this the Parisians noticed, and appreciated on its merits. Her speed and finish in movements, effortless lightness in turns, gracefulness—delighted the public. The young Bogdanova's success was complete; her parents' years of work and trouble were finally rewarded; the beautiful future of flattering dreams began to turn into reality. Imagine what a triumphant celebration this day was for a Russian family living in a foreign land; imagine all the expectations, timorous hopes, doubts, and dangers which agitated their souls before this. The triumph compelled them to turn to their distant homeland in their minds and joyous dreams, the land which in these hours was incessantly discussed by the audience delighted with the Russian artist: she did her homeland honour. Who would not understand the tears of joy with which the happy family perceived everything that day?

Soon afterwards the Parisian newspapers referred to Nadezhda Bogdanova's début in *feuilletons*; reviewers saw in her a first-class talent, mentioning her in connection with illustrious dancers; they predicted a glorious future for her, despite disagreements over her teacher St-Léon and *La Vivandière*. 'In Miss Nadezhda Bogdanova,' wrote the *feuilletonist* of *La Presse*, 'we noted agility, speed and precision of expression; everything she did was marked by cleanness and vivacity... If one judges her by this ballet then one must say, using a musical term, she manages better in *presto* than in *adagio*. The grace of this merry, light and lively dancer one would sooner call playful (*espiègle*) than passionate: without changing her nature, she can develop in herself much that is charmingly original. She reminds one neither of Elssler nor Taglioni nor Carlotta Grisi, but Cerrito somewhat, a quality no doubt transmitted to her by her teacher St-Léon. She animates her mime, and gives vital sense to each position. Her success was never for a moment in question; incessant cries of approval accompanied her every step.'

We could repeat this characterization now. But having seen Cerrito, we must note that we don't find any resemblance between the two artists any more. Miss Bogdanova has developed that individuality which the Parisian *feuilletonist* predicted.

'The applause of the public,' noted another paper, 'it seems, told the artist, "You are still young and lovely, your early talent will develop quickly if you work hard. The present is already beautiful for you, the future will be magnificent—you have only to want it".' Roqueplan himself shared this opinion, adding his approval to the exclamations of other spectators who were encouraging the débutante.

It was impossible, however, not to observe that the French punctilious boastfulness was unable to restrain itself, as for example in the *feuilletonist* of the newspaper *Pays*, who, recognizing remarkable talent in Miss Bogdanova, added: 'She is Russian only in name; she received her choreographic training in Paris; her professor was the Frenchman, St-Léon, who presented her to the public himself,' etc. According to this, our dear artist is obliged to France for everything, even her talent, which, one must suppose, was inspired in her by the air of France. We prefer not to display similar boastfulness, depriving Paris of the honour of training our compatriot; Paris definitely developed her taste, and gave true direction to her art. But neither shall we discredit the honour of her work before Paris, her childhood years, linked to need and deprivation, years which prepared her for her Parisian triumphs; we do not want to deprive her father of his due, who divined and worked to develop her talent, nor omit paying honour to the celebrated Fanny Elssler, who has delighted us all, and who by her just praise, one could say, decided the young and inexperienced dancer's fate; nor do we want, finally, to deny honour to ourselves, and so we recall Miss Bogdanova's initial triumph in French society, where she appeared directly from Russia, without having had a single lesson from her French professors. . . .

XI

The day after her début Miss Bogdanova was obliged, according to custom, to appear in class. Here St-Léon congratulated her on her brilliant success and announced that he had agreed with Roqueplan to permit her some three additional appearances on stage in the ballet, and then for her to cease making appearances for a whole year: 'You ought not to exhaust your strength,' he added, 'but get stronger, so then you will fully justify the hopes of the public; a new ballet will be prepared for you, and you will learn it next year.'

Such news ought to have saddened the artist, who had awaited her début impatiently and after her first success hoped to triumph on the stage. But she had no choice but to agree with her experienced professor, and resign herself to his decision. So again day followed day—monotonous days in continuous labour and study; they differed from the previous ones only by the pleasant recollection of her début, and with it greater hopes and greater confidence. In accordance with St-Léon's promise, Miss Bogdanova was still to be seen in public, in *La Vivandière* and then, on 1 December, in the Hungarian Dance of *Vert-vert* (of Leuven and Mazilier, music of Deldevège [*sic*] and Tolbecque).

Notices of her second début were more enthusiastic than before: 'Her successes,' said one newspaper, 'are so rapid that with difficulty one realizes she was a débutante just this winter, albeit she was charming even then. This artist has now scaled the steep heights of choreography. A few more months of effort, work, and devotion, and she will compete with any of our celebrities of dance.'

She danced her latest ballet with the famous Parisian dancer Priora and triumphed beside her. 'With Priora a rival appeared,' wrote Jules Janin about *Vert-vert*, 'who contended with her for the palm of excellence and for the triumph itself—it was the tender daughter of the North, the charming snowdrop (*perceneige*) from the sparking snows of Russia, the sister of Maria Taglioni, Nadezhda Bogdanova. . . .'

The next year, 1852, Miss Bogdanova appeared in the ballet of the opera *Le Juif errant* (music of Halévy, words by Scribe). Much space would be necessary to convey all the enthusiastic newspaper reviews.

After a triumphant imperial performance, our young artist received a gift from Louis Napoleon, a brooch which signified his recognition of her talent.

After that it remained for her patiently to wait out the year, and inaugurate her career amidst celebrations and striving. At last the promised new ballet was ready. St-Léon wanted to set to teaching it to his student when unexpected changes in the management were made. A celebrated artist returned from Spain and appeared again at the Opéra, which caused St-Léon to resign. Mazilier was reappointed in his place. Various backstage intrigues ensued, and the new ballet was given to the new dancer, while an insignificant, intentionally shortened role was left to our aggrieved artist instead of the leading part. Despite this, she pondered this role so carefully, and gave it so many individual details and such a graceful place, that she diverted the audience's gaze from the principal dancer. This was her true triumph: talent and innocence won out over envy and intrigue. Here, perhaps, she took revenge without knowing it—nobly, having in mind nothing but art. The ballet was *Orfa* (by Trianon and Mazilier, music of Adolphe Adam), first produced at the beginning of January 1853. The press notices unquestionably gave Miss Bogdanova the palm of excellence, even Jules Janin, who had looked on her first début with more reserve than the others. 'In truth,' he noted, 'she is the daughter of those very snows which nurtured Taglioni.'

Thus the Russian artist, with her talent, attracted the notice of the Parisian public, whose high opinion she had won with her first appearances. Immune to intrigue, she was engaged as a first dancer at the Grand Opéra. This broadened her field of activity but did not change her way of life, which, as before, continued quietly and modestly in the family circle, among concerns about her art. The same modesty and caution distinguished her in the theatre from other dancers and compelled their respect. Roqueplan valued her good qualities, which were indisputable, approved the silence by which she conquered all backstage intrigues, and showed her special esteem. He even ordered a separate dressing-room for her and had it decorated at state expense. That was unprecedented.

In the same year Miss Bogdanova appeared with great success in the ballet within Niedermeyer's new opera, *La Fronde*, and in Mazilier's new ballet *Aelia et Mysis*. In reviews from this time the successes at the conservatoire of her musician brothers, violinist and pianist, began to be noticed. 'You see,' a *feuilletonist* observed, 'one fine day the entire artistic family will return to Petersburg. But for another two or three years at least we can take pleasure in the talents and the glory they have acquired in France.'

XII

The spring of the following year Miss Bogdanova was invited to the ballet in Vienna. Magnificent celebrations were being prepared there on the occasion of the Austrian emperor's wedding. Roqueplan, expressing particular favour towards the artist, gave her three months' leave.

Having signed a contract for Vienna, Miss Bogdanova prepared the ballet *Giselle*, from which she had already danced excerpts in provincial theatres during her tour of Russia. It was pleasant for her to recall her childhood years with this ballet, and to triumph in it again, more fully, on one of Europe's first stages before a select public. These three months passed unnoticed for the triumphing artist: performing the same ballet did not bore the Viennese public, as recently it has not tired us.

One novelty about this must not be overlooked: Miss Bogdanova danced in Vienna among live trees and flowers placed on the stage; in the second act of *Giselle* the flowers were picked and thrown, and after the performance were quickly passed into her admirers' hands.

Though she appeared in only one ballet, she delighted the spectators with other dances, among which the newspaper reviews placed the mazurka in special relief; they also mentioned our artist alongside Elssler and Taglioni, which was the greatest possible praise.

As Vienna was admiring her with delight, Jules Janin in Paris reminded the French about her, the French who just then were looking upon Russia and everything Russian most unfavourably: 'At this hour, as I speak, a child of Russia, oh heaven! A Russian, oh great gods! A Nadège, oh grief! A Bogdanov, abomination of grief, is dancing in Vienna, in Austria, at the celebrations of the young emperor and his young princess, sparkling with all the graces of life and the crown! She was much loved at the Opéra in Paris, this little Nadège; she was found lively, alert, elegant, and very pretty, with a touch of Carlotta Grisi and something of Guy-Stéphan. Alas! She had to leave, she has left! She is borne away—she too—by the impetuous blast of war, and see! She dances there in a neutral country, half French, half Russian; and if she could hold peace in her fists, her two little hands would quickly be opened. She has danced the role of Giselle to unanimous applause, which Carlotta Grisi made a marvel worthy of La Sylphide.'

At the beginning of August, on the Russian artist's return to Paris, crowned with new laurels, Janin wrote about her again in the same spirit: 'This week again (oh the futility of the *eastern question*!) you figure that this insolent little Bogdanoff, with her rosy cheek and her name in "-off", a daughter of Russia, a *Russian*, o heaven! She dares, she has dared to return! She is with us, at this hour she has passed, audacious child, through the Arc de Triomphe d'Etoile, between cannons and cannonades, guns and gunfire, such music and such drums! She comes, unhappy little creature, with light and nimble gait, to knock at the doors of the Paris Opéra! And—horror upon horror—her brother, a Bogdanoff, who should be called Pierre or Paul, or even Nicolas Bogdanoff—had he not dared, the Cossack! to approach, bow in hand, our Conservatoire of Music, and the slack Frenchmen—did they not award a gold medal to this Russian violin, to this Nicolas (Alexandre) Bogdanoff? This is the end of the world, I tell you. Behold, we have come round to a veritable invasion.'

At this time the Bogdanov brothers finished their conservatoire course in triumph, and received gold medals in recognition of their talents. The family's new joy and the celebration were indescribable. Another goal had been achieved by strength and persistent labours. Just when Russia began to struggle with its enemies and surprise them with her courage, bravery, mercy, and all the qualities which inform a true Christian soldier, two modest youths and their young sister, alone in a country hostile to their homeland, proved triumphantly that Russia produces diverse talents who must be recognized even by people who despise a Russian name, talents enjoyed even by people who unjustly condemn Russia as the *land of barbarians*. This was a flattering role for our artists; and it is flattering to us to see them as representatives of Russia amidst our enemies of the past.

XIII

These pleasant family hours were more than once disturbed by unpleasant moments, disappointments, and low spirits. The Grand Opéra was renamed the Imperial Theatre, one of many changes. Roqueplan had to retire, and the theatre closed for three months.

The new director signed a contract with Miss Bogdanova for another year, but soon revealed his own ways of favouring other dancers and dismissing the Russian artist. He did not want to honour his contractual obligation to produce a new ballet for her, and responded very indelicately to Miss Bogdanova's reminders and requests. Each reader may judge whether this was honourable. Meanwhile the Parisian public, regardless of her small roles, constantly showed its good disposition and concern for our compatriot's talent. The newspapers continued to declare her a first-rate talent and lavished praise in their notices. They observed that she lacked roles, but that she was not to blame for this. Consider for example this notice from *La Patrie*: 'One may congratulate the theatre management on a talent like Miss Bogdanova. The public receives this dancer with great pleasure. At the time of her first début three or four years ago she was very strong despite her youth. Since then, by constant work and striving, she has established herself in the first rank. In Vienna she enjoyed the most brilliant success. Indeed, it is just to place her in a prominent light (*c'est justice, que de la mettre en evidence*).' But in that new era they wanted to give her not prominence, but no exposure whatever. The Théâtre Italien in Paris invited her to appear for several months, but the [Opéra] management did not allow it. One can imagine the awkward moments our dear aggrieved artist must have endured in this uneven fight. She was left to await the expiration of her contract and to place her hopes in her homeland, towards which the feelings and thoughts of our young artists strove.

1855 passed most unpleasantly for them, aggravated by the difficulties of a Russian family among a people who, celebrating non-existent victories, shouted immoderate cries against Russia in their newspapers. It was hard to listen to this every day, at each step, to be silent and languish, ignorant of the homeland's fate. But they waited until autumn and received permission to leave Paris. At that time they received a lucrative offer from London, but they rejected it out of hand and started out for Russia. Despite much unpleasantness, they departed grateful to the city and country where they spent their youth, where they had received a definitive education and secured a place in their profession. For five years, modestly and timidly, they had walked upon this ground; now they left it, self-assured, following the true path and looking happily to the future. They also took many pleasant memories, never to be erased: yes, and what artist finds unpleasant the place where fame was first won, where with faltering heart he listened to incorruptible applause for his talent and was comforted by young, bright hopes? Paris had not received these Russians indifferently: the press regretted that she was to be parted with them, and expressed the wish to see them again, among their favoured guests.

In November the artistic family stopped for several days in Berlin. Here Nadezhda Bogdanova danced Giselle in the royal theatre and made the same impression as in Vienna six months earlier. The Berlin newspapers unanimously sang her praises. But our artist found a special celebration of her talent and art in Warsaw, famous for its ballet. Warsaw received her at each appearance with lively enthusiasm, and recognizing her rare talent, rewarded her not only with curtain calls and applause, but with expensive gifts. She left the most pleasant recollections with ballet lovers in Warsaw: even now they speak of her with delight. And Nadezhda Konstantinovna recalls Warsaw's kindness towards artists as well.

One must note that the praises we mention are not arbitrary, but taken from published notices, scrupulous and therefore always in agreement with each other: let no one look in these pages for a deliberate panegyric.

XIV

At the beginning of 1856 our artistic family arrived in their native land to be among their compatriots. They wanted to dedicate their talents to us,

1. Charles Didelot, from an engraving by Reimond

2. Jules Perrot

3. Cesare Pugni, Composer of *The Pharaoh's Daughter* and *The Little Humpbacked Horse*

4. Sergei Nikolaevich Khudekov

5. Marie Taglioni, probably as Flora in Didelot's *Zéphire et Flore*. Lithograph by Chalon and Lane, London, 1831

6. Adam Pavlovich Glushkovsky in the ballet, *Raoul de Créqui*

7. Alexandre Mikhailovich Gedeonov, Director of Imperial Theatres, 1834–58

8. Timofei Alexeyevich Stukolkin

9. Marius Ivanovich Petipa

10. Mathilde Kshesinskaya as Esmeralda, in a revival of Perrot's *Esmeralda* in 1899

11. *The Pharaoh's Daughter*, Prologue: The Interior of a Pyramid. Decoration by Kukanov for a revival of the Ballet in Moscow in 1892

12. Interior view of the Amphitheatre at Olga Island, Peterhof

13. Ekaterina Ottovna Vazem

14. Virginia Zucchi

15. Carlotta Brianza, creator of the role of Aurora in *The Sleeping Beauty*

16. Pierina Legnani, creator of the role of Raymonda

17. Decoration by Lambin for *Raymonda*, Act I, Scene 2

18. Sketch of a decoration by Golovin for *The Magic Mirror*, Act IV, Scene 7

19. Vatslav Nijinsky, creator of the role of the Favourite Slave in *Le Pavillon d'Armide*

having passed through a difficult school far from home: it can be hoped that Russia, like a mother, will receive her children warmly and lovingly. Indeed, could they be refused this maternal welcome, this warm greeting, after honourably upholding a Russian name everywhere, after spending their time industriously in a foreign land, being full of fresh and youthful strength, yet not distracted by the enchantment of foreign glory, wanting instead to link themselves with the Russian soil which was part of them? We know that the large Russian family will find it heart-warming and good to cherish and comfort this family, will rejoice in their successes, and take pleasure in their talents. Their strength, their tender youth and constant diligence are our guarantee of their furthest possible development. The Petersburg public has already worthily and justly appreciated Nadezhda Konstantinovna Bogdanova's talent. All Russian people rejoice in Russian talents. What more is there? In our time can any hostile circumstance be found which is not foreign? No, that is impossible! Our monarch has said that a walk of life is open for us in everything good and beautiful. Long since have we loved our home life and our glory, and under no circumstances would we not receive Russian talents with the warmest sympathy. We hope from the soul that their number will increase, and that they will be, in our general view, on a level with foreign talents. The time has already passed for Russian talents to regret having a Russian name, names now esteemed in all lands. In our midst let them now be esteemed yet more. And let us point this out to the Bogdanov artistic family as a sign of sincere welcome. Let us encourage them and wish them the greatest, fastest, and happiest successes, with which, of course, their well-being must be linked. They have worked hard and suffered much; let them now work hard and take much pleasure in it.

VI
The 1860s

INTRODUCTION

At the beginning of the 1860s, the Petersburg ballet had the look of business as usual. Arthur St-Léon, invited to Russia as balletmaster of the Moscow company, had been promoted into Perrot's place on the latter's departure.[1] He was now first balletmaster in St Petersburg, and Carolina Rosati, who had joined the company in 1859, was the principal ballerina.

Already in the first years of the decade, however, an extraordinary situation was evolving which would affect the Petersburg ballet for the rest of the century. The company had often had more than one choreographer at a time, but in the 1860s two resident choreographers of approximately equal talent (if not reputation) vied fiercely with one another for official recognition and public favour. Where previously one had always enjoyed precedence, there was now a genuine rivalry for the first time.

St-Léon's competitor was Marius Petipa, who after assisting with the productions of *Paquita* and *Satanilla* before Jules Perrot's arrival, had observed the latter's work for ten years and had begun to try his hand at original composition.[2] In retrospect, *A Regency Marriage*, *The Blue Dahlia*, and *The Parisian Market* seem of little consequence when compared with *The Sleeping Beauty* and *Raymonda*, though *The Parisian Market* was also produced in the city of its title and was revived in St Petersburg in the 1890s. We may nevertheless assume that the making of these early ballets stirred in Petipa an awareness of his creative abilities. By the end of the 1860s Petipa had become the chief balletmaster of the imperial theatres—his place at the top of the hierarchy undisputed—and he would remain so until his retirement in 1903.

The rivalry of Petipa and St-Léon had at least three important consequences for the 1860s. First, it extended to principal ballerinas and led

---

[1] On St-Léon, see Ivor Guest, ed., *Letters From a Ballet-Master. The Correspondence of Arthur Saint-Léon* (London, 1981), and Guest's *The Ballet of the Second Empire 1847–1858* (London, 1955) and *The Ballet of the Second Empire 1858–1870* (London, 1953); V[era Mikhailovna] Krasovskaya, *Russkii baletnyi teatr vtoroi poloviny XIX veka* [Russian Ballet Theatre of the Second Half of the 19th Century] (Leningrad and Moscow, 1963), pp. 63–89; Yurii Iosifovich Slonimskii, *Mastera baleta* [Masters of the Ballet] (Leningrad, 1937), pp. 135–[67].

[2] On Petipa see, e.g., *Memuary Mariusa Petipa, Solista EGO IMPERATORSKOGO VELICHESTVA i baletmeistera Imperatorskikh teatrov* [The Memoirs of Marius Petipa, Soloist of HIS IMPERIAL MAJESTY and Balletmaster of the Imperial Theaters] (St Petersburg, 1906), tr. Helen Whittaker as *Russian Ballet Master. The Memoirs of Marius Petipa* (London, 1958); Krasovskaya, *Russkii baletnyi teatr vtoroi poloviny XIX veka*, pp. 211–336, tr. in part by Cynthia Read in 'Marius Petipa and "The Sleeping Beauty"', *Dance Perspectives* 49 (Spring 1972), pp. 6–56; Slonimskii, *Mastera baleta*, pp. 201-[82], tr. Anatole Chujoy in *Dance Index* VI (1947–8), 100–48; A[nna Markovna] Nekhendzi, compiler, *Marius Petipa. Materialy, vospominaniya, stat'i* [Marius Petipa. Materials, Recollections, Articles] (Leningrad, 1971), German tr. as *Marius Petipa. Meister des klassischen Balletts. Selbstzeugnisse, Dokumente, Erinnerungen*, ed. Eberhard Rebling (Berlin, 1975).

to the rise of factions in the audience, producing an outbreak of raucous balletomania. Second, it brought recognition to Russian ballerinas on an equal or near-equal basis with foreigners, who continued to be hired. In this regard the lines of partisanship between Petipa and St-Léon are not consistently drawn. Petipa championed the Italian Rosati and St-Léon the Russian Muravieva; Petipa later advocated his Russian wife Maria Surovshchikova and St-Léon a number of foreigners. Third, the rivalry produced two of the most successful ballets ever created for the imperial repertoire: *The Pharaoh's Daughter* and *The Little Humpbacked Horse*—one serious, the other comic, one by Petipa, the other by St-Léon, one for a visiting celebrity, the other for a Russian ballerina.

Having served fifteen years in the imperial ballet, which to all appearances he had adopted as his permanent home, Petipa no doubt sensed that he would sooner be recognized as St-Léon's legitimate peer by a splendid coup than through long and faithful service. So when the moment came, he seized it. The chronicler Alexandre Pleshcheyev recounts what happened:

On 18 January 1862, at Rosati's benefit, Marius Ivanovich Petipa produced the grandiose ballet in 3 acts and 6 scenes, *The Pharaoh's Daughter*, which has graced our stage to the present day. St-Georges wrote the scenario for this ballet, full of dramatic movement, and Cesare Pugni its music.

Concerning Petipa's work on the Petersburg stage, this ballet has played an important and decisive role: soon after its production Marius Ivanovich was named second balletmaster, and with St-Léon's departure, he remained the only one.

Rosati herself expressed the wish that Petipa produce a new ballet for her last benefit. It was believed backstage that the ballerina had passed her forty-first year (according to biographical accounts she was thirty-six), and she had decided, on leaving Petersburg, to bid farewell to the stage forever. In the waning of her career, Rosati wanted to be seen in a new and effective role.

Petipa, who was in Paris at the command of the theatre direction, received from St-Georges the completely worked out scenario of *The Pharaoh's Daughter*. Meanwhile, Andrei Ivanovich Saburov [Director of Imperial Theatres] for some reason got angry with Rosati, towards whom in his heart he felt great sympathy, and declined to mount the ballet despite the fact that the scenario and music, which Pugni had begun to write, were ready. Rosati called on Petipa and they went to the director together.

'When will you order rehearsals of the new ballet to begin?' Rosati asked.

'No . . . Not at all . . . there is no money . . . no time to produce *The Pharaoh's Daughter*!' Saburov answered.

'But I am entitled by contract to a new ballet and you must honour my request.'

'I cannot . . . do you hear, I cannot! It is impossible . . . But then, tell me, Petipa, could you produce a large ballet in seven weeks?'

'Yes, I shall try, and probably succeed.'

Perplexed by such an unexpected answer, the director knit his brow, ordered St-Léon to cease rehearsals and leave the production of the new ballet to Petipa. In six weeks everything was ready . . .[3]

[3] Aleksandr [Alekseevich] Pleshcheev, *Nash balet (1673–1899). Balet v Rossii do nachala XIX stoletiya i balet v S.-Peterburge do 1899 goda* [Our Ballet (1673–1899). Ballet in Russia to the Beginning of the 19th Century and Ballet in St Petersburg to the Year 1899], 2nd edn. (St Petersburg, 1899), pp. 192–3. The persistent claim that Petipa created *The Pharaoh's Daughter* in six weeks must be taken literally—that is, Saburov's permission to proceed with rehearsals may have come six to seven weeks before the first performance. Conceptual preparations would have taken longer, and we know from newspaper announcements not six weeks but four months before the first performance that the scenario of the new ballet was ready, Rosati was to star, and Petipa was to be its choreographer. See, e.g., *Severnaya pchela*, 1861, No. 211 [23 Sept.], p. 806. As of this announcement the first performance of *The Pharaoh's Daughter*, which was to be in four acts, was set for sometime in December. Saburov's antipathy to spending money on a ballet may have been related in some way to his surpassing love of opera, of which Vazem writes (*Zapiski baleriny Sankt-Peterburgskogo bol'shogo teatra*, p. 38).

9

Libretto of *The Pharaoh's Daughter*

Vernoy de Saint-Georges and Marius Petipa

THE PHARAOH'S DAUGHTER
Grand Ballet in Three Acts and Nine
Scenes, with Prologue and Epilogue
Scenario by Mr St-Georges
The Ballet is produced on the local stage
by Mr Petipa
Price: 30 copecks silver
Saint Petersburg

PROLOGUE

SCENE I

A desert in the east. It is night. The sky is clear and filled with stars. To the right a group of palm trees. To the left a fountain amidst some greenery. At the back of the stage a pyramid, in which a door has been hewn opening to the desert. On the horizon mountains are visible, and a road along which a caravan is making its way into the valley.

Tableau 1

A caravan of Armenian merchants approaches to the sounds of a march. Black slaves lead camels loaded with wares; young female slaves and bayadères, some on camels, others on foot, accompany their masters, merchants taking them to be sold into Egyptian seraglios.

The caravan stops for rest in the shade of the palm trees.

Slaves, having pitched an elaborate tent and scattered rich oriental carpets and pillows around on the ground, begin to prepare a meal.

The travellers finish their prayers, then sit down to eat.

Tableau 2

New voyagers appear at the top of a hill. They come down into the valley slowly, apparently wearied by their trip.

They have but one camel. Their baggage is light, like that of contemporary tourists.

They are the young Lord Wilson and his servant John Bull.

The latter, it seems, is much more tired than his master, who, seeing the pyramid, gazes at it in delight.

Dragging himself along with difficulty, John Bull leans on the camel. At the sight of the resting Armenians, he directs his master's attention to them, and happily anticipates the prospect of enjoying their dinner.

Lord Wilson walks up to the merchants and asks to join them. They receive him with pleasure, invite him to sit down, and offer him fruit; meanwhile slave girls pour him a drink.

John Bull scurries to lay his hands on a few scraps from dinner, wolfs them down, then tears a tankard of wine from a black slave's hands and returns it, drained, to the slave's chagrin.

At their lords' order the young bayadères dance before them, at first calmly, but stimulated by the wine the slaves have brought, they get excited and abandon themselves to spirited, bacchic passions. John Bull, also flushed by the wine, mixes in with the bayadères' dances, but they laugh at his antics.

At the end of the dances peals of thunder are heard: the sky is radiant, like the northern lights. The blast of a frightening simoom rises up and disturbs the desert sands. The merchants stand up, frightened. The caravan falls into disarray, slaves hurry to pack away the baggage. The simoom increases to great ferocity. A whirlwind blows away the tent which had sheltered the travellers. The palm trees break and fall down. A great sea of sand blows on the horizon. Everyone seeks refuge and finally, in a group, they run to take shelter in the pyramid.

A true gentleman-tourist, Lord Wilson looks with tranquil curiosity at this frightening spectacle and settles down to sketch it in his album, to the amazement of his servant, who begs him to save himself. John Bull tries to flee, but a burst of wind throws him to the ground, and despite his best efforts he cannot get back on his feet.

The storm intensifies so much that it is impossible to stay out on the desert. Lord Wilson sadly makes his way to the entrance of the pyramid, accompanied by John Bull, whom the simoom knocks off his feet. He slips along to the very doors of the pyramid, entering with Lord Wilson.

SCENE 2

An interior view of the pyramid. At the left, a statue of the Pharaoh sitting on a granite throne. At the back of the stage, stone sphinxes of warriors. Around the walls stretches a long line of mummies. In the middle of the stage is a niche for the principal mummy, around which all the others are arranged, as if a retinue. The niche is gilded and decorated with colours, and in general is distinguished by its special magnificence.

Tableau 1

Having overcome their fright, the merchants settle down with their fellow travellers in the protection of these ancient vaults to rest and to smoke opium from long chibouks, which their slaves prepare for them.

Tableau 2

Presently Lord Wilson and John Bull enter. The latter, at the sight of the soft pillows prepared for one of the merchants, lies down on them and refuses to yield his place to anyone. Smiling, the merchant lets him stay there.

From the back of the dark hall appears an old man dressed in rags, the watchman of the pyramid. He is astonished at the unexpected visit of such a multitude of visitors to this dwelling of the dead.

An inquisitive tourist, Lord Wilson gives the old man some gold coins and asks him about the mummies. The old man gladly satisfies his curiosity, then, stopping in front of the chief mummy, bows before it, and showing Lord Wilson its royal band explains that it is the resting place of the Pharaoh's favourite daughter.

Having finished conversing with the watchman, Lord Wilson walks over to the merchants and asks to smoke some opium.

'Take care! See what opium can do,' the leader of the caravan answers, pointing to one of the smokers, plunged into an anxious dream.

'It makes no difference!' retorts Lord Wilson, taking the chibouk from the sleeping man, 'I want to be delirious, like him.'

Following his master's example, John Bull also smokes. Gradually both Englishmen and all the others fall into a deep sleep.

Soon light clouds descend and obscure the sleepers.

Then various fantastic events occur, the fruit of dreams stimulated by agitated feelings aroused by the opium.

A bright ray of light illuminates the chief mummy's face, the face of a beautiful girl. There is a peal of thunder, the mummy comes alive, and steps down from the wall. It is Aspicia, young and beautiful, the great Pharaoh's favourite daughter.

Returning to consciousness she remembers her past, looks around in amazement at her subjects, and is horrified at the sight of the gloomy dwelling. But the girl's coquettishness comes to the fore. Aspicia looks into the metal mirror attached to her sash, and finding herself as beautiful as ever, is comforted.

This miracle is only the first of many.

The old watchman throws off his rags and is transformed into a handsome youth, into the genie-guardian of the pyramid, with a gold wreath on his head and a sceptre in his hands. At a wave of his staff everything comes to life—the mummies encircling Aspicia are animated with the brilliance of youth. Warriors in full armour come down from their niches. The pyramid is illuminated with fantastic lights.

Girls surround Aspicia and bow to her, when she notices the sleeping Englishman, and finding him handsome, points him out to her girlfriends. They share her delight. Lord Wilson, disturbed and anxious in his dream, turns and reaches out to Aspicia. She places her hand on his heart, then lifts it away and presses it to her own.

Observing Aspicia's fascination, the genie of the pyramid bears her off on the clouds.

Then everything disappears in the darkness; thick clouds cover the stage, in the midst of which an inscription in bright letters is seen: 'A Dream From the Past.'

End of the Prologue

ACT I

SCENE 3

A hollow in the forest, surrounded by dense tree growth and illuminated by bright sunlight. In the depth of the forest a hilly road is visible, which goes across a little bridge thrown betweeen two cliffs. At the right, an arbour of wild flowers and bench of moss.

Tableau 1

At the rise of the curtain, trumpets and horns are heard echoing in the forest. The Pharaoh and his daughter, the beautiful Aspicia, next in line for the throne, are lion hunting.

The young princess appears, accompanied by her women and several hunters. Across Aspicia's shoulders she wears a quiver; in her hands she holds a golden bow. Her girlfriends dance a martial dance, in which their charming mistress also takes part.

Tired and wearied by the intense heat, Aspicia wants to rest in the arbour. The bench is covered with gold-embellished purple cloth. The princess, surrounded by the girls, lies down.

The young Nubian Ramzea, Aspicia's favourite slave and confidante, sits down on the bench, resting the princess's head on her breast. The huntresses form a graceful grouping around them.

One of the slaves sits at Aspicia's feet and plays a lute shaped like a small hand harp. To the quiet, melodious sounds of this instrument the princess and her companions fall asleep, wearied by the hunt. Soon even the musician, letting fall the instrument, dozes off.

Just then a monkey appears in one of the trees beneath which the princess and her companions are sleeping. Clinging to the branches, it comes down to Aspicia. Awakened by the rustle of the leaves Aspicia starts, and taking up her bow, aims at her foe . . . The monkey, sensing danger, jumps up on to a tree, and after several leaps and amusing faces, disappears.

Meanwhile a bee begins to hover around Aspicia, having flown out from the calyx of a fragrant aloe. It circles around the princess, taking her for a splendid flower, and tries obsessively to reach her enchanting mouth, as if it intended to draw sweet mead from it.

Aspicia's friends, awakened, try to drive the unpleasant insect away from their mistress.

Two Egyptians are seen in the distance. One of them is young, good-looking, and dressed in a bright costume. The other is humorous and clumsy. These are Lord Wilson and John Bull, living in a fantastic world as if it were reality, the result of opium hallucinations; Wilson is now Taor, Bull is Pasifont. They intend to make an appearance at the Pharaoh's court as seekers of adventure. As usual Pasifont seems tired, and an insatiable glutton, he sits down to eat fruit and the other foods with which his travelling satchel is filled.

Taor, noticing the girls' alarm, hastens to free Aspicia from the worrisome bee.

Aspicia, astonished by the stranger's unexpected appearance, is bewildered. Taor, enchanted by her beauty, cannot take his eyes off the charming princess. Thinking that he has seen this captivating beauty before, he approaches her. To the music which played when Aspicia touched her hand to his heart at her appearance in the pyramid, he now places his hand on the young princess's heart, who is unquestionably enraptured with him.

But he has hardly yielded to the most eager delight than a frightening noise is heard in the depths of the forest. Horns and trumpets sound; hunters run on stage. Aspicia comes to her senses. Taor hastens to withdraw, dragging Pasifont behind him, to the glutton's extreme displeasure, for he had not yet managed to finish his breakfast. The hunters tell Aspicia that an angry pride of lions is roaming the forest. A lion's roar is heard.

Frightened, the girls rush off in all directions. Aspicia also flees, but soon returns in horror. Indicating that a lion is chasing her, she hides in a thicket of the forest. The lion runs in, leaps, and follows in Aspicia's tracks.

At the noise, Taor enters and asks Aspicia's friends what has happened, but in their fright they hardly answer. Suddenly he notices the princess racing desperately across the bridge. In pursuit, the lion clears the bridge in one leap and stops, looking greedily at its prize.

Taor, quickly taking up a bow left behind by one of the hunters, takes aim at the wild animal, accurately fires an arrow at it, and runs to Aspicia's assistance. Meanwhile the mortally wounded lion tries to jump, but having lost its strength, falls over the precipice.

Taor carries down the unconscious princess in his arms.

The women surround Aspicia, trying to revive her.

Tableau 2

Trumpet fanfares announce the Pharaoh's approach.

He appears on a military chariot drawn by negroes and surrounded by hunters and warriors. He looks indignantly at the low-born Taor, who has made bold to support with his own arms the daughter of the great sovereign of Egypt.

Aspicia quickly presents Taor as her rescuer. 'He saved me! His courage delivered me from the clutches of this wild animal,' she says, pointing to the arrow-pierced lion which several hunters are carrying in.

The Pharaoh's wrath changes into liveliest gratitude. He permits Taor to kiss his hand and expresses the wish to take him under his august protection. Then the Pharaoh's chariot is brought in, which he and his daughter ascend. Aspicia makes a friendly sign to Taor, to his delight. The great sovereign's procession sets out for the palace of the rulers of Egypt.

Taking on a haughty air, Pasifont looks at the Egyptian nobility around him with a dignified air as he walks behind his master, who is seated on a palanquin of honour.

SCENE 4

The palace of the Pharaoh in Thebes. A splendid hall, opening out on to a garden, with porticos along the sides. In the back of the stage a large platform on which servants are are laying a magnificent table. Caryatids holding lamps are standing around the entire hall. A broad staircase, covered with a rich carpet, leads to the platform where the Pharaoh's throne is located.

Tableau 1

Slaves are bustling about the hall, preparing for a brilliant celebration.

Tableau 2

Taor and Pasifont enter, Taor overjoyed that he can see Aspicia, the object of his passionate desires; with anxious looks, full of love, he searches for her around the hall.

Tableau 3

Aspicia appears, surrounded by young girls and slaves.

She goes to Taor and offers him boxes filled with precious gifts and gold as a sign of gratitude for saving her life. Taor declines them. His enamoured heart does not thirst for precious stones; he is hoping to win the happiness of his life. Taor and Aspicia begin a declaration of love.

Pasifont trembles with fear at the sight of his master's boldness, but he guards the lovers and soon warns them of the Pharaoh's approach.

Tableau 4

A triumphal march is played. The Pharaoh enters, surrounded by a brilliant retinue of courtiers, nobility, warriors, black slaves and white. He ascends his throne.

Tableau 5

A courtier of the King of Nubia, accompanied by nobles, approaches the sovereign of Egypt. He kneels before him and announces the arrival of his king, who has come to ask for Aspicia's hand.

The princess is horrified; Taor tries to mask his despair.

The Pharaoh gets up from his throne and goes to meet the Nubian king. The latter appears with a numerous retinue of nobles and warriors in splendid costume. The Pharaoh offers him his hand and presents Aspicia, pale and agitated.

Struck by the young princess's beauty, the Nubian kneels before her, offers her his hand and heart, and swears to be faithful. To all appearances, the Pharaoh listens to his proposal with satisfaction.

Meanwhile Taor, unnoticed, draws a dagger to strike down his rival, but Aspicia's pleading glance stops him and tempers his anger.

The Pharaoh promises to grant the youthful king's wish, and while he is signing a papyrus, presented by his ministers, which establishes friendly relations with the Nubian king, Aspicia goes over to Taor and tells him, quietly: 'Do not worry! I will never belong to him.'

'But to whom then?' Taor asks.

'To you,' Aspicia answers, and taking a gold ring from her finger, gives it to the young man, who hides his joy with effort.

The ruler of Egypt descends his throne and orders festivities to begin, celebrating his favourite daughter's escape from death.

Dances are performed (*grand pas d'action*), during which a plot is devised between Aspicia, Ramzea, and Taor.

'In a few minutes you will belong to my rival,' Taor whispers to Aspicia. 'What can I do?' asks the princess. 'Flee!' answers Taor, 'If you truly love me, then decide.' 'Agreed,' answers Aspicia, 'but this hall is so large that we could never get out of it unnoticed.' Taor shows her a key that Ramzea has given to him. 'This is a key to a secret door in that far wall,' she says. 'A barque is waiting for us on the Nile . . . Let us flee!'

Undecided for a moment, Aspicia finally promises to fulfil Taor's request.

After this *pas*, the caryatids draw near, carrying baskets of flowers on their heads. They form various graceful groupings and dance, and at the end children appear from their baskets, who finish the dances with them.

The Pharaoh and the Nubian king walk to the table. During this procession their path is strewn with flowers. In the hall incense burners are lighted with aromatic grasses. All the lamps are lighted at once. Everyone bows before the ruler of Egypt.

After the Pharaoh sits down at the table with the Nubian King, dances begin anew at the front of the stage, during which Aspicia's abduction is accomplished.

The young Ramzea veils her mistress, again beset by indecision. Taor begs her, and taking her by the hand, leads her towards the secret door! Finally Aspicia agrees. Looking to all sides, the lovers, fearful of being noticed, approach the exit. Ramzea guards them. Pasifont opens the door, and letting Taor and Aspicia pass, disappears after them.

But one of the slaves reports the princess's flight.

Angered, the Pharaoh orders Ramzea brought to him. Tearfully, she approaches her master.

The Pharaoh threatens her with his wrath if she does not disclose where his daughter has fled, but Ramzea does not know.

The Pharaoh commands his entourage to find Aspicia. Guards hurry to carry out their sovereign's order. The Nubian king, noticing the secret door, quickly follows along the fugitives' path. The Pharaoh is in a terrible rage. Frightened noblemen leave in pursuit. General confusion.

The end of Act I

ACT II

SCENE 5

A room in a fisherman's cottage on a bank of the Nile. At the left is the entrance, at the top a window with several steps, and at the right a door leading into an adjacent room. Nets and other fishing gear are hanging on the walls.

Tableau 1

Fishermen and women run merrily into the cottage. Girls carrying grape-vines press grape juice into mugs, and give the mugs to the fishermen. The fisherman and his wife who live in the cottage act as host and hostess, and entertain their company. The villagers start to dance.

There is a knock at the door to the left. The host hurries to open the door.

Tableau 2

Dressed in simple country clothes, Taor, Pasifont and Aspicia enter and request hospitality. The fishermen receive them with pleasure, and the fisherwomen circle around Aspicia and offer her fruit and drink; the peasants continue the dances which had been interrupted.

Taor and Aspicia, drawn into the festivity, mix with the crowd of peasants and perform a lively character dance with them. At the end of the dance, as night falls, the fishermen prepare to leave to fish by firelight. Some take up tridents, others burning torches. The girls carry out the fishing nets. General activity of the peasants with tridents and torches, during which they bid farewell to one another and withdraw.

In gratitude for the hospitality, Taor offers to help his host at his work. He says farewell to Aspicia, but having stepped away from her goes back again, presses his beloved to his heart, and only then follows his agreeable companions.

All retire except Aspicia.

Tableau 3

The young princess at first listens to the gradually fading sounds of the fishermen's steps, then, realizing that she is alone, begins to worry... Her fear increases because night has fallen and the fisherman's hut is lit only by a single lamp and the mysterious moonlight, which shines through the open window.

Looking around the cottage, Aspicia hastens to close the door through which the fishermen left, then runs to the door opposite, horrified at the sight of a person entering, covered in a long cape which conceals his face.

Tableau 4

Coming in, this person locks the door behind him, then throws off his cape, and Aspicia recognizes her fiancé, the Nubian king. He kneels before her, wanting to calm her, but the princess, paying him no heed, orders him to leave.

The Nubian declares that nothing will make him agree to give her up to his rival, and that he has come to take her back. Aspicia asks him to abandon his suit, explaining that she cannot love him, that her heart belongs to another whom she has sworn to love forever. This admission arouses the Nubian's fury; he rushes at Aspicia, trying to seize her.

Fleeing him, Aspicia steps up to the window, and pointing out to the audacious Nubian the river coursing beneath it, resolutely declares that she will jump into the water if he makes bold to pursue her any further.

The Nubian first backs away, but doubting that Aspicia would carry out such a threat, he rushes at her again. Without hope, raising her eyes to heaven, the girl sends her last farewells to everything dear to her on earth, and jumps into the river.

Shattered by this deed, the king lets out a wild cry. Hearing this, his attendants run in. He points to the open window, and despairing, wants to leap after the girl, but his courtiers restrain him. Just then Taor and Pasifont return to the cottage.

Seeing them, the embittered Nubian orders both to be seized and threatens Taor with revenge for Aspicia's abduction.

SCENE 6

The Nile

Aspicia descends peacefully from the surface to the bottom of the river; the deeper she goes, the more a new world opens up around her. A magnificent coral grotto and brilliant, diamond-like stalagtites rise up in the underwater kingdom. In the middle of this splendid grotto majestic Father Nile, with his grey beard, sits on a jasper throne adorned with sea plants. He holds a coral-studded gold trident. Naiads, undines, nereids, and various other subterranean divinities surround him.

Noticing the body of the young girl floating down through the waters, Nile takes the unconscious Aspicia into his embrace, studies her face and turns his gaze to the princess's golden necklace. Seeing the name 'Aspicia' inscribed on it, he recognizes the daughter of Egypt's mighty sovereign.

Venerable Nile bows deferentially to the Pharaoh's daughter and charges the naiads to care for her. They take her into the azure grotto.

Then the Nile orders a festival in honour of their guest. At his sign, [dancers representing] the rivers Guadalquivir, Thames, Rhine, Congo, Neva, and Tiber, their tributaries and rivulets, flow in and perform character dances of the countries through which they run. Dances of the naiads and the undines follow (a *grand ballabile*), among whom the charming Aspicia appears in the costume of an underwater fairy. After a general dance, Aspicia, wanting to know her beloved's situation, begs the Nile's permission to see Taor. The god of the river gives his consent.

At the wave of his trident, the image of Taor appears now at the top of a cliff, now in the transparent spray of a waterfall. Aspicia, delighted, tries to reach the image dear to her, but every time she does it disappears, making affectionate gestures.

Then the girl falls at the god's feet, begging him to return her to earth, to the one so dear to her heart. Father Nile is chagrined by this request, and seemingly explains to Aspicia that he has no power over the dead; but yielding to Aspicia's pleas, he agrees to fulfil her wish.

At his command a huge mother-of-pearl seashell, bedecked with flowers and lotus leaves, rises from the grotto. With the naiads' help Aspicia gracefully takes her place in this flowery nook. A rotating pillar of water lifts the shell above the arches of the grotto, while Aspicia, in a farewell gesture, strews flowers on the naiads and undines who follow her course. Nile stands among them, pointing her way back to earth.

The end of Act II

ACT III

SCENE 7

The Pharaoh's garden. On the right, a colossal statue of Osiris.

Tableau 1

Seated on a granite throne, surrounded with wives, courtiers and slaves, the sorrowful and anxious Pharaoh sternly questions augurs about Aspicia, his only daughter and successor to his power and wealth.

The high priest opens a large book resting on the backs of kneeling black slaves. He invokes the gods and pronounces incantations while other priests and priestesses perform an inspired, fantastic dance around him.

But the oracle is silent, and the chief augur, bowing to the Pharaoh, declares it impossible to explain what has happened to the princess. With a threatening look, the Pharaoh rises from his throne. Everyone kneels, fearing the sovereign's wrath.

Tableau 2

The Nubian king enters, but without telling the Pharaoh about his daughter's death, announces that he caught Aspicia's abductors. Guards lead in the prisoners. Taor walks proudly and fearlessly, but his servant Pasifont, in irons as is his master, trembles at the sight of the wrathful Egyptian sovereign, to whom he cannot even raise his eyes.

The Pharaoh commands Taor to reveal where his fugitive daughter is hiding. Taor answers that she is missing, that he is saddened by the loss, and that he is prepared to sacrifice his life if only to see Aspicia again, unharmed and happy. The Pharaoh, paying no heed to the young man's ardour, demands that he return his daughter. Taor repeats that he does not know where the princess, beloved to him, is hiding.

'Then see how I punish the disobedient,' the Pharaoh cries.

At his lord's command, the high priest brings out a large basket with flowers, carried by two other priests; they place the basket in front of the Pharaoh's throne. Another two priests lead in a black slave with his chest bared, and force him to kneel in front of the basket.

The high priest begins his incantation. A peal of thunder is heard. Darkness covers the stage; then a pale, mysterious light is seen. All kneel. The flowers in the basket begin to move, from which crawls an emerald-coloured snake. It sways beautifully in the air, uncoils and recoils several times, and looks at those present with its bright, ruby-like eyes. It is a serpent of Isis, one of the Egyptian gods. At a sign from the high priest, the sacred reptile sinks its fangs into the black slave's breast, and having bitten him, calmly slithers back into the basket and hides among the flowers. The unfortunate slave, mortally striken, falls and dies in terrible convulsions.

The priests carry the basket out and place it on a granite pedestal.

At the sight of such a horrible spectacle, Taor controls his fear with effort, but summons his courage and preserves his calm. Pasifont falls face down and stays there, motionless. The Pharaoh turns to Taor: 'You saw what an awful death befell this slave,' he says, 'the same will happen to you if you do not reveal where Aspicia, whom you abducted, is hiding.'

Just then the merry tune of a village march is heard, gradually approaching from a distance.

Curious, everyone stands up.

From the depth of the stage appear peasant men and women with flowers and oak branches. Ramzea, running to the Pharaoh, informs him that his beloved daughter has returned.

Tableau 3

On their shoulders the peasants are carrying a sedan chair decorated with flowers and leaves, on which Aspicia is seated, surrounded by a merry crowd of young peasant girls. Greeting her parent from afar, the princess alights from her flowery throne and rushes to her father, who embraces her joyously, kisses her, and cannot look enough at his favourite child.

He asks what happened to her. Aspicia, frightened at the sight of the Nubian king, takes refuge in her father's embrace. In a lively narrative she tells of her adventure in the fisherman's cottage and accuses the king of forcing her to jump into the Nile.

'Is this true?' the Pharaoh asks the Nubian king.

The king finds it difficult to answer. The princess calls on the gods as witness and asks the Nubian to do likewise. Fearing heaven's wrath, he cannot refute Aspicia's accusation.

The Pharaoh, convinced that the Nubian king made an attempt on his daughter's life, declares a breach between the two realms and destroys the treaty of friendly relations. The embittered Nubian threatens the Pharaoh with revenge, but the latter, scorning it, orders him and his retainers to leave.

General confusion. The Nubian makes to leave, threatening the Egyptians, who in turn, with insults, accompany the Nubians out of the hall.

Meanwhile Aspicia, despondent, leans on the fateful basket which contains the serpent. Ramzea runs to warn her of the danger, and tells her of the frightful scene of the slave's death. Struck by her story, Aspicia looks anxiously for Taor, and notices him between two priests leading him to his execution. (The Pharaoh did not forgive Taor the abduction of his daughter and ordered him to be executed with his accomplice Pasifont, whom bodyguards forcibly carry to the place of execution, as he, out of fear, is unable to move.)

Aspicia falls at her father's feet, begging him to forgive Taor, but the Pharaoh pays no attention to her. Despairing, the princess explains that she will die with her beloved. The Pharaoh is unmoved. The sovereign's pride is drowning out the voice of his heart.

Meanwhile the sorrowful procession approaches the place of execution. Pale, with dishevelled hair, Aspicia, seeing her father's inflexibility, runs up to the basket with the sacred serpent, prepared to thrust her arm into it. The frightened Pharaoh runs to his daughter and restrains her. Touched by Aspicia's heroic self-sacrifice, he forgives Taor and Pasifont.

Freed, Taor falls at his beloved's feet. The Pharaoh raises him up and declares him his daughter's husband. Aspicia and Taor, deliriously happy, rush to the Pharaoh, who receives them into his embrace. The entire court congratulates the young people and their generous monarch.

General rejoicing. Egyptian men and women, with little metal discs at their elbows and in their hands, rush in and begin a character dance (*pas de crotales*), in the course of which various picturesque groupings are formed.

Drawn to the crowd's celebrating, Aspicia takes part in the dances. The bacchic merriment ends with ethereal groupings, during which the stage is covered in clouds.

Scene 8

Instead of the Pharaoh's luxurious gardens, we see the interior of the pyramid again, as in the prologue. Lord Wilson, John Bull, and the Armenian merchants are asleep. The bronze statues of the Pharaoh and the mummies are resting in their niches. The old man covered in rags, the guard of the pyramids, is dozing in a corner.

Scene 9

The top of the pyramid opens and is illuminated by a fantastic light, in which is represented a magnificent

APOTHEOSIS

In the heavens all the gods of Egypt encircle the principal divinities, Osiris and Isis. Beneath them are grouped the Pharaoh, Aspicia and the entire world of ancient Egypt.

Just then Lord Wilson, John Bull and all the others in the pyramid awaken.

THE END

Printing permitted, provided that upon publication the number of copies legally agreed upon would be delivered to the Censorship Committee. St Petersburg. 15 January 1862. Censor V. Beketov.

10
'The First Performance of the Ballet *The Pharaoh's Daughter*'
Sergei Nikolaevich Khudekov

The success of *The Pharaoh's Daughter* was immediate and extraordinary. 'The new ballet is the last word in balletic art,' one critic wrote, 'The magnificence of the production exceeds all description.'[1] 'Since Perrot we cannot recall such a huge success, and in fact the work of Messrs St-Georges and Petipa is marked by that magnificence, that grandness to which the famous choreographer accustomed us.' On the beauties of the decorations, this reviewer continued,

> we cannot begin to expound; to appreciate them fully they must be seen. The groupings were unusually picturesque and the dances distinguished by their characteristic quality. The groupings of the caryatids with baskets of flowers on their heads are particularly effective, as also the groups with the little metal plates during the Bacchic festival of the third act.
> Nor shall we analyse all the dances, but let us note in general that Mr Petipa produces them intelligently and always thinks them through carefully, with the talents of the dancers in mind, so as never to tire them but to bring their superior qualities into relief; in a word, he is an excellent balletmaster.[2]

The principals, including Rosati, Petipa himself as Taor, and Timofei Stukolkin as Pasifont were also praised, together with the soloists who performed the divertissement of the rivers.

In his first major composition, Petipa called upon the grandeur and pageantry which were to serve him well for the remainder of his career. These would vary in accordance with the circumstances, as earthly celebrations or utopian visions from a fantastic world, but they were part of a formulaic conception of danced drama which *The Pharaoh's Daughter* first begins to reveal. Moreover, particular devices of this ballet were used again, by Petipa and others, without concern for the borrowing. Nikia, the heroine of *La Bayadère*, dies from the bite of a poisonous snake concealed in a basket of flowers; the Lilac Fairy, like Aspicia, journeys in a barque of mother-of-pearl; and Amata's vision of Lucio in *The Vestal* is modelled on the vision which Aspicia has of Taor in the kingdom of the god of the Nile.

By the end of the nineteenth century, *The Pharaoh's Daughter* (like *The Little Humpbacked Horse*) had attained the status of a classic. The esteem it enjoyed,

---

[1] Z, 'Theatrical and Musical Chronicle', *Russkii mir* [Russian World], 1862, No. 4, p. 104.
[2] M. R., '*The Pharaoh's Daughter* (Ballet in 3 acts)', *Syn otechestva* [Son of the Fatherland], 1862, No. 18 [20 Jan.], p. 139.

pointed up by a revival of the ballet in 1898, seems to have bestirred Sergei Nikolaevich Khudekov to write his recollections of the première. 'The First Performance of the Ballet *The Pharaoh's Daughter*' was published in the newspaper *Peterburgskaya gazeta* on 15 October 1898. It is the shorter of his two such reminiscences (the earlier and longer, translated below, pp. 250–75, was of *The Little Humpbacked Horse*). For decades Khudekov had been the editor of the *Peterburgskaya gazeta*; he was also Petipa's collaborator on a number of ballet scenarios, and the author of the extensive *Istoriya tantsev* [The History of Dances], 4 vols. St Petersburg/Petrograd, 1913–18.

The Pharaoh's Daughter was the first important choreographic work produced by Marius Petipa. After this ballet its author's reputation as a talented choreographer was firmly established. In truth *The Pharaoh's Daughter* enjoyed a colossal success. I remember the first performance, which took place in January of 1862 on the Sunday before Butter Week, after which it was given every day throughout Butter Week; it was a huge box-office success, and speculators in tickets made huge sums. The scenario of *The Pharaoh's Daughter* was written at the management's request by the French writer Mr St-Georges, to whom 3,000 francs were paid for this labour. It is remarkable that this huge work was produced in only *six weeks*. This required a significant effort from the balletmaster, the composer Mr Pugni, and the decorators.

The necessity and the haste of production was brought about by the circumstance that this ballet was mounted for the beautiful dancer Rosati, who was dancing in Petersburg for the last time and wanted to take advantage of her contractual right to dance in a new ballet at her benefit performance.

Despite the haste, the ballet succeeded magnificently: Marius Petipa won fully-deserved laurels, which proved he could properly compete with his rival of that time, Mr St-Léon, who had given neither him nor his young wife any opportunity whatever. Mr Pugni's music overflowed with melodies which were choreographed into various dances and quadrilles; the decorator Roller excelled himself—for the perspective of the hall in the palace of the Pharaoh he was awarded the title of academician.

This ballet was given a hundred times and was considered a favourite for benefit artists because having *The Pharaoh's Daughter* on the bill would produce perpetually full houses. Many artistes have danced the leading role, which demands, besides choreographic art, plastique and excellent mime. But the best among them was of course Mlle Rosati, who created the role of Aspicia. An imposing figure of striking appearance, with classical, regal gestures—these came together in Mlle Rosati, who unfortunately danced this ballet only eight times. After a brilliant success, she left the stage and retired to Paris, leaving a host of admirers in Russia. One of them sent a little cask of caviar, which 'the Pharaoh's Daughter' adored, to Mlle Rosati in Paris every week practically up to the time of his death. In addition, the ballerina was known for her thoroughbred pug dogs, perhaps the finest in the world, distinguished by their smallness and the extraordinary ugliness of their 'doggy faces'. In Paris the artist maintained a pug dog breeding kennel, sustained by the best specimens sent to her from Moscow by the same old balletomane, who put on the dandified airs of a much younger man.

After Rosati, the role of Aspicia was given to Marius Petipa's wife. This young, pretty dancer, if she did not surpass the creator of the role, was not inferior to her either, if for no other reason because she had youth and beauty on her side. These two were dangerous rivals for Mlle Rosati, whose shoulders bore only the baggage of past successes. Although 'Balletomane' [Konstantin Skalkovsky], author of the book *Ballet*, assures us that Mlle Rosati had a special success in the dance with the lute, he gleaned this information from an inaccurate source. There was no such dance in *The Pharaoh's Daughter*. He probably confused it with the 'pas de sabre', with the dance in which Rosati in fact distinguished herself.

Rosati was replaced by Marie Petipa. The first appearance of this attractive, graceful artiste in the pyramid, the slow entrance of the mummy called back to life, her gaze fixed on one point, and then her astonishment at the sight of the sleeping, handsome man—this all produced endless applause.

This became Marie Petipa's crowning role, and not one artiste after her had such a tumultuous success in this ballet. Mlles Dor, Vergina, and Vazem all danced it, but they were a long way from Marie Petipa's Aspicia, famous for her purely classical plastique, femininity in poses and movements, and the extremely fine expression and liveliness of her mime.

She strikes a pose, and an artist could paint a picture, she makes an imperious gesture, and for the sculptor's chisel it would be a marvellous, well-formed model. That was Mme Petipa, who always enjoyed great success, though she was never distinguished by mastery of choreographic difficulties, and the twists and turns for which contemporary artiste-virtuosi are so greedy. Lacking a natural gift, becoming mistress of two or three turns on one leg is accomplished by labour. Indeed, though many years have passed since the day our ballet lost Mme Petipa, and many celebrities and semi-celebrities of all kinds have come since then, of foreign importation and our own vintage, the like of Mme Petipa has not been seen again. She did not astound anyone with her dances, but enchanted everyone immediately. The Italians nowadays astonish with their dances; the spectator is exactly as in the circus looking at the acrobats working just beneath the roof. With gasping breath and sinking heart he watches the belaboured dancer—will she finish her *pas* or not?! . . . Will she fall down or not, will she break her neck or not?! Then she finishes her *pas* and the public forgets her. Mme Petipa was not like that on stage. Binoculars were trained on her even at times when she was not dancing. Mme Petipa lived her stage situation, and for that reason chose sensible ballets with narrative content and dramatic mimed scenes, not incoherent ballets such as they began to produce in later times. Precisely for that reason, in a majority of the dances specially composed by her husband, a certain thought or point was being expressed. These were, so to speak, 'dances in the action'. The 'pas de sabre' in *The Pharaoh's Daughter* was such a dance, in which Mme Petipa, alone of all who performed it, enjoyed so great a success that after her performance a loud roar rose up; the public was choked with delight, demanding several encores of the dance.

Among the artistes who danced in the first performances of *The Pharaoh's Daughter*, Mlle Madaeva was especially distinguished in her personification of the river Neva. She had to repeat this Russian dance, ending with a broad wave of the arms and a low bow from the waist, three or four times. Madaeva's first name was Marthe, but she was called Mathilde or Motinka.

Mlle Kemmerer I created the role of the fisherwoman, Mlle Radina I danced the 'pas des crotales' with solid success, and finally Mlle (Maria) Sokolova, with her astonishing, ethereal *ballon*, distinguished herself in the 'pas des cariatides' together with Mlle Kosheleva, who whirled so rapidly that it seemed she might fly either into the orchestra or, in her dizzying spin, beyond the boards into the director's box.

Mr Golts created the role of the Pharaoh; he played this role several hundred times and considered it his very favourite. He was grand and magnificent in it. His weighty, tall figure, with sharply outlined, round gestures, was ideally suited to the representation of Egypt's ancient sovereign. Many artists followed Mr Golts as the Pharaoh, but no one ever replaced him. Inimitably funny was Mr Stukolkin I, who created the role of the Englishman's servant. The Englishman himself was played by Marius Petipa, who later yielded this role to Mr Lev Ivanov, who created the role of the fisherman at the first performance.

Mr Lev Ivanov (now a balletmaster) could well be the only artist still in the theatre who took part in the first performance of *The Pharaoh's Daughter*. The last Mohican in this wonderful Pleiades of former artists remains Mr Felix Kshesinsky, who thirty-six years ago artistically created the Nubian king. He shall appear again in this role [in the forthcoming revival].

S—kov

11

Libretto of *The Little Humpbacked Horse, or The Tsar-Maiden*

Arthur Saint-Léon

St-Léon answered *The Pharaoh's Daughter* with *The Little Humpbacked Horse*, a ballet in which he stressed contrasts with Petipa's work and aimed at greater audience appeal. Its characters were not Egyptian nobles but Russian peasants, its tone not serious but farcical. Beneath the surface, however, there were reminders of *Pharaoh* in the cloud-covered scene-changes near the beginning, in the undersea kingdom and the divertissement of nationalities. Concerning outward magnificence, the reports of *Humpbacked Horse* were also similar to those of *Pharaoh*:

> The production of the ballet was splendid: nine excellent new decorations, costumes after drawings by the artist Charlemagne, machines by Roller, a magic fountain of real water with changing colours, chemical devices by Mr Shishko, accessories by Mr Gavrilov—all these satisfied the most exacting requirements. The music was very characteristic, melodious, and correlated fully to the story. The performance of the ballet was irreproachable: all the participants brought off their roles with enthusiasm, and the leading character, Mlle Muravieva, simply performed miracles.[1]

While nothing about *The Little Humpbacked Horse* would stand in the way of its success, one observer of the first performance described what proved to be St-Léon's principal failing—the lack of genuine dramatic interest and large-scale architectural planning to go with his undisputed facility as a creator of dances. After recounting the scenario in detail, the critic of *Son of the Fatherland* concluded:

> What you see is a design, fully corresponding to contemporary requirements, the richest material for the *richest* of divertissements, but not for a ballet. Mr St-Léon's speciality, incidentally, is the divertissement with dances of different character; with particular relish he reproduces the dances of various nationalities. Often he manages to impart character to them, a true stamp. The thought of introducing a Russian element into the ballet was not a bad one, but a more poetical shading had to be given to it, the dances had to be more distinctively grouped, and not all adapted more or less to the same rhythm, that of the famous Russian dance, the trepak . . . Beyond that almost all the dances—of the animated frescoes, the nereids or rusalkas—fit into the same form, and lack originality . . .

[1] 'Petersburg Notes', *Golos* [The Voice], 1864, No. 334, 4 Dec., p. 3.

Mr St-Léon's greatest blunder was that he somehow moved the leading character into the background—the Tsar-Maiden, or, more correctly, the first ballerina Mlle Muravieva, who doesn't dance very much and appears only in the third scene. . . . We might have done without the outrageous dances of the crayfish and even the ruff and the carp. . . . In places Mr St-Léon's experience and talent as a balletmaster come through, but I prefer his earlier works from a purely choregraphic standpoint.[2]

Like Didelot's *The Captive of the Caucasus*, St-Léon's *The Little Humpbacked Horse* presented a Russian story with methods perfected in earlier, non-Russian works. Like *The Captive*, it was produced at a time when Russian dancers were in the ascendent and patriotic feelings were running high. The devotees of Russian art who lobbied St-Léon to produce *The Little Humpbacked Horse* knew they were dealing with a cosmopolitan figure but were hardly in a position to complain, and nothing suggests they were tempted. Here at least was a visiting celebrity who spoke Russian fluently and was willing to set a Russian folk tale (unlike Petipa, who spent most of his life in Russia, never mastered the language, and avoided Russian topics in his ballets). Since national character in ballet is a matter of perception, Russians could see the Russian in *The Little Humpbacked Horse* as they chose: in the subject-matter, external features or profound elements of style, mixed or undiluted with traits attributable to another nation or culture.

After *The Pharaoh's Daughter* and *The Little Humpbacked Horse* Petipa and St-Léon learned the difficulties of repeating their success. Their next attempts in the same vein, Petipa's *The Beauty of Lebanon* and St-Léon's *The Golden Fish*, were epigones which both failed.

THE LITTLE HUMPBACKED HORSE
or

The Tsar-Maiden
Magic Ballet in Four Acts
and Nine Scenes

by Mr Saint-Léon

Music by Pugni

Saint Petersburg

At the Typographer F. Stellovsky

1864

DRAMATIS PERSONAE:

Pyotr, *a peasant*
Danilo, *his son*

[2] M. R., 'Theatre Chronicle', *Syn otechestva*, 1864, No. 291 [5 Dec.], p. 2302.

Gavrilo, *his son*
Ivan, *a simpleton also Pyotr's son*
The Tsar-Maiden, *daughter of the Moon, sister of the Sun*
The Little Humpbacked Horse (*a genie which appears in the form of a horse*)
The Khan of the Kirgiz-Kaisaks
Mutcha, *the khan's retainer*
Shadzha,
Isfez,   } *the khan's grooms*
Kushkar,
The Queen of the Nereids
One of the khan's wives
A Russian merchant
His wife
A retail trader
A country woman
A Jew
A Kirgiz warrior

Peasants who dance at the market—Kirgiz, Tartar, Chuvash, Tungus, Samoyed, Bulgar, Kurd—nereids, various ocean fish, oysters, crayfish, shells

ACT I

Scenes 1 and 2
The Russian Bazaar

The stage represents a square near the village of Krasnovodsk; on the left Pyotr's homestead, which is located at one corner of the square. Part of the homestead, facing the audience, is a stable with an awning. In the distance, Pyotr's fields; they are visibly wasted away, and the young shoots on them are beaten down.

The settlement is located at the foot of the Urals. It is after lunch, and the merchants are sleeping near their carts. Pyotr's three sons are snoring beneath the eaves of their shed. Little by little other merchants come in and wake up the sleepers, including the three brothers. A Jew then appears to buy up the best of the goods on display; but the Russians respond rather coldly to his zeal and mock him in every way they can. The simpleton joins the people and also chaffs the Jew. Pyotr appears, and reproaches his sons for playing pranks when they have readied nothing for the bazaar. His sons hurry to fetch sacks filled with grain: they have nothing else to sell. Other peasants display fancy produce; they marvel at Pyotr's sorrow, for his land is now completely spent, when once it brought forth excellent harvests. Pyotr responds that his sons Danilo, Gavrilo, and Ivanushka are to blame because they work so little at farming. 'I have decided to punish them,' he says, 'and will give them neither hearth nor home until the fields are plentiful, as they were before.' 'Do not grieve, Pyotr,' says one of the merchants, 'you will recover. And until then there is no reason to quarrel with your sons. Certainly just now there is no point in it, because evening has come and a celebration is about to begin.' After this conversation young people take up various games and entertainments, mostly dances. With nightfall Pyotr's neighbours disperse, and he and his sons remain by themselves.

The old peasant begins to scold his sons again, especially Danilo and Gavrilo. He does not scold Ivanushka because a fool is not capable of much anyway. Having warned them once more that he will punish them for being remiss in their farming duties, Pyotr goes home. The brothers are dejected. They try to figure out why their land has dried up so badly despite their efforts—and they cannot. Ivanushka observes that evil spirits must be doing mischief; his brothers laugh at him. Meanwhile, after much debate, they agree to stand guard at night in the fields to learn what is bringing them such devastation. As they are about to leave a furious storm arises. Girls who were walking out late run to Pyotr's homestead, seeking shelter from the storm under the eaves of his shed. Danilo is first to reach them; he offers to escort them home, and they leave. Gavrilo, whom the rain has drenched to the skin, feels that their new sentry duty is beyond him, and leaves for home. Ivanushka the Foolish is left alone; he is merry and perfectly contented with himself.

Midnight comes, and the crowing of a cockerel is heard. Suddenly in the gloomy darkness Ivanushka notices that something white is skimming across the fields. He rushes off to encounter this strange phenomenon and discovers it to be a white horse with a shining mane, stamping the young shoots and breaking loose the grains with its hooves. Not at all frightened by it, Ivanushka stops, and as the horse runs past him he seizes it by its golden mane, jumps up on its back, kicks the animal's side with his powerful legs, and in a fury seizes it by the neck.

Meanwhile thick clouds obscure the sky; from time to time flickering lightning illuminates the horse-spectre, which quickly flies off into the heavens carrying the intrepid Ivanushka. At last the storm quietens; the sky gradually clears, and a new decoration represents an uninhabited place in the Urals. The magic horse comes back to earth and Ivanushka sees before him a small, odd human figure with a humped back and a horse's head. 'Ivanushka,' says this creature, 'Your courage amazes me. You were not frightened when I lifted you up to the clouds. I want to reward you for this: ask me whatever you wish.' 'My father will not feed nor house me because our fields are being destroyed and nothing will grow on them,' says Ivanushka. 'I have promised to make every effort to bring the fields back to their former plenty, and now that I know that this confounded white horse is responsible, I have decided to kill it straightaway.' 'That is a stupid idea,' the little horse answers, shaking his head. 'To deal with him, silly Ivanushka, is not within your powers, but I want to show you how much I value your courage and zeal, and so I say to you: return home. You will find a pair of the finest horses in your father's stable; They are my gift to you. I too shall be in the stable in the form of a little humpbacked horse: pay me special heed. When you need something, just touch your little horse's hump and your every wish will immediately be fulfilled. Go home now, and fear nothing: I shall save you from all misfortune and disaster.'

Ivanushka, of course, is very happy over this. At that moment the strange creature changes into a little humpbacked horse; Ivanushka climbs up on it and speeds off into the skies.

The cloud which had obscured part of the stage fades away and the previous decoration appears again. Fatigued, Ivanushka goes up to his father's house. The little humpbacked horse walks with him. Ivan lies down on the ground and quickly falls asleep, holding on to his little horse's reins. Suddenly he awakens; the strange adventure to which he was party comes back to him; it seems like a dream now, and he tries to find in it something resembling reality. He taps the hump of his little horse, and immediately the odd creature who gave him the little horse appears before him. The creature asks: 'What is it that you need?' Reassured that he had not dreamed the strange adventure, Ivanushka lies down again, and falls into a deep sleep.

Danilo is returning from a lover's tryst. To do so unnoticed he crawls through a window, but Gavrilo, who is awake, explains that he saw two fine horses in their stable which Ivanushka the Foolish brought from who knows where. Danilo is astonished, seeing in the animal crib a pair of spirited, golden-coated horses, which he thinks are invaluable. 'Here,' he says, 'is the opportunity to get rich ourselves and to make Papa rich too. Let us take these horses and sell them to the wealthy Kirghiz khan!' Gavrilo agrees, but Danilo finds it difficult to get the horses out of the stable because Ivanushka is sleeping right at the stable gate. Gavrilo explains that at the back of the stable there is an old door through which they can lead the horses.

The little humpback awakens Ivanushka and tells him that his brothers have stolen his horses and taken them to sell to the Kirghiz khan. Ivanushka despairs. 'Do not grieve,' the little horse responds, 'we shall follow them directly.' And they set out in pursuit of the brothers.

End of the first and second scenes

SCENE 3

Inside the richly appointed halls of the Kirghiz khan.

The khan of the Kirghiz Kaisaks is surrounded by his wives, retainers, and grooms. To divert the khan, his favourite wife dances. Suddenly there is a loud noise. Guards have seized three foreigners arguing among each other over a pair of magnificent horses. The khan wishes to know the details of the situation. The foreigners are led in; they are Pyotr's three sons, already familiar to us, each of whom is asserting his right to the horses Gavrilo and Danilo found in their father's stable. The khan questions the brothers, who get confused and cannot answer. But the little horse, transformed into an old hunchbacked Jew, makes his way into the khan's palace and walks up to Ivanushka. 'Lord,' he says, bowing respectfully to the khan, 'the horses at issue in this dispute belong to this young man; I myself sold them to him and can easily prove it. My receipt is probably in Ivanushka's pocket even as we speak.' Ivan is astonished by these words; he checks his clothes but cannot bring himself to reach into his pockets. At his feet he finds a meticulously detailed account on a piece of paper. 'Don't keep searching for nothing,' the little horse tells him: 'the receipt is there; you just dropped it.' And picking up the paper from the floor, the little horse presents it respectfully to the khan.

Danilo and Gavrilo are terrified; they rush for the doors to escape, but are stopped by guards. 'Don't be frightened,' Ivanushka tells them. 'If it pleases his lordship to buy the horses from me, I shall share the money equally with you.' He extends his hand to them, and all three rush to embrace, still in total ignorance. What happy accident brought them out of their difficult situation, and whence this receipt? At that moment the khan orders Mutcha to inspect the horses; he does and returns a moment later, claiming that never in his life did he expect to see such superb horses. The khan pays generously for them: he commands that five sacks of silver be heaped on Ivanushka, and that the horses immediately be led to his stables, of which they shall be an adornment. But the little horse, still in the guise of an old hunchbacked Jew, approaches the khan again. 'Condescend,' he says, 'to permit your slave to say a word . . . I know my horses very well . . . Had they killed these two young men, it would have served them right. I sold them to Ivan only because I knew that he alone can manage them and because it would be best to entrust the horses to him.' 'Now then—really?' says the khan, 'I agree! Ivan will stay with me and become my stable master . . .' Overjoyed by this honour Ivan immediately shares the money with his brothers. Everyone leaves and Ivan is given the uniform of a stablemaster, which he puts on immediately.

The little horse turns to Ivanushka. 'I have steered you clear of misfortune,' he tells him, 'but you could get into some other trouble and need my help. Take this little whip, and when you want me, snap it and I shall immediately appear to you.' Having said this, he disappears.

Mutcha, envying the favour shown to Ivanushka, surrounds him with his comrades and forces him to drink some wine... Somewhat emboldened, Mutcha observes that pretty women are the only thing missing for an absolutely good time. He turns to the Persian frescos which cover the walls of the hall... 'Ah,' he says, 'why couldn't these charming creations come alive?...' 'Well, let's not settle for that,' Ivanushka responds calmly, and snaps his whip. The little horse appears. 'Can you not,' Ivanushka asks him, 'do something to bring these beautiful women to life?' And he points to the frescos which cover the walls. 'I can,' answers the little horse, 'but only on the condition that nobody touches them; otherwise they will immediately disappear.' Mutcha and his comrades, hearing this, cannot restrain their laughter. But their mirth gives way to amazement when they observe the frescos come to life, come down from the walls, and join them.

DANCE OF THE ANIMATED FRESCOS

The dances of these supernatural creatures cause indescribable delight in the khan's retainer and his comrades. They rush to the frescos with their arms outstretched, but they return to the wall. The khan suddenly enters; he is most agitated and explains that he has had an unusual dream. 'No other mortal,' he says, 'has ever seen a woman as beautiful as the one who appeared to me in this dream.' Having glanced at the frescos which cover the walls, he shrugs his shoulders and says: 'They are beautiful women, but none better than the woman who at present fills my heart is to be found in the entire universe. Without her I will have no happiness on earth!' With these words Mutcha, recalling the strange power of which Ivanushka had given such obvious proof, takes the khan aside and explains to him that Ivanushka is disloyal, and that if he wished, he could obtain the beauty of whom his sovereign dreamed. 'Is such a thing possible?' the khan asks enthusiastically, and turning immediately to Ivan, says 'Ivan! Set out this minute to get me this pearl of beauty...'

Ivanushka is taken aback, and tries by all possible means to get around the order. 'Obey!' the incensed khan shouts at him. 'Only this young princess can melt the ice of my old age and restore a youthful ardour in me. If you don't get her for me, I shall consider you a traitor and order you punished by the whip.' Ivan is much grieved by this. The khan's nobles scoff at him. The khan repeats his command and gives him a few moments to think about it; then he withdraws with his entire retinue.

Alone, Ivanushka remembers the little horse; he takes his whip and waves it in the air. The little horse appears. In tears Ivanushka runs to him . . . 'Ah,' he says, 'a great misfortune has befallen me! The khan, our ruler, has seen a woman of extraordinary beauty in a dream and has fallen in love with her. He doesn't know who she is, nor where she lives, yet he has ordered me to bring her here without delay. What can be done? If I do not fulfil his order, I shall be whipped without mercy.' 'Do not grieve,' the little horse answers, 'you will not suffer the whip! . . . Go to the khan and tell him that you will show him the most beautiful women of different lands on this wall, and that it remains but for him to point out which one has so tenderly wounded his heart. Leave the rest to me.' With these words the little horse disappears. Alone, Ivanushka is thinking about how to explain this to the khan . . . and just then the khan comes back. 'Now then,' he asks, 'are you ready to carry out my order?' 'Everything shall be as you command, sire,' Ivan answers, 'but before I can find the woman to whom you have given your heart, I must know who she is. Be so good as to watch this wall closely; on it I will show you the most beautiful women of all peoples and realms, and when you see the woman who appeared in your dream, point her out to me, and I shall get her for you.' Ivanushka raises his magic whip to the wall, where images of the women of various countries appear: a bronze-skinned Indian of South America, an Algerian Jewess, a daughter of Asian India with a lotus blossom in her hands, a black Egyptian, a white-faced daughter of the Swiss Alps, and so forth . . . The khan, greatly excited, at first does not spot his ideal among these women . . . Suddenly, O joy! 'It is she!' exclaims the khan, 'it is she! . . .' A beautiful girl appears on the wall, bathing her feet in the bright azure waters of the ocean . . . In a moment the vision disappears. 'Very well,' says Ivanushka, 'this beauty will soon be yours.' He snaps his whip; the little humpbacked horse appears from beneath the earth. Ivanushka climbs on its back and they fly off into the air.

Scene. End of the first act

ACT II

SCENE 4

The Abduction

The stage represents a fantastic locale, a charming island which looks coquettishly out on to the peaceful bosom of the blue-watered ocean.

At the rise of the curtain, Ivanushka, astride the little humpbacked horse, is coming down to earth from the airy expanses. 'This is it,' says the little horse, 'the favourite haunt of the Tsar-Maiden, Sister of the Radiant Sun and Daughter of the Bright Crescent Moon, whom you must abduct and take to the khan; let us hide until our opportunity comes.' They hide. The ocean, which had been completely calm, becomes agitated, and from its waters emerge rusalkas who come ashore. They are preparing to meet the Tsar-Maiden, who appears soon afterwards. She glides along in a huge, dolphin-shaped seashell, comes ashore and disembarks, accompanied by the rusalkas who make up her court.

GRAND BALLABILE

After the dances, the queen of the rusalkas reminds the Tsar-Maiden that night is falling and it is time for them to leave. But the Tsar-Maiden is captivated by the magic island and wants to stay. The queen of the rusalkas insists, and the Tsar-Maiden is walking behind her when suddenly, at the little horse's order, a fountain appears from which pour streams of a thousand shades. The Tsar-Maiden, struck by this miracle, runs to the fountain to admire her reflection in its waters, which change colour at her approach. Ivanushka and the little humpbacked horse walk up to her; at a sign from the horse Ivanushka rushes at her, while the little horse drives off emissaries of the sun and the moon, who have appeared to prevent the Tsar-Maiden's abduction. The rusalkas surround the Tsar-Maiden in a tight-knit group. Ivanushka, who cautiously fights his way to the back, with a quick movement seizes the Tsar-Maiden by her tresses . . . The rusalkas flee in fright; the Tsar-Maiden begs Ivanushka to let her go . . . 'Fine,' says Ivanushka, 'I shall let you go, but not before I take you to the khan, my sovereign, who is expecting you at this very moment.' The Tsar-Maiden tries to resist, but faints. Ivanushka carries her in his arms, and sits down with his precious burden in a chariot which takes them to the khan's realm.

End of the fourth scene

ACT III
Scene 5
The Return to the the Khan's Palace

The khan's deadline to Ivanushka to return with the Tsar-Maiden has long since passed, and still there is no word of him.

A profound melancholy comes over the old khan, his passion for and his sufferings over the unknown beauty having increased in like measure. He is so embittered that his retainers tremble at his very look; his gloomy spirit is reflected in his wrath at those around him; in vain his favourite wife tries to entertain him with amusements and dances, for they irritate him all the more. One by one the couriers return whom the khan has dispatched to find Ivanushka, reporting that despite their best efforts nowhere have they found a trace of him. Ivanushka's two brothers are led in; they likewise have managed to learn nothing about him. With this news the khan flies into a rage and is at the point of bringing his wrath down on each of them.

But suddenly he and his entire court are struck by some benumbing force; it seems that everything around them is spinning, and they all quickly fall into a deep sleep. The little humpbacked horse appears. In his tracks Ivanushka carries the unconscious Tsar-Maiden. Next to the khan's divan another luxurious couch appears, on which Ivanushka places her. At a signal from the little horse, Ivanushka snaps his whip about: the khan and his retinue awaken. Seeing the Tsar-Maiden beside him, the khan can hardly believe that his dream is coming true. Excited, he stands up, begins to pace back and forth, gives his courtiers a command, and they bring in expensive gifts for the Tsar-Maiden, which he intends to give her when Ivanushka wakes her up.

At last the Tsar-Maiden awakens; the khan and his court bow down at her feet, but she is indifferent both to the gifts and the honours being rendered to her. Only one feeling stirs her soul: the awareness that she is in bondage.

'Forgive me,' the khan says to her, 'The justification for my behaviour is in serving the fiery love which you have instilled in me. I cannot live without you, and I place my khanate at your feet. Henceforth everyone here will obey you completely, queen of my heart—you are the owner of all you survey here; I myself shall be your first and most devoted slave.' The princess listens indifferently to these passionate outpourings; she is not attracted by the magnificence of power. To amuse the grieving beauty as best they can, Ivanushka's brothers Danilo and Gavrilo attempt to play something for her on their fifes; but their music lacks order and harmony. The little humpbacked horse gives Ivanushka a sign and he, taking a fife from one of his brothers, begins to play it. Hearing Ivanushka play, the Tsar-Maiden becomes animated and imperceptibly begins to dance. The khan's joy exceeds all measure, and he expresses his delight in various ways: he falls at the Tsar-Maiden's feet, reaches out to her and begs her to share his khanate. The princess cannot overcome her repugnance of the khan, but to conceal it somewhat she says to him: 'Very well, I will be your wife, but only on one condition . . .' 'Speak,' the khan breaks in, 'your slightest wish will immediately be fulfilled!' 'Ah,' says the Tsar-Maiden, 'when we were speeding over the ocean, I lost my mother's ring:

order it to be returned! And if in three days you get it for me, I will be yours!' 'Agreed!' answers the khan, who then says, turning to Ivanushka: 'You can do anything; find me the Tsar-Maiden's ring. I will expect you back in three days . . .' Unhappy with the khan's new command, Ivanushka bids him and his courtiers farewell and disappears with his faithful little humpbacked horse.

End of the Fifth Scene

TABLEAU
SCENES 6 AND 7

The stage represents part of the Arctic Ocean. In the middle a whale is seen, half thrust out above the water.

Ivanushka and the little humpbacked horse tread heavily along the waves and stop near the sea monster. Ivanushka is frightened to death. The little horse tries every means to cheer him up, then goes over to the whale and says: 'You must help me find a ring which has been lost in your dominions; it belongs to the Tsar-Maiden, whom you are obliged to obey in everything.' Then the little horse descends with Ivanushka to the bottom of the sea. There they meet the queen of the rusalkas, who is charged to entertain Ivanushka and the little horse during their search; at the whale's command, all the inhabitants of the sea set out to find the Tsar-Maiden's ring. A ruff finds it and brings it to the queen of the rusalkas, who gives it to Ivanushka.

SUBTERRANEAN BACCHANALE

End of the Sixth and Seventh Scenes

ACT FOUR
SCENE 8

In the khan's palace the Tsar-Maiden languishes, as before, in the tedium of enforced captivity. The khan, trying every means at his disposal to amuse her, takes her to a charming oasis in one of the canyons of the Urals near a Tartar hamlet. Ivanushka's time is up, and the khan cannot hide his unease . . . Night falls, and there is no messenger in sight. Suddenly Ivanushka appears in a crowd of the khan's courtiers; he is radiant with happiness, for he has retrieved the Tsar-Maiden's ring. There is no longer any barrier to the wedding of the khan and the beautiful girl, but she still refuses to marry a man of such venerable age. Suddenly her face brightens and a smile of delight appears on her lips. 'Ah! If only you really loved me!' she says to the khan, 'Our complete happiness might soon be at hand. But you are old and in this state you do not suit me as a husband. I know how to restore your youth. You have but to bathe in a cauldron of boiling water, and then—in a kettle of milk. All your wrinkles and illnesses will be eradicated by the boiling water, and the milk will return you to your first youth and bloom.' The khan professes readiness to conform to the Tsar-Maiden's wishes. Two cauldrons are brought in, but the sight of the water, seething with white froth, strikes an irrepressible fear into the khan, and he cannot bring himself to endure the trial. He is alarmed by the thought that he might die. Noticing Ivanushka, he turns to him. 'Listen,' he says to him, 'you are able to perform such improbable miracles—you go into the cauldron first, and if you emerge from it unscathed, then I will give you everything that you ask and will myself go next.' After taking counsel with the little horse, Ivanushka boldly jumps into the first cauldron and then the second, and is transformed into the handsomest of young men. Astonished by this transformation, the khan hastens into the cauldron; but no sooner does he sink into the boiling water than he loses his life.

The Tsar-Maiden gives her hand to Ivanushka, and the people, who witnessed this miracle, unanimously proclaim him their khan. For his perfect happiness Ivanushka lacks only his father and brothers, but no sooner has this occurred to him than they appear. The greybeard Pyotr rushes to embrace his son, presses him to his heart with great emotion, and apologizes to all three of his sons. 'So then, Ivanushka,' says the little humpbacked horse, 'I have rewarded you as I promised, and you will no longer require my services. It is time for us to bid you farewell!' With these words he vanishes.

A general dance of representatives of all the peoples of the Russian empire.

SCENE 9

APOTHEOSIS

Permitted by the censor, St Petersburg, 1 December 1864.

12

'The Petersburg Ballet During the Production of *The Little Humpbacked Horse* (Recollections)'

Sergei Nikolaevich Khudekov

In 'The Petersburg Ballet During the Production of *The Little Humpbacked Horse*', Sergei Khudekov gives us much more than an account of the first performance. It is an important retrospective on the Petersburg ballet of the 1860s (despite Khudekov's persistent references to the 1870s) penned by an eyewitness. It is also a warm, even nostalgic essay which takes us backstage, to the rehearsal hall, to the buffet during the intervals, to the business office of the theatres, and into the auditorium, where we are introduced to the manic balletomanes of the time. This rare piece of work was published in the *Peterburgskaya gazeta* in two instalments, on 14 and 21 January 1896. *The Little Humpbacked Horse* had been revived by Marius Petipa the preceding month.

The revival of *The Little Humpbacked Horse* at the Maryinsky Theatre has brought back a great swirl of memories of the Petersburg temple of Terpsichore [thirty years ago]. It was a treacherous stage for artistes then, with its petty intrigues, its joys and sufferings, and the gaggle of balletomanes, of diverse point of view and social station, which encircled it.

I was present at the first performance of *Humpbacked Horse* and to this moment recall that evening, which still vividly conjures up before me the 'golden age' of ballet in Petersburg. Yes! That was its golden age. Ballet was patronized in every social sphere; virtually no stratum of society was indifferent to it. Ballet occupied first place on the stage of the Bolshoy Theatre. Some people lived exclusively by their interest in ballet, and took seriously everything which touched upon this art.

And now? Different times and different ways!

'Things are far from what they once were!' sigh balletomanes of those days who have lived beyond their time. Everything has degenerated, everything is diminished! 'Courtiers' then were not limited to worshipping ethereal fairies; they loved the art itself, they pondered the secrets of choreography; besides, a galaxy of amateur fanatics existed then, for whom the sight of half-clad nymphs and naiads elicited no sinful thought whatever. No, they were enthusiasts of a pure art which provided elevated aesthetic pleasure!

On the occasion of this revival of *Humpbacked Horse* I shall limit myself to recollections of the time when the Russian stars Muravieva and Petipa flourished, when the Petersburg ballet shone in utmost splendour by its superior choreographic artists.

The Bolshoy Theatre

Opposite the Maryinsky Theatre to this day stands an imposing building which harbours a mass of legends closely connected with the development of theatrical art in Russia. Choreography was especially lucky there. Like meteors, a pleiades of stars, foreign and Russian, flew through it; Petersburg applauded Taglioni there, and Fanny Elssler, Ferraris, Cerrito, Carlotta Grisi, Rosati, Grantzow, Dor, Salvioni, and many other visiting celebrities. Native Russian talent also developed and flourished there which went on to storm the whole of Europe: Muravieva, Lebedeva, Petipa (I speak of not so distant times), Vazem, Vergina, Sokolova, Kemmerer, and others. There too balletmasters—Perrot, St-Léon, and Marius Petipa, unfading to this day—deployed their skills, shone by their art, and perfected themselves. None of these creators in ballet had careers strewn only with roses: in this temple of Terpsichore they suffered their measure of thorns as well.

In the 1870s [i.e., the 1860s] ballet was very fashionable. The theatre management specially patronized this type of spectacle. It was given three times a week—Tuesday, Thursday, and Sunday; on the other days Italian opera was offered at the Bolshoy. With the arrival in Russia of Patti and other singers of the most outstanding and expensive kind, ballet began to get 'squeezed', as balletomanes then expressed it. Instead of three performances, ballet came to be given twice, and finally once a week, on Sunday. Ultimately, to the chagrin of all genuine lovers of the choreographic art, the ballet was transferred from the Bolshoy to the Maryinsky Theatre, where it alternated with performances of Russian opera, the vital signs of which were barely noticeable at the time. National opera was scorned.

The balletomanes loudly protested this alleged abasement of ballet. Their complaints appeared in newspaper articles, where it was argued that henceforth Russian ballet, the pride of the Northern Palmyra, must undoubtedly decline. The Maryinsky stage was not suitable for ballet! It was too wide and not deep enough and therefore obviously impossible for ballet, which required various effects of decorations and machines. The balletomanes repeated this time and again, but their voice was not heeded.

On the other hand, the machinists of the Bolshoy Theatre rejoiced at the transfer. Roller, the late chief machinist of that theatre, told me that to whatever degree the depth of the Bolshoy stage was suitable for decorative effects, it was so much the more impossible for mechanical apparatus. There was always standing water in the basement, where most of the machinery was located, and one had to marvel at the patience and stamina of the Russian 'evening soldiers' who for inconsequential pay spent almost six continuous hours up to their knees in water shifting various machines. 'Every day I had to bring in several extra *poods* [a unit of measure] of tallow for the greasing of damp cables,' Roller repeatedly complained to me, walking about the stage in his never-changing, ever-dirty, cockaded service cap.

Roller, however, was perpetually complaining. If the balletmaster asked him for a special mechanical device in the course of a new production, the obstinate machinist-academic's answer was always the same: 'Zis eez impossible!' It would turn out, however, that the impossible was made possible by order of the Director of Theatres, and the machinery department of the Bolshoy Theatre was always distinguished by its innovations.

Balletomanes of the 1870s

If the 1840s and the 1860s had their distinctive people and *littérateurs*, then why could the 1870s not have had their distinctive balletomanes?

They did!

In the 1870s there was still unspent money around from the redemption of government bonds, and railway fever was at its height; an orgy of concessioneering drew a mass of people to Petersburg who thirsted for a fast profit, and hand in hand with profit went equally fast living, as regards both money and life. In those days every youth with fat and generous pockets considered himself obliged to attend the ballet and 'court' one of the muse's pupils. It was a fashion which lasted until regiments of painted Frenchwomen appeared in Petersburg. Even nice old gentlemen were not averse to courting ballet girls.

I vividly remember one such man, who always sat in the first row of the stalls. He was a living cadaver! His flat, glassy eyes, without the slightest trace of fire—the fire had been squandered on courtesans—now wanted, it seemed, to drop out of their sockets... His voluptuously hanging lower lip constantly mouthed something. His knees trembled, and with a coloured foulard kerchief he incessantly wiped away large beads of sweat which glistened on his skeletal head. He never missed a performance; he died of a stroke an hour after returning from the ballet. There were not a few old lechers and balletic butterflies then, whose wings were burned by artistes more than once but who nevertheless remained stubbornly true to their idols, deserving a better fate, trying to get their feelings reciprocated.

Most of the balletomanes, of course, were drawn from the *jeunesse dorée*, with that youthful passion, that fever which does not admit the slightest philosophical reasoning. Even if an entire fortune were not staked, then most of it was, provided that success with a dancer came of it; and if one did not succeed at first it did not dampen a courtier's ardour, but rather fanned the flames of unsatisfied passion even further, whereupon more money for gifts, dinners, and flowers was poured into this sieve of the Danaides.

There were many such people. In the Bolshoy Theatre time and again one had occasion to eavesdrop as young men, with charactertistic fire and flippancy, discussed the legs of dancers, so-and-so's pleasing build, the points of their trotting horses, and scoundrel creditors tired of waiting for their money . . . On ballet nights the auditorium of the Bolshoy Theatre would always present an especially lively scene . . . The balletomanes of the 1870s had money, and they flaunted it without the slightest care, thinking naught for the morrow. And all this was done, in their words, 'for the maintenance of art.'

The ballet had immense vitality. Thanks to this, it provided the public with passing delights at the same time choreography itself was developing and being perfected. The ballet profited from the public's enthusiasm; artistes strove, studied, perfected themselves, wishing to advance. They realized that a move from the coryphée to soloist not only affected their material situation, but also the public's view of them. A dancer's shares would immediately increase in value on the balletic stock exchange, where reputations were quoted in keeping with an artist's talent and her status on the stage. In this, however, I think that any given instance today is no different from the 1870s: evaluation is made according to established levels of accomplishment.

The Battle of the Factions

Balletomanes entered the auditorium of the Bolshoy Theatre as if it were their home. Each had his place in the stalls and knew in advance whom he would meet and who would sit next to him. It was like a big family, but one not always on friendly terms. Factionalism was the perpetual lot of the ballet public. Its cause lay in a tradition according to which two stars were engaged in the ballet at the same time—rival first ballerinas. Adherents of one or the other dancer used every means to extol their favourite to the detriment of the other party's idol.

This combat of balletic Montagues and Capulets frequently turned to laughter. It never came to bloodshed—what suffered most were the pockets of our admirers of the various rusalkas and nymphs. At the time *Humpbacked Horse* was produced, two parties were active: some balletomanes went mad over Marfa Muravieva, who created the role of the Tsar-Maiden; others bowed to the talent of Maria Surovshchikova Petipa (first wife of the balletmaster Marius Petipa and mother of Marie Mariusovna Petipa).

It was a protracted and terrible battle. One day, after *Humpbacked Horse*, they would give Muravieva five bouquets and two baskets of flowers, and after the next performance of *The Pharaoh's Daughter*, seven bouquets and three baskets would come out of the orchestra for Maria Surovshchikova.

'And what have you got today? Well then? Show us!'

'You think *you* won?' burst out the Petipists. '*We* gained the upper hand!'

But the Muravievists were not to be caught napping, and at the next performance of *Humpbacked Horse*, Muravieva carried away, in an unwieldy official carriage, ten bouquets, a lyre of flowers and a huge wreath with the inscription 'Unforgettable Tsar-Maiden'. Apropos this point: Muravieva never travelled in her own carriage, as suits a prima ballerina; she used an official 'large coach' for a very simple reason: one bowed to her colossal talent but not her beauty, with which she was not especially well endowed by nature.

And so the game of gift-giving continued. To Muravieva ear-rings, to Petipa ear-rings with a star; after this to the former a diamond crescent to be worn on the head, as befits the Tsar-Maiden, daughter of the Moon, and so forth. The crescendo swelled in this manner to the last day of Butter Week, when the balletomanes bade farewell to their favourites until the following season. On this day the programme was always mixed, that is, made up of the best excerpts from different ballets in which both ballerinas and all soloists had danced during the course of the winter. The balletomanes prepared for this performance a month in advance, the Guelphs carefully concealing from the Ghibellines what ovations they were prepared to make for their idols.

On this day the auditorium was transformed into two genuinely hostile camps. Tickets were obtained only with a struggle; speculators got substantial sums, and the general public came to this performance not for aesthetic pleasure but to witness the noise, shouting, and uproar which accompanied the appearance now of this artist, now of that. It was the only day of the year when the warring parties could compare their powers directly, when the prima ballerinas danced together on the same day and when all soloists, without exception, received gifts and flowers. At

Muravieva's entrance she was applauded exactly five minutes, and in the interval she was called out twenty to thirty times.

Then began an act of a ballet with Petipa. The frenzy of her adherents knew no limits. The balletomanes Alexander Pavlovich Ushakov, K—ov (who was lame), Apollon Grinev, and others pulled watches out of their pockets and at Maria Petipa's appearance applauded exactly *seven minutes*, to allow the boast that the public preferred Petipa, as it applauded seven minutes at her entrance—that is, two minutes more than for Muravieva. The same held true for calls. For one—twenty; for the other, twenty-five.

Into this Butter Week performance also went a rivalry of gifts. But here the Muravievists got the upper hand. Their presents were always more costly and more numerous, thanks to the chief Muravievist, Al. Sap—ov, celebrated at that time for his wealth, who spared nothing in supplying Muravieva with gifts, adding a substantial sum from his own pocket to the monies collected by subscription.

Such were the balletomanes of the 1870s!

One can judge the antagonism between the Muravievists and the Petipists by this anecdote, which brought everyone a laugh, though the balletomanes took it very seriously. The vaudeville *Zehn Mädchen und kein Mann* was enjoying great success at the Alexandrinsky Theatre. The translator of this vaudeville introduced some couplets into it striking for their topicality at the same time. Thus, the following was sung:

> Muravieva, Petipa
> Will not dance it through this *pas*!

The adherents of Maria Surovshchikova Petipa took offence at the fact that the name Muravieva was placed first and Petipa second; very persuasively they requested the performer to sing:

> Petipa and Muravieva
> Will not dance through such a number.

[In these verses, Khudekov slightly alters the dancers' surnames to create puns, making of Muravieva's followers the 'ant's' party, and Petipa's that of '*balagan* acrobats'.]

The couplet was sung on the Alexandrinsky stage in two variants. Both Petipists and Muravievists were satisfied.

Balletomanes, Claqueurs, and Fanatics

The art of clapping in the ballet was raised to the status of creative work. The balletomanes assured us that applause had its own poetry and charm.

According to them, it could only have value and significance if one knew *how* to clap. 'One must begin at the proper time, pause and take a breather at the proper time, and at the proper time shout "brrravo!"' In other words, lack of skill could spoil the whole business. Inopportune calls for an artist can arouse the displeasure of the public, with which it is afterwards difficult to cope! 'It is a science all its own! You must seize the moment when the whole crowd will follow your lead unquestioningly, as if they had all been electrified at once.'

So spoke the *men of the 1870s*!

Apollon Afanasievich Grinev knew how to seize these moments in a ballet; he was a great master of the craft. If he took it upon himself to 'support' an artiste, he really kept his word. In the interval he would run around to all the young balletomanes requesting 'support', to be expressed in energetic clapping.

Finally the moment came.

'Quickly, now, give it everything!' Grinev would say, bubbling over with emotion. 'Fire! Don't spare the palms of your hands!' he cried out, fidgeting in his seat. His method of clapping was unique. He placed both hands together and then struck them against one another so deftly that they produced thick sounds, like pistol shots. He would show off this ability. Half-rising from his seat at the end of the first row on the left side, he would 'work' to the utmost, his face in a sweat, looking around the audience and nodding at the gallery, towards which he made a sign when it was time for its occupants to clap.

'A trill! A trill!' [he would say, as if conducting an orchestra] 'And then quieter, and quieter, and be silent! Rest for a moment, and give it everything again!' One could clap for half an hour like this; but if a person got carried away, his hands could get swollen after only a minute. This is how Apollon Grinev taught new arrivals, young balletomanes he enlisted into his 'party' and initiated into the secrets of the claqueur's art.

Alexandre Pavlovich Ushakov excelled in this field no less ably; with his furious applause he either brought his neighbours in the stalls to the verge of despair or carried the entire hall along with him. If he wanted to call someone out for a bow, he called her! His obstinacy was astonishing; he would clap almost alone, shouting out the name of the danseuse. He did not quieten down until the artist being summoned had flown out from the wings after her variation. In the intervening moments Alexandre Pavlovich would become some kind of fanatic, a whirling dervish in a frenzy. He ignored everyone else, stamping his feet, clapping his hands and shouting 'Bravo!' simultaneously. Often, when giving a lecture on the fascinations of ballet, he would so galvanize his neighbour in the fourth row (on the edge of the middle aisle of the stalls) that even this neophyte would begin to applaud the dancers like a stalwart gentleman totally familiar with the secrets of the choreographic art.

'Now you'll see,' he says to this neighbour, a stranger to him, 'now Maria Petrovna will dance! She has such "ballon" it scares me to think about it.'

'I see! And who, if I may be so bold, is Maria Petrovna?'

'You have to be an idiot not to know Maria Petrovna!' Ushakov exclaims in great excitement. 'It is Sokolova! She easily brings off *entrechats sept royal*! . . . She is the only soloist of this type which we have! Ballon? The mind boggles! But then, do you understand what an *entrechat sept royal* is?'

Ushakov's neighbour does not know, and must either take offence or hear out his new instructor of choreography. And for his part Ushakov, unaware that he has offended someone he doesn't even know, continues his lecture, calling by name all the dancers who came out on stage.

'Ah—that's she! . . . Look! That's the "geography map"! Coryphée Z. We call her that because her neck is snow-white and all her veins show through to the skin exactly like rivers on a map! . . . And that is Efremova . . . So then, bear up now, I beg of you!' And Ushakov has begun to clap irrepressibly, fulfilling his duty to this soloist.

By the end of the performance the stranger has reconciled himself with the oddities of Ushakov the fanatic balletomane, has become his friend, and attends the ballet ever more often, apparently interested in it, while everywhere exchanging bows with Ushakov as he would with an old friend. Ushakov recruited such friends into his legion with almost every performance.

Alexandre Pavlovich managed so to instil this passion for ballet into a most serious person, professor of the mining institute N. I. K—ov, that for several years running the silver-haired gentleman virtually never missed a ballet performance. He got acquainted with artists, went to their homes of an evening, having mastered the detailed terminology of the choreographic art. In the company of the light-winged, he boasted a knowledge of *rondes de jambes*, pirouettes, battements, every type of entrechat, glissades on pointe, etc., etc. N. I. K—ov combined minerology with choreography in that he loved to give artists little rings of stones, the first letters of the names of which made up a dancer's name. Thus for one of them, 'Maria', he once in my presence gave a ring made of Malachite, Amethyst, Ruby, Izumrud [emerald] and YAshma [jasper]. After that, ringlets with stones corresponding to the letters of their names appeared on many dancers' dainty little fingers.

A dancer would turn to N. I. K—ov at evening tea with a request to name 'her' minerals. And if it was impossible for her jeweller to get the right stones, the professor would supply such unobtainable ones as jacinth, heliotrope, or others.

He went to the ballet not to make amorous advances, no! He simply liked this world and loved choreography, receiving aesthetic pleasure from it, and admiring a smartly and elegantly performed classical variation. He freely admitted this.

The Buffet at the Bolshoy Theatre

The old Paris Opéra, now burned down, had its Café du divan, located towards the end of the first opera arcade. Habitués of the Opéra—artists, literary people, musicians—would retire there for an exchange of impressions after a performance. A similar exchange of ideas among balletomanes of the 1870s took place in the celebrated 'smoker' of the Bolshoy Theatre. Side-stepping through the narrow entrance, between the counter and the wall, balletomanes during intervals pushed their way to this gathering place which they themselves had created.

This smoking-room had two sections; in the first, a store-room, stood a small table covered with a tablecloth of doubtful whiteness, beyond which a stairway of three to four steps descended into the main theatre kitchen, where the cooking was done on masquerade days. In this kitchen, with its perpetually smoking stove, the cream of the capital's youth gathered during intervals to have a cigarette, and when the opportunity arose, to down a flagon with their friends, that is, to drain a bottle of Röderer, fashionable at the time. In those days one could meet millionaires there, owners of the most considerable fortunes. Some of the most tightly stuffed pockets on people with some of the most empty heads assembled there, and vice versa—magnificent minds with empty pockets. Although that time is not long passed, it is likely that few people still alive recall the smoker, and occasionally at that, where in the soot of tobacco smoke lively arguments were conducted about the achievements and the talent of this or that dancer.

In this smoke-filled room, taking no notice of its discomforts, men were reborn as balletomanes. They loudly spoke their minds, shy before no one. But only their own kind gathered here—privileged balletomanes, privy to backstage squabbles and the secrets of the ballet. Very young officers rubbed shoulders with generals; foppish civilians encountered silver-haired diplomats. They argued about ballet; testimonials to artists rang out here, and the future reputations of lady dancers were made. Dancers' talents were advanced, sometimes of course exaggerated beyond all measure and merit, and the success of a newly produced choreographic work was determined.

Interval conversations could be very sharp; on such occasions those arguing remained in the smoker even after the action had recommenced on stage, continuing to prove the bankruptcy of an adversary's viewpoints concerning the merits an artist, or a newly produced dance, or a simple variation composed for a dancer.

The 'exchange of ideas' concerning one dancer lingers in my memory. 'Hah!' noted X, one of the balletic regulars. 'She dances well . . . but she has paws instead of feet, and instead of shoulders, a skeleton . . . in general she . . . is not that . . . well, I would hardly be seduced by *that* body . . .'

This remark produced a veritable storm in the smoker. The most unflattering epithets were directed at this gentleman. 'So that's it, you assess the merits of a dancer the same way you analyse an article or inspect a horse! Is that really why you come to the ballet? . . .' 'Of course!' answered our balletomane, collecting his wits after a second round at the bar. 'What is ballet? Indeed! . . . An exhibit of beautiful women . . . a flower bed in which everyone can pick the flowers of pleasure . . . I can't conceive of ballet otherwise . . .'

The fanatic Alexandre Pavlovich Ushakov, who was present on that occasion, did not see the matter that way. First he called the person talking an idiot, probably by right of fellowship: they both were nearly identical graduates of the same educational institution. Then Ushakov lectured him on choreography, forcing him to renounce the epithets he had uttered against the dancer and to make an apology before everyone there . . . After this, of course, corks flew and champagne flowed . . .

Working-class balletomanes went to smoke in the room next to the box-office, where I. G. Nikiforov, who served as chief cashier for nearly twenty-five years, received them cordially. There conversations about ballet proceeded in a more sanguine fashion; only Apollon Afasanievich Grinev was an exception, who loved to mix with 'the powerful of this world' and who always waxed passionate in the evaluation of artists' talents.

Grinev would come into the box-office and after two or three of his energetic phrases the room would become too small. He got excited, which fact was proclaimed by his novel manner of carrying on an argument.

'Now then!' he would ask someone, 'How do you find Petipa in *Pharaoh*? Excellent, no?'

'Yes, excellent!' would come the answer.

'How so excellent? Not true! She is magnificent!'

'Yes! Indeed, magnificent!'

'You don't understand! She is not magnificent . . . she . . . is divine!'

'But that's still not true! Not divine, she . . . she . . . is simply the ideal . . . the incarnation of elegance . . . of beauty!'

It seems that if a respondent could testify for the dancer with the full range of the most enthusiastic praises, he would not come up against the ever-excited, ever-seething Grinev, who was of a type long remembered by balletic artists who knew him.

There was another 'type' which never missed a ballet performance. He was a person of limited means who came all the way from a poor neighborhood in the Vyborg district to the Bolshoy Theatre on foot. He was the martyr of choreography. He arrived at the beginning of a performance and left only when the lights were put out. The balletomanes knew and loved him, but despite their best efforts could not teach him to applaud. 'What are you doing with your paws spread out like a fan?' G. asked. 'You'll never get anywhere doing *that* . . . Make your hands like boxes and bang away to your heart's content! . . .'

For years on end the same servant worked in the smoker. He was a fixture there. Someone nicknamed him 'The Nose', a Gogolesque appellation which stayed with him and which he used to his advantage. Every New Year he considered himself obliged to give out cards which read:

Happy New Year from
'N.O.S.E.'

For this of course he received an extra rouble note.

Various Trends in the Ballet

During the production of *The Humpbacked Horse* two utterly divergent views of the choreographic art were placed in sharp relief. These trends were formed, as it were, out of the prevailing circumstances. Two first dancers were invited to the ballet at the same time, and each had an attendant balletmaster sworn to her advocacy, obliged to revive old ballets and mount the new fruits of his inspiration for her.

After the departure of Perrot [Khudekov here uses the word 'death' in error], who left us *Faust*, *Esmeralda*, *Caterina*, *The Naiad*, *Armida* and others, the post of balletmaster passed to the choreographers St-Léon and Marius Petipa. St-Léon mounted ballets for Muravieva, Petipa for his own wife. Their rivalry was great. According to the management's policy, one grand ballet calling for large expenditures was produced each year, and one small ballet for the benefit performance of the second ballerina! Each balletmaster thus produced a large choreographic work every other year. Clearly their competition was tremendous. They mustered little love and tender feeling for one another, and in fact secretly endeavoured to

1st Performance of The Humpbacked Horse   261

irritate each other. A schism divided the ballet company; some artists were delighted with everything St-Léon did, others gave Petipa's gift its due. Testimonials for and against one or the other balletmaster reached the press, which also formed two camps.

I was never an adherent of St-Léon's talent. I recognized him as a master at composing separate dances and especially at mounting variations for a dancer which were crafted to her gifts and genre. Never did I consider him a master of composing an artistic whole. Apparently aware of this defect, St-Léon chose stories bereft of content for his ballets, in which the ballerina never had to show and develop her talent as a mime. In his ballets, pantomime and story—the soul of a complete choreographic work—were always missing. They were simple divertissements, collections of dances little connected organically with the action. In his work artists danced not of logical necessity, but at a magic genie's command. If it pleased the balletmaster for them to dance, they simply danced.

In St-Léon's ballets, without exception, the first danseuse never had a mimed scene. They claim that St-Léon did this because he could not demonstrate pantomime. Producing ballets without narrative content lowered him a great deal in the estimation of choreographic savants. An opera without a story and narrative movement turns into a concert; in the exact parallel, a ballet without content becomes a divertissement and is not a complete theatrical work, even if produced in costume with successful dancers. Such ballets will pass unnoticed in the history of the art and will never shine like wondrous pearls, as the ever-young, poetical ballet *Giselle*.

As soon as he was taken from the ranks of first dancers to become a balletmaster, Petipa immediately developed another view of ballet. He was an adherent and direct successor of his teacher Perrot, who first would seek a theme, endeavouring to make his ballet a complete, carefully conceived dramatic work in which artists could shine in both dances and mime. That is why Perrot's, *Faust, Esmeralda*, and others live on to this day.

Marius Petipa went even further. Drawing on traditions worked out by the balletmasters Coralli, Mazilier, and Perrot, the novice choreographer never abandoned strictly classic fundamentals, which permit nothing to exceed the limits of elegance. It is evident from the ballets he produced in that period that Petipa chose stories accessible to his public, suitable for arrangement into mime. Being an excellent mime himself, he taught it to artists in masterly fashion. Besides this, he took care to correlate his dances with the action, never permitting them to be dramatically unmotivated appendages which might caress the eye but were otherwise unjustified. His so-called *pas d'action* are sensible and readily understood.

They are not like those *pas de deux* or *pas de quatre* imported by the Italians where an artist puts before the public no more than an examination performance of choreographic difficulties. *The Pharaoh's Daughter, The Beauty of Lebanon, Le Roi Candaule, La Bayadère, Roxana, Zoraya* and others—all are ballets whose stories could successfully be transferred to the dramatic stage.

The balletmasters St-Léon and Petipa, working at the same time, were also sharply distinguished by the nature of their own talents. Each put his mark on a new choreographic work, and in doing so said something about the tendencies of his talent.

St-Léon composed groupings and large ensemble dances only because he had to. He could not deal with large groups of people: his artistic taste was not developed well enough. His *morceaux d'ensemble* were lifeless and pallid. As for Mr Petipa, he is a great master of arranging masses of dancers as it suits him. Every grouping he composes is just asking to be put on to an artist's canvas. Everything is elegant and harmonious, each grouping in strict agreement with aesthetic principles, outside which a single superfluous stroke risks disrupting the harmony of the whole.

On the other hand, St-Léon was able to compose those correct, rhythmic movements which in balletic parlance are called classical dances—variations. Soloists said that they were always comfortable dancing variations he devised.

The public of that time loved and esteemed Petipa's narrative ballets according to the merits of each work, preferring them in general to the choreographic vinaigrettes St-Léon concocted. Nowadays attitudes have changed: external sharpness and brilliance are essential, not interior content. But surely is it not better to combine *one with the other*? On a model stage the means are always there to consider the suitability of a ballet scenario. Of course it would be better! The public and art itself would gain.

The Staff of the Ballet Troupe

Muravieva and Petipa were the leading dancers at that time. They had utterly different talents: *Muravieva*, a student of Hugué, was the ideal *terre à terre* dancer, that is, an artist not blessed with ethereality and lightness; despite that, the speed of her leg movements and her ability to follow the beat adroitly and elegantly however diverse the structures of the sounds—were truly astounding. The precision of her execution, the first sign of a fully-formed talent, was above all praise. Muravieva did not flaunt the technical feats for which Italian ballerinas are praised today; instead, she astonished the public by the correctness of her dances, correctness acquired in the Petersburg school thanks to its accurate instruction in classical technique.

She embroidered patterns about the stage with her legs, wove a lace of the finest quality; her dances were filigree work of the highest standard! They spoke of Muravieva in such terms. She danced as if trifling, playing, as if she derived great pleasure from it. To execute clearly and sharply her *variation taquetée* was to freeze into poses in time with the music without showing the slightest fatigue. She was a true representative of the strictly classical style, and never resorted to that 'finger-and-toe' virtuosity to which contemporary ballerinas are susceptible. But alas! Muravieva delighted the entire hall with her correct dances, but enchanted no one.

The gift to charm was inherent in her contemporary and rival, Maria Petipa. This artist, whose maiden name was Surovshchikova, was a creation, in the fullest sense of the word, of her balletmaster husband. He realized that his wife could never shine as a first-class executant of dances, as she lacked the requisite natural gifts; among her problems laziness played a part. Wishing nevertheless to advance her cause, Marius Petipa directed his energies at developing her plastique and mime.

He achieved colossal results. Beautiful by nature, with a charming smile and wonderful eyes, Mme Petipa was transformed by him into the personification of grace. All of her, from head to toe, the remarkable roundedness of her arms, was an Aphrodite perfectly crafted in marble! Each gesture, pose, and attitude recalled the figures of ancient Greece! And how graceful she was when performing the dances he especially composed for her!!

The dance with the dagger in *The Pharaoh's Daughter*, or the *pas*, 'La Charmeuse', from *The Beauty of Lebanon* brought the public to a state of frenzied ecstasy. There were no technical difficulties in these undemanding dances, but their execution breathed passion, charming languor, and a fascination which yields to description only with difficulty.

Maria Petipa did not astound with her variations, but so enchanted the public that her wonderful image was imprinted on the spectator's imagination for a long time.

Soloists

In this golden age of ballet, each soloist was distinguished from her colleagues by some special quality. The following soloists linger in my memory:

Alexandra Nikolaevna Kemmerer I. She performed ballerina's roles in Petersburg and Moscow, but reverted to the rank of soloist before the end of her career. The qualities of a classical dancer and a performer of character dances came together in this artist. Mlle Kemmerer I had no rivals in Spanish dances and generally in any dance where energy, burning, youthful passion, life, and fast *temps* were required.

Maria Petrovna Sokolova. Light, ethereal, 'turned-out' (that is, with legs correctly set) . . . and that's that. Her manner of performance was lifeless; in every ballet she danced but one variation where her elevation was brilliant, but in a career of many years she never performed a single character dance.

Maria Alexandrovna Efremova. A tall blonde. She danced correctly but without animation, as if she were performing some arduous labour. Wicked balletomanes said of her: 'She's no dancer, she's a log runner.' On the other hand, her enthusiasts called her 'pointes of steel'.

Lyubov Petrovna Radina. An artist in every respect. She was in her element everywhere, in both classical and character dances. She was promoted for her charming performance as the Spirit of the Valley in the ballet *Théolinda*. Naturally beautiful, with a mischievous smile, she performed those wild, unbridled dances Petipa composed for her with incomparable brio. Her manner of performance, however, was always correct, without breaking aesthetic rules. Apropos the latter: aesthetics are still practised with excellent success on the stage of the Maryinsky Theatre. In the old days it was said of unbridled artists: 'She is a *street dancer*, not an *artiste of the ballet*.' Radina was a favourite of the entire public.

Anna Ivanovna Prikhunova (wife of the dancer A. Bogdanov). An excellent, strictly classical soloist. She was famous for the fact that no one had ever seen her smile. She went through her *petites batteries balancés et fouettés* with unruffled sang-froid, danced as if angry, and even came out for calls reluctantly.

Matilda Nikolaevna Madaeva. Her smile lit up the room. For elegance, the best swift small feet in the ballet; her *coup de pied* was enchantment! In this artist resided much femininity, which effortlessly attracted the hearts of the audience. Her talent was not especially brilliant, but winning in the extreme!

Vera Alexandrovna Lyadova. Always enjoyed great success. Her dear little cancan in *Graziella*, for the utter absence of anything unseemly, was marked by elegance and gracefulness of movement. Not one movement exceeded the limits of propriety, so even the cancan was acceptable on the classical stage in her performance. Later she left the ballet for operetta, which, thanks to Lyadova—La Belle Hélène and La Périchole—made a lasting home for itself on the stage of the Alexandrinsky Theatre.

All Petersburg attended her funeral: those services are probably remembered to this day by the many literary people who took part in them during the bearing of the deceased's remains along Millionnaya Street.

[*Anna Dmitrievna*] *Kosheva*. A dizzying artist, who could knock you over. In the Dance of the Caryatids from *The Pharaoh's Daughter* she performed her variation of successive *jetés en tournant* so quickly that you expected her to break up into little pieces striking the column supporting the Director's box, next to which she stopped at the end of her dance. When her variation came, the balletomanes often made wagers: 'Will she make it or not?' She never failed to make it.

Later a large number of the most talented soloists came forward: the impeccably classical Shaposhnikova, the airy Amosova, the passionate Simskaya II, and so forth. Of these I will not speak; it would take us too far afield from the first production of *Humpbacked Horse*.

The public goes to ballet to admire ethereal, light-winged creatures and to delight in the ever-smiling and life-affirming servants of Terpsichore on stage. 'How merry she is!' the viewer thinks, applauding a dancer. But rare is the spectator who is initiated into the secrets of the balletic science. A young girl needs much perseverance, dedication, and labour to become a dancer-soloist of even moderate attainments.

The daily sufferings begin with matriculation into the school at the age of nine or ten. At that age the little creature is being earmarked for a treacherous calling by neither aptitude nor love of art. People desirous of getting into the school have always been numerous, but in those days succeeded only 'by patronage'. The head of the school, Pavel Stepanovich Fyodorov, all-powerful in the theatre world at that time, accepted students for the most part only by patronage. Parents would obtain a letter of recommendation from someone of exalted rank, and the school was enriched by a new pupil of the muses the same day. No heed was paid to height nor build nor beauty nor talent; every little girl with a famous recommendation was accepted.

With her entry into the school the tortures of the future choreographic starlet began. Standing her heel to heel, a teacher sparing no pains trained the little girl to place her feet in various complicated positions. Instruction—a 'class'—was conducted in all the strictly formulated rules of art. Meanwhile, the new child was leaned up against the barre and bent over in a pose to train the future dancer how to keep one leg perpendicular to the other.

As in singing, so in dance, the pledge of success lies in elementary training. A voice is correctly placed, and the artist controls it easily; the feet and body are correctly trained, and after that a career depends on the dancer herself, depending on her continued involvement in her art. To develop their choreographic abilities, artists not infrequently inflict torments and agonies upon themselves. Taglioni, after a two-hour daily lesson with her father, would faint from exhaustion. They would undress her and run for a doctor's help to bring her around. Her hour of triumph in the evening never came cheap!

The famous English ballerina Nathalie Fitzjames, in order to acquire *ballon* and airiness, invented a special method of training her legs to the horizontal and its perpendicular at the same time. She lay down on the floor, face down, extended her legs horizontally with the body and then ordered her chambermaid to sit on her back. She stayed in this position for several minutes, stretching her limbs. These self-inflicted torments brought her to London's temples of glory.

Even at the end of her schooling the danseuse cannot breathe freely, saying: 'Whew! It's over!' No, if she wants to maintain her ease and agility of movement and to remain a soloist, she must practise every day, she must continue the choreographic ritual, lest a few days' rest be paid for with several months of hard work. Lightness and flexibility of limb, which so delight the public, are quickly lost. The fate of a dancer is eternal battements in dancing class. In the manner of the wandering Jew, 'Dance, dance!' is continuously repeated in her ear.

The Salary of Ballet Artists

Things were different in the old days! One cannot but marvel at how the servants of Terpsichore existed then. Salaries at the time of *Humpbacked Horse* were extremely paltry. Like the civil service, the ballet had a hierarchy of pay grades, with graduated steps from the lowest to the highest rank.

I have a list of the names of all the ballet artists in 1873.

The balletic ranks of women's personnel were as follows:

(1) Ballerinas/first dancers. (2) Soloists. (3) Coryphées and (4) Corps de ballet.

At that time there were two first dancers, that is, who performed principal roles in ballets: Mme Vazem (graduated in 1867), who received 1,143 roubles per year in salary plus 25 roubles for each performance; Vergina (graduated 1868) received 1,143 roubles in salary plus 5 roubles for each performance.

The *soloists* numbered 11: Kemmerer I (1858), Kosheva (1856), Madaeva (1858), Prikhunova (1860), Radina I (1854), Amosova (1869), Simskaya II (1872), Shaposhnikova (1868) and Shchetnina (1859). They received from 700 to 1,143 roubles per year in salary, and per-performance fees of from 5 to 25 roubles. Then there were 52 *coryphées*, who received from 300 to 600 roubles.

1st Performance of The Humpbacked Horse 267

In the line of succession after them came 44 members of the *corps de ballet*, who received from 174 to 240 roubles.

Two women dancers were listed as mimists: Mlles Troitskaya and Nikulina, at 400 roubles per year.

Male personnel consisted of 11 dancers, of whom only one, Mr Johanson, received 5,000 roubles in salary plus 10 roubles for each performance, and the rest, namely Golts, who had served since 1822, Gerdt I since 1860, Lev Ivanov from 1850, received the highest salary, an invariable 1,143 roubles, with per-performance fees not exceeding 15 roubles.

Three comic dancers were listed on the roster: Stukolkin at 1,143 roubles plus 10 roubles per-performance, Troitsky at 1,143 roubles plus 7 per-performance, and Bystrov at 500 roubles. There were 15 male coryphées with a salary from 240 to 400 roubles; there were 25 men in the *corps de ballet* with salaries from 174 to 240 roubles.

The régisseur at that time was Marcel; his assistants were Messrs Efimov, Gelau, and Dyadichkin, whose jubilee was so triumphantly celebrated in the Maryinsky Theatre a few days ago. At that time, filling the same position as today, he received only 270 roubles per year in salary.

By what means the several women dancers lived who received 14 roubles, 50 copecks per month it is not difficult to guess; but how male personnel lived on that amount is positively incomprehensible. This is a riddle almost impossible to solve, all the more since ballet personnel did not have a minute of leisure during an entire season. It was impossible for them to make a spare copeck on the side: in the mornings rehearsal, in the evenings either a ballet or some opera divertissement.

The exceptions, of course, were artists who gave dancing lessons privately, especially those provided with official posts. These people took life easily compared to their comrades working only for art.

A few days ago I asked a retired artist who lived on 1,143 roubles and per-performance payments how she felt about those times. 'Life was good!' she answered, 'no worse than now! It was a "golden" era of ballet then. A crowd of aristocratic youth paid court to us, and oh! how strictly we behaved, maintaining our purity. It was difficult for men in those days to get us to reciprocate their feelings, but if someone held true to the path of lengthy courtship, he could be sure that his heart's beloved would be "faithful all her life"! The ballet in our time was famous for this!' the former soloist boasted.

'How many of us got married then!' And she began to count: 'Listen! Kemmerer, Madaeva, Muravieva, Kantsyreva, Prikhunova I, Kosheva, Vasilieva, Vergina, Sokolova I, Zhenya Sokolova, and others, all of them now the most esteemed and revered mothers of families which stem from old, aristocratic courtly lines . . . Really now, was it not a *golden* age of ballet?'

Balletic 'Conversation'

Balletomanes of the 1870s had their own vocabularies, a special terminology to express various balletic activities. Instead of 'to court' they said 'to hit on'. The expression 'to look through binoculars' was never used. When the neighbour of someone who was courting a dancer noticed her standing behind the first wing, he would say, pointing her out, 'Look! There comes yours! . . . *lay to* your binoculars.' 'There's no patting or pinching this one, but have at this one with everything you got . . . Have a go!' Balletomanes talked to each other this way.

The theatre school likewise placed a special imprint on balletic personnel. Students developed their own jargon, their own manner of conversing, which often was difficult for the uninitiated to comprehend. Here is what remains of my small vocabulary from those times:

'Ooh . . . That nasty man!'

'I love to . . . on a troika.'

'I have a big solo in the new ballet.'

'You lucky dear! I only have a small one . . . no *adagio*, no coda . . . just a walk-on . . . one bit. I will complain to Pavel Stepanovich . . .'

'And in my solo there is even a *sujet* (that is, pantomime).'

'Well . . . I'm busy in other ways.'

'And nasty Marius dismissed me! He says: "You have a nothing brain on your head! You," he says, "out of head!"'

'Ah, girls, you wouldn't believe it!' This was a favourite phrase of all ballet types.

'They promised to add some day-perks (that is, per-performance fees).'

'Look, girls, Manechka doesn't dance, she knits with her feet . . . ah, you wouldn't believe it! . . . She had her leg muscles surgically removed! . . .', the balletic sisterhood would go on.

'Sweet little soul', 'dovey', 'my sweet' or 'dearest'—favoured epithets by which balletic people spiced their conversation both with good friends and with persons they hardly knew.

'Oh stop, you don't mean it . . .', etc., etc.

'And do you feel the same way about him?' . . .

'Oh that's just silly . . . I despise him!'

The Production of 'The Little Humpbacked Horse'

St-Léon understood that stories with a predominant fantastic element are essential for a ballet. Thanks to this the balletmaster was able to give free rein to his thoughts and to introduce whatever dances came into his head. Where there is a fairy tale, he reasoned, anything is permissible! If a mazurka crosses one's mind in a ballet which involves red Indians, put it in!

He did exactly that.

The ballerina Marfa Muravieva, who had achieved worldwide celebrity, was at St-Léon's disposal in those days. She had danced with colossal success at the Grand Opéra in Paris, which then as now dispensed diplomas of 'Magnitude, First Class' to its stars. *The Little Humpbacked Horse* was produced for Muravieva.

Following the unsuccessful *Pearl of Seville*, and the more successful short ballet *Fiametta*, which succeeded thanks to its wonderful melodies, given out almost to excess, it was St-Léon's turn to produce a new ballet for Muravieva's benefit performance. For a long time he searched for a story. One Moscow writer sent him a scenario for *The Frog Princess*, but St-Léon declined it, not wanting to mount *anything Russian*. The choreographer was cosmopolitan from head to toe. He acknowledged no particular national tendency in art, to which he would devote his whole career once he abandoned his favourite instrument, the violin, which he played like a first-class virtuoso.

St-Léon suffered nothing Russian because he recognized nothing artistic in it. But soon he had to abandon this belief. Fate itself decided on the production of a ballet with a story from the inexhaustible world of Russian stories.

Among Muravieva's most passionate admirers, and as it turned out, among St-Léon's personal friends, were the occupants of a letter-designated box called 'the infernal *loge*', below and near the director's box in the Bolshoy Theatre. This subscription box was called 'infernal' because it was upholstered in bright red material, which reflected on all those who sat in it, as if in Hell. The admirers of our celebrated Muravieva sat there: A. Sapozhnikov, M. Lopukhin, A—t, P—ii, and others. To these incorrigible balletomanes was born the happy thought of bringing forth upon the stage a Russian fairy tale. Lopukhin was the soul of the project; chiefly with his help the entire company decided on Ershov's tale. They prevailed upon the obstinate St-Léon a long time before he finally agreed to produce a ballet with a Russian story.

The clever foreigner at last reckoned that if the *Russian* ballet succeeded, his somewhat insecure position as balletmaster of the Petersburg stage, resulting from his previous failures, would be strengthened. He did not err in this calculation, certain that his friends in the infernal *loge* would support him, friends obliged to publicize their own brain child. At last the première of *Humpbacked Horse* took place, and this first attempt at a Russian ballet must be considered successful in large part, if not completely.

In the press (*The Voice* at that time carried full accounts of ballet performances), reviewers did not lavish special praise on the balletmaster. Nor was the all-powerful music critic of the French and Italian companies, Mavriki Rappoport, known as 'Moses' behind his back, *satisfied* with *Humpbacked Horse*. He dignified himself with the title of music critic and thought much of himself, whereas in fact he understood precisely nothing about art, for his own involvement as an artist was practically nil.

Muravieva danced, so to speak, *idling away her time*! She was not given one mimed scene. This fault was laid at the balletmaster's feet, although he had done this intentionally. Realizing that Muravieva was a quite mediocre mime, bereft of plastique, he chose those parts of Ershov's tale where the danseuse would not have to act. This was intentional and justified.

Quite unfavourable notices of *Humpbacked Horse* appeared in some newspapers. This (it was written) was a spectacle worthy of a *balagan*, not the Bolshoy Theatre. The cockerel's cries in the first act, the old khan's falling to the floor, Foolish Ivanushka's clownish antics, and finally, the oddly produced role of the 'humpbacked horse' itself, which obliged the artist to skip, leap and stamp his feet in place at all times in imitation of a real horse—all this gave St-Léon's enemies and ill-wishers abundant ammunition. 'A *balagan*, a *balagan*, but not a ballet!' his detractors pounded away. As a show for children the ballet had a huge success regardless of these criticisms, and in 1874, when the ballerina Vazem was ascendant, it reached 150 performances.

The First Performance of 'The Little Humpbacked Horse'

The balletomanes awaited the first performance as they would a sacred feast. They took an interest in rehearsals, and, unseen by the theatre management, gained access to the hall of the Bolshoy Theatre, stealing into the depths of the boxes where they evaded guards and ushers, people very amenable to gifts.

Finally the première came; it was Muravieva's benefit performance.

Muravieva herself did not produce any particular impression in the role created for her, the Tsar-Maiden. Nowhere could the artist shine, as there was nothing substantial in the role; she could not create a single enchanting image to be preserved in the viewer's memory, as she was hampered by the

1st Performance of The Humpbacked Horse 271

poverty of the material the balletmaster created for her. Any coryphée or girl of the corps de ballet 'near the water' could perform the Tsar-Maiden with similar success. The ballet would not suffer in the least because of this, nor would the wholeness of the audience's impression be disturbed in any way. There is positively nothing in the role of the Tsar-Maiden for a ballerina to do, as Foolish Ivanushka and his horse are given pride of place; Ivan's whip has more to do than the ballerina.

In the fountain scene, where the Tsar-Maiden makes her first appearance, Marfa Muravieva had no success at all. The variations St-Léon created for her were composed in the Muraviesque spirit, that is, with *temps à l'école taquetée*, in the school of Fanny Elssler and Ferraris. Thanks only to the correctness and the precision of her dancing, Muravieva was successful with her variations.

One might add that Muravieva did not resemble most of today's Italian ballerinas who live by the motto, 'the more difficult the better'. Members of the contemporary Italian school are insipid because they lack aesthetics, plastique, and elegance. The whole focus of this new school, which comprises a confused *nihilism* in choreography, is directed not at the development of elegance and a sense of beauty in dances, but at tricks of the legs and all manner of idiosyncracy, which belong in *féeries* or the *balagany* at the edge of town, but not on a self-respecting stage.

That is one reason why the nihilist-Italians, with inflated reputations alleging them to be *colossi* of choreography, do not appear on the stage of the Grand Opéra in Paris, the only stage after Petersburg by right and according to its merits bearing the title, 'Academy of Dancing'.

Yet Marfa Muravieva'a talent was very special! She might go out on stage, and someone in the audience seeing her for the first time would shrug his shoulders and say: 'Is this she . . . the famous Muravieva? . . . it cannot be!' This artist's small figure was not distinguished by its physical beauty. Nature made her for some purpose quite other than to shine in the treacherous calling of dancing. Small of stature, she was also very long-waisted; like Taglioni's, her hands hung below her knees, which, however, is considered a virtue in dancers able to make rounded movements of the arms. Muravieva's feet were not fine, and her powerful calf muscles showed through her tights. Her face was oblong with a long nose, a sharp profile and quite broad cheek-bones; her eyes were soft, velvety and tender. Muravieva was utterly incapable of doing her hair and paid little attention to her looks. If they call Virginia Zucchi 'Rag Doll' for her awful *coiffure*, or, rather, for her absence of *coiffure*, then we must call Muravieva a dancer with licked-down hair; she particularly cultivated the smoothing of her hair.

Whatever her natural defects, she astonished the crowd. She would begin to dance and be instantly transformed. Her physical defects diminished and the spectators, electrified by her dances, would become her true and lasting admirers. Such was the power of her talent! Muravieva subdued the crowd, which straightaway adored her, although adorers in the balletic sense she never had. Many admirers assembled around her though none pursued 'reciprocity' or sought after her heart. Once offstage, this artist was unable to awaken a balletomane's ardour.

Muravieva had great success in *Humpbacked Horse* after dancing the comic *pas* in the khan's headquarters. Following the spiritedly executed mazurka she quickly rose up on pointe, and with arms raised, danced a *kamarinskaya* to the accompaniment of Wieniawski's magical violin. In *Humpbacked Horse* the latter had nearly as great a success as the artists of the ballet. He shook his long black hair and somehow, from above, he would strike all the strings with his entire bow to play the mazurka at a fast tempo. In his hands the violin became a living being. Muravieva said that dancing to Wieniawski's violin she 'floated away', feeling that her dances 'were inspired'. She often said that her success in this *pas* was due solely to Wieniawski. That, of course, is an exaggeration. Even without Wieniawski she would have been incomparable, though the power of this wonderful violinist could no doubt influence the dancer during any dance.

After the *kamarinskaya* the public's enthusiasm knew no bounds: 'Bis! Bis!' cried the audience. Wieniawski began again, and Muravieva repeated her solo. Barely touching the boards, the famous ballerina flew around the stage . . . then quickly raising up on pointe, in an instant she froze and grew calm . . . to make a transition back to the quiet *temp* of the *kamarinskaya*, and the whole crescendo which subsequently quickens this *temp*. She finished her dance with a broad sweep of one arm and a low bow from the waist, Russian style. A shout went up even after the encore. Muravieva had to repeat this dance several times at the first performance, with ever-greater success each time.

That concluded the artist's triumph in this ballet, if one does not count the classical variation in the last act, after which the Tsar-Maiden stood on a shield and was carried around the stage to a loud and lively march. Later on members of the audience, caught up in the spirit of this march, clapped in time with the music until the the dancer was let down from the shield. When the role of the Tsar-Maiden was performed by Madaeva, whose first names were Matilda Nikolaevna, the balletomane G. got very excited; he could not restrain himself, turned to his neighbour, and cried out loudly, audible in the last rows of the stalls: 'Now give it your best! You see, it's Motinka being carried on the shields!' And he began to clap in time with the music; after him, by reflex action, the public joined in, adding their rhythmical applause to his.

The pre-eminent role in *Humpbacked* was that of Foolish Ivanushka. It was initially assigned to Stukolkin, indisputably our best comedian, but fate decreed otherwise: Stukolkin injured his leg, and the role passed to Troitsky, an artist who at the time was just beginning his career. By playing Ivanushka, Troitsky immediately became a prominent member of the company. This role advanced his career, despite the fact that he performed it in a dry and lifeless manner, not as it ought to be played according to the original fairy tale.

A role performed at the première of a work is reserved for that artist until the end of his or her career. In Troitsky's performance, Ivanushka was not that good Russian lad who in spite of his silliness had his wits about him. Mr Troitsky was a kind of knight and hero, frequently playing for superficial effects of doubtful comic content; cracking his whip, he accidentally hit himself in the leg, or his neighbour in the shoulder, and then huddled over in pain and hopped about. He rode astride turtles and crayfish, urging them on, drawing cheap laughs from the balcony and from children. Mr Troitsky's long, gaunt figure, borne on stage exactly as if he were stiff as a poker, did not correspond with the character of the well-fatted 'foolish lad' at all. Nevertheless, to the end of his career Mr Troitsky was the only performer of this role, thanks to the system in effect at that time. In Moscow I saw Mr Geltser in the role of Ivanushka. Here was the true hero of Ershov's tale; he created a type which old Muscovite balletomanes probably remember to this day.

Lev Ivanov and Alexei Bogdanov played Ivanushka's brothers; both these artists were later balletmasters, Bogdanov a very unsuccessful one. Their father was played by old Golts, who had danced the mazurka in the first performance of the opera, *A Life for the Tsar*. Golts got so carried away in *Humpbacked Horse* that at one performance, when his sons invite him to join them in the trepak, he cried out through the entire hall: 'Right you are, boys, don't be timid! Look lively! Carry on!' Then he launched into the dance. The public, unaccustomed to spoken words in a ballet, was delighted by this unexpected outburst, and gave the veteran artist a loud ovation. Subsequently when this scene came around, the regulars always waited for Golts to be on the verge of crying out, but no, he never did it again; he had been reprimanded because 'chattering with one's tongue' does not befit a ballet artist—that's what legs are for!

Kshesinsky played the khan; he projected the role of the weakened old voluptuary excellently; as a tasteful artist who took his art seriously, he never resorted to caricature.

The Russian Dance in the first act, with a little kerchief carried in the hand, was performed by Mlle Madaeva. Waving the kerchief, she glided exactly like a swan. Here was a soloist who enchanted everyone with her wonderful, alluring smile. She regularly repeated this dance by public demand.

In the second scene, the khan's wife was played by Mlle Kosheva. Nowadays this scene is performed with animated statues. In the first performance, they were called 'animated carpets', because in a Kirghiz khan's tent there could be no frescos hanging, only carpets. The animated figures on these carpets were performed by the soloists Mlles Vasilieva, Prikhunova, Sokolova, and in later times Amosova, the latter famous for the fact that she was deaf and nevertheless kept perfect rhythm; she watched the beat, never taking her eyes from the conductor's baton. The conductor at that time, Mr Papkov, knew this and happily followed the artist's movements, helping her to keep the beat.

It was Lopukhin who gave St-Léon the idea for the divertissement of the last act, danced by the different nationalities who populate Russia. Most of these dances were successful; the Ukrainian Dance, which Mlle Lyadova danced with Alexei Bogdanov, was especially pleasing. The originality of her costume, the graceful shaking of her apron, and finally the unexpected kiss at the end called forth a storm of applause. This dance was always repeated, and Mlle Lyadova, as if to call attention to it, always kissed her partner in an especially juicy fashion, audible throughout the entire hall.

The Lettish Dance in the performance of the good-looking Legat also pleased. The Cossack Dance, where Lev Ivanov (and later Pavel Gerdt) executed a series of cabrioles to the right and left with a tapping of heel to heel, was also repeated from the first performance on.

Mr Picheau performed the role of the humpbacked horse, which throughout the entire ballet must have a broad smile on his face, which showed Picheau's snow-white teeth. He often complained to me that he thought it unworthy to represent a horse. 'But what can be done?! It must be; I need the per-performance fees for my huge family.' And to his death he never yielded up the role of the horse, although he continuously varied the obligatory marking time in place, resting now on the right foot, now on the left, imitating a horse 'champing at the bit'. 'Nothing can be done! One must create some *kind of stallion*!' he would say, putting on his costume, covered with heavy brass harness ornaments.

The first performance came to an end. Muravieva was called out at least thirty times. She came out with the creator of the ballet, St-Léon. The success of the new choreographic work was secured, thanks to the novelty of its story, although Muravieva did not like to dance in it. St-Léon supposed that a similar success awaited every other ballet taken from the world of Russian fairy tale. He was, however, soon disappointed in this. Pursuing easy laurels, he returned to a Russian theme again after the unsuccessful *Bride of Wallachia*. He produced *The Golden Fish*, and this ballet failed triumphantly. Neither the praises of friends nor the brilliance of the decorations helped; the ballet quickly sank into oblivion.

Since that time the Petersburg stage has long awaited a new choreographic work based on a Russian fairy tale, from that inexhaustible source of fantasy for both the author of scenarios and for the balletmaster. It is an all but untouched realm, overflowing with the most poetical images which await realization by the Terpsichorean priestesses of our balletic world.

S. N. KH—KOV

VII
The 1870s

INTRODUCTION

When St-Léon died in 1870, Marius Petipa in effect became the first balletmaster of the imperial theatres. In 1872 the German ballerina Adèle Grantzow, perhaps the most important of St-Léon's 'discoveries', danced in Russia for the last time and left a vacancy on her departure filled by Russian dancers until Virginia Zucchi's début on the imperial stage in 1885. For most of that thirteen-year period, Ekaterina Ottovna Vazem was the first ballerina of the imperial theatres.

Vazem was a *terre-à-terre* dancer with a brilliant technique, which helps explain her admiration for St-Léon, the creator of extraordinary dances for her like-gifted elder colleague, Marfa Muravieva. In her *Memoirs* Vazem praises St-Léon at the expense of Marius Petipa, about whom, possibly because of differences of temperament or the collective irritations of a long professional collaboration, she has few complimentary things to say. Yet few other people could write with such authority about Petipa and the Petersburg ballet of the 1870s, virtually the first period ever in which the company flourished without foreign celebrities. Vazem twice surveys Petipa's ballets, once in a discussion of the balletmaster and again when describing her career after the last foreign ballerina had left.

The following extracts are Vazem's description of Petipa from Chapter 4 (pp. 62–7, 74–6) and her review of the ballets in which she performed from 1874 to 1880 from Chapter 11 (pp. 161–72) of *Zapiski baleriny Sankt-Peterburgskogo Bol'shogo teatra, 1867–1884* (Leningrad and Moscow, 1937).

13
From the *Memoirs of a Ballerina of the St Petersburg Bolshoy Theatre, 1867–1884*
Ekaterina Ottovna Vazem

Chapter Four. The Balletmaster M. I. Petipa

It was my lot to work with St-Léon no more than two years. He was, if I may express it thus, devoured by his rival, the balletmaster Marius Ivanovich Petipa.

Resting on his European name, St-Léon behaved in a very haughty way. He probably thought that his authority as balletmaster in Paris was sufficient guarantee of success in Petersburg.

Petipa presented a very different figure. Having arrived in Russia at the end of the 1840s as a completely unknown dancer, he strove at all costs to make a career, and thus to please the broad mass of public, the theatre management, and mostly the imperial court. At the time of my entry on to the stage Petipa, after one-act trifles, had produced one genuinely solid grand ballet, *The Pharaoh's Daughter*, and tried feverishly to sustain the public sympathy which arose after this production. He tried in every possible way to ingratiate himself with the balletomanes, among whom a numerous party had formed around him. It was also the party that worshipped his wife, the ballerina Maria Surovshchikova Petipa, for whom her husband mounted most of his ballets. Its members were labelled 'Petipists'. It professed enmity towards the 'Muravievists', devotees of the ballerina Muravieva, St-Léon's favourite and the leading performer in most of his ballets. This enmity was open and even spread to the press, where reviewers strove to extol one camp and disparage the other, often forgetting all sense of fairness. Petipa's friends used every suitable occasion to underscore the superiority of this balletmaster to St-Léon, who displeased them. This could not help having an effect on the theatre management. And it could not have come about without the influence of the balletmaster's wife. Be this as it may, one fine day, the expiration day of St-Léon's annual contract with the direction, that contract was not renewed, and Petipa, who had enjoyed a brilliant success with his grand ballet *Le Roi Candaule* not long before, remained the sole artistic director of the Petersburg ballet, a post he filled until the beginning of the present century.

The genre which Petipa cultivated was much more to the taste of the Petersburg public than that put forward by St-Léon. In the epoch being described Petipa's principal concern was to give ballets *à grand spectacle* with an engaging, at times absorbing dramatic story and brilliant mise-en-scène. For the latter he deftly managed to wheedle very sizeable sums of money from the management. These choreographic dramas were liked, of course, more than St-Léon's dramatically quite naïve ballets. With respect to choreographic classicism, however, Petipa's compositions yielded much to St-Léon's. Although a true adherent of classical ballet, as, of course, were all choreographers of his time, Petipa knew classical dance far more superficially than St-Léon, many *pas* that St-Léon used being completely unknown to him. With the departure of St-Léon, those fine, 'minute' *pas*, which required genuine filigree work from a dancer, forever vanished from our stage. Variations in a composition by Petipa were monotonous and, I would say, 'cruder' than those of St-Léon.

But he was a true master at composing mass dances and groupings for the corps de ballet, and also character dances and mimed scenes, as he himself was an excellent character dancer and mime.

The second indisputable minus of Petipa compared with St-Léon was his non-musicality. He did not feel music, its rhythm; therefore dances he composed very often turned out to be extremely awkward for the performer, and soloists had either to request that he change them or to remount their number themselves 'on the sly', since Petipa could not remember everything he had done in the course of a long ballet. In addition, the balletmaster was never confident of the success of his productions, although he worked them out at home, and changed them time after time in the process of rehearsing. This made necessary a very large number of rehearsals, many more than St-Léon required. At rehearsal Petipa was often fussy and inclined to be nervous, and this nervousness was invariably transmitted to the artists. Nor was Petipa very strong in the assessment of the artistic individuality of soloists and in the ability to utilize them fully; in any event he yielded to St-Léon in this respect. Ordinarily he asked the first dancers what variation they wanted to dance. When it came to me, I always answered that it made no difference, but that in the different acts the dances be varied in accordance with the rhythm of the music and the *pas* being performed.

Finally, our balletmaster could be reproached for his extreme lack of education, because of which most of his ballets were teeming with blunders and absurdities. In this respect, however, the public was very undemanding, to the point that if the performance were effective the rest made no difference to it ...

In their choreographic content Petipa's ballets were in essence just as much divertissements as the ballets of St-Léon, only here the divertissement was more strongly stitched into the fabric of the story. The balletmaster's (or 'balemesa', as Petipa referred to himself in his makeshift half-Russian) favourite device was to stretch out the action to some sort of celebration, especially a wedding, there to have wide latitude for the production of various classical and character dances.

It is acceptable for us to call classical ballet 'Petipa's ballet', setting it off against the ballets of later innovators in choreography, beginning with Alexandre Gorsky. Strictly speaking, however, that designation is incorrect. Petipa was by no means the creator or founder of a particular school of choreography. During his career he put forward on stage artistic principles worked out by other choreographers, his predecessors and contemporaries such as St-Léon, from whom he was not shy about borrowing everything which suited him in his own work.

He always tried to follow artistic tendencies in ballet, familiarizing himself attentively with all the latest news in the field. In my time he went to Paris and other European cities in the spring of every year to get acquainted with the *dernier cri* of balletic fashion, carefully searching out new effects and devices, which he transferred to our stage the next season. Then, with the rise of interest in Italian ballet with its leg-breaking feats, Petipa began to develop the style of the Italians in us as well.

His main effort as regards this foreign material consisted of adapting it to local conditions.

With the departure of St-Léon the entire ballet repertoire came to be made up exclusively of Petipa's ballets or of ballets of others, revived by Petipa with varying degrees of revision.

[Vazem now reviews Petipa's principal ballets, grouping them into successes and failures.]

Individual failures, however, ought hardly to discredit Petipa's indisputable merits. In the course of a season he mounted no fewer than two large ballets, not counting one-act ballets and separate dances. In light of such work it would be difficult to expect nothing but masterpieces from him. In general the ballet repertoire was assembled in an interesting way, and concerning the quality of execution there is nothing more to say. Other European ballet theatres could not bear comparison with our ballet; the Petersburg ballet at that time stood at an extraordinary height. If ballet performances did not always bring in the takings expected of them, the reason was mainly that the public was constantly drawn away from the ballet, which trod the boards of the same Bolshoy Theatre as the Italian opera, which had a number of outstanding vocal talents on its roster. The Italian opera was the most fashionable theatre in Petersburg then. To make up for it benefit performances in ballet attracted a huge, smart crowd of spectators to the theatre, and above the box office on these days 'Full House' was invariably to be seen.

In his relations with the company Petipa was calculating to the point of insolence. If to some artists he was always polite and courteous, often lavishing compliments on them with purely French, mawkish sweetness (one must realize that among my colleagues were individuals who counted grand dukes and various other powerful persons as 'friends', people on which the career of a balletmaster wholly depended), then with 'small fry' he conversed with the peremptory tone of a commander, often not sparing his words. 'Listène ma belle' was Petipa's favourite expression to coryphées and women of the corps de ballet. 'Yew danse like mai cook' would be heard at rehearsals, and so forth.

Abnormally touchy, Petipa guarded his authority as balletmaster most jealously and did not relish being contradicted. He got very annoyed if artists criticized his work. Secondary artists always remained humbly silent, but not infrequently there were clashes with first dancers. He was not overfond of me because I often contradicted him, although he esteemed me as an artist. It must be said in fairness, however, that if after numerous arguments he was fully satisfied that he had been wrong, he was always prepared to admit it. I danced through many of Petipa's ballets and always met with his complete satisfaction and gratitude for my work. Petipa greatly furthered my career, producing effective numbers for me, and (I say this without boasting) I also contributed in no small way to the success of his productions through my participation. In any event, the artistic results of our collaboration proved to be most favourable.

Petipa has been accused of loving gold; they said he took bribes from artists for new productions and for good 'parts'. I do not know how true this is. If Petipa could indisputably be accused of anything, it is only that he loved, in Griboyedov's words, 'to take care of his own'. Owing exclusively to his efforts, his first wife Maria Surovshchikova Petipa, a weak dancer, attained the rank of ballerina. Having a balletmaster for a father is the only explanation for the career of his daughter Marie Mariusovna Petipa, who went on stage without passing through a course in the theatre school, and who for a long time filled the post of first character soloist despite the fact that there were stronger and more appropriate people for this post in the company. He pulled all he could out of his second wife also, the dancer Lyubov Leonidovna Savitskaya, daughter of Leonid Lvovich Leonidov, a tragedian famous at that time in the Alexandrinsky Theatre; she was suited only for mass dances in long skirts. A very crude and loose-tongued lady, Savitskaya quite often quarrelled with her husband at rehearsals, during which she would at times shower him with the most vulgar abuse.

In private life Petipa was a typical Frenchman, outwardly polite, false and artificially jovial. He was distinguished by a great weakness for the fair sex, started up love affairs with whomever came along, from society women to theatre seamstresses, and was extremely proud of his conquests.

Chapter Eleven. My Subsequent Service

The middle of the 1873/4 season marked the beginning of my own repertoire; before then I had only danced old ballets produced for other ballerinas. At the beginning of 1874 I appeared for the first time in a grand ballet produced by Petipa especially for me—*The Butterfly*. The story of the ballet, written by the Parisian ballet librettist St-Georges, was not, God knows, especially engaging. At the time this ballet was produced, Petipa probably took into account that a ballet *The Butterfly*, to the same scenario but with music by Offenbach, had been performed successfully in Paris by the ballerina Emma Livry, who was burned on the stage of the Grand Opéra during a dress rehearsal of the opera *La Muette de Portici*. Minkus wrote the music for *The Butterfly* here; Petipa produced very many dances for this ballet, and in general they were interesting. Here, incidentally, in the 'Dances of the Butterflies', I had a variation to the music of a waltz by Venzano, which at the time enjoyed great popularity. The celebrated Adelina Patti sang it at the Italian opera in Donizetti's *Linda di Chamounix*. In this variation, which began with *temps* requiring elevation, I made two *pirouettes renversées* on pointe and stopped, as they say in ballet, *à la seconde* (in second position). This was new. Before that Dor did the same pirouettes in *Le Corsaire*, which astonished everyone, but only on demi-pointe, which was much easier. The variation ended with my jumping on one pointe, which no ballerina had ever done before. The balletomanes immediately christened this *pas* the 'Vazem Variation', similar to the Ferraris, Dor, Grantzow Variations, etc., already in existence. As regards participants in the performance, *The Butterfly* was well produced. Golts, Lev Ivanov, and Alexandre Bogdanov played its acting roles, and all the best soloists performed the dances, led by Madaeva, Radina, Kshesinsky, and of course Gerdt, who excelled in his ethereal variation in the 'Dances of the Butterfly'. The character dances were very effective: Persian, Malabar, and especially the dance of the Circassian women, performed by the corps de ballet armed with lances.

The following season Petipa's ballet *The Bandits* was given for the first time for my benefit performance; it was of purely 'passing' interest, devoid of story and choreographic content. Its *pièce de résistance* was the concluding divertissement, entitled 'The Allegory of the Continents', which was totally unrelated to the story of the ballet. It was doubtless suggested to the balletmaster by the *féeries* which began to come into fashion on western stages at that time, especially those of Italy and Paris. In this artistically dubious 'allegory', numerous representatives from five continents promenaded in front of the audience and then performed various national dances in constantly changing stage illumination. For me, a classical ballerina, Petipa found nothing better than the number, 'Europe—cosmopolitan'. I came on in a costume which represented an odd combination of the dress of various nationalities, and danced in succession, with brief pauses, a Spanish cachucha, a German waltz, a French cancan, an English gigue, and finally a Russian dance. I remember I was greatly troubled as I had to dance the cancan alone, which as everyone knows is always performed by a couple. I got out of this difficulty by deciding to perform it imagining I had a cavalier, and cancanned this solo to the merry sounds of the allegro from the Suppé's famous overture *The Queen of Spades*. Before that I had never danced a cancan in my life, but I was told it didn't come out too badly. The public insisted on an encore but I did not consent, considering successes like this below my artistic dignity. I encored only the last dance—the 'Russian'.

The ovations of our public after this 'cosmopolitan' were a most revealing verdict about the level of its artistic taste. I was quite disturbed that I had performed a marvellous classical variation in this ballet to Auer's violin solo, my apparent success in which, for all the dance's solid virtues, could not compare with what came to me for this 'cosmopolitan' nonsense.

In my regular benefit performance the next season I appeared for the first time in *Le Roi Candaule*. For me this was a serious test. The part of Queen Nizia was, as I have pointed out, composed for the ballerina Dor, famous for her technique, with whom I now had to contend with my own abilities. As was said and written at the time, I passed this test with honour, at least in purely choreographic respects. Most difficult technically for the ballerina were the 'dances of Venus' in the second act, and especially the 'pas Dor', which I have already mentioned for its unaesthetic character. At first I hesitated—the utterly unpleasant impression Dor left on me alienated me from it. I called this to balletmaster Petipa's attention, asking him if he wouldn't substitute another number for this one.

'Yes, Madame, I know that thees *pas* eez not verree pretty,' Petipa answered, 'but if yew don' danse eet, all artistes think yew can't . . . I advise you danse eet.'

I did, and managed the difficulties perfectly. My other dances in *Candaule* presented nothing remarkable. In general I wasn't very sympathetic to this ballet and seldom appeared in it. It always seemed boring to me because of its endless mimed scenes. The public also responded quite coldly to *Candaule*. My main partners were Kshesinsky in the title role and Lev Ivanov as Gyges. For all his talent and stage experience, Kshesinsky was not a very successful Candaule. He was too harsh and unpolished for the image of the powerful Lydian king. In any case, during the revival of this ballet many years later when Gerdt played Candaule, this role was incomparably more impressive and vivid. In my time Gerdt was limited to the dances in the second act of this ballet.

As far as I can remember, ballet criticism, making pronouncements about my appearances and noting my mastery in dance as such, often reproached me for insufficiently expressive and persuasive mime—that is, for dramatic coldness. It is not for me to analyse the rightness of such judgement—the pluses and minuses of every artist are more evident to onlookers. The theatre administration, however, did not seem to share the reviewers' opinion. The best proof of this was my being assigned, late in 1876, the purely mimed leading role in the Italian opera's production of Auber's *Fenella* [also known as *La Muette de Portici*, in which the leading female role is traditionally performed by a ballerina].

In the ballet my next new part was that of the bayadère Nikia in *La Bayadère*, produced by Petipa for my benefit performance at the beginning of 1877. Of all the ballets which I had occasion to create, this was my favourite. I liked its beautiful, very theatrical scenario, its interesting, lively dances in the most varied genres, and finally Minkus's music, which the composer managed especially well as regards melody and its co-ordination with the character of the scenes and dances.

I associate with *La Bayadère* the recollection of a clash with Petipa at rehearsal. In the third scene the bayadére Nikia performs a dance with a basket of flowers in a comparatively slow tempo. For this number they made an oriental costume for me with filmy trousers and bracelets on my legs. Petipa composed this dance of *batteries*, so-called 'cabrioles', the throwing out of one leg to meet the other. I pointed out to him that such *temps* were in complete disagreement with the music and the costume. How, in fact, could one perform cabrioles in wide trousers? As usual, the balletmaster began to argue. He thought that I was contradicting him out of stubbornness. After a protracted argument we agreed that Nikia's dance would be made up of smooth movements and plastic poses, and would have the character of a dramatic scene, as if it were a choreographic monologue addressed by the bayadère to her beloved Solor. Thus I prevailed, but Petipa, apparently, held a grudge against me.

We came to rehearsals of the last act. In it Solor is celebrating his wedding with Princess Hamsatti, but their union is disrupted by the shade of the bayadère, murdered at the bride's wish so that she could not prevent them from marrying. Nikia's intervention is expressed in a *grand pas d'action* with Solor, Hamsatti, and soloists, among whom the bayadère's shade suddenly appears, visible only to her bridegroom. I danced the 'shade', and for my entrance Petipa again produced something absurd, made up of delicate, busy little *pas*. Without a second thought I rejected the choreography, which was 'not with the music', nor did it match the general concept of the dance. For the entrance of a shade who is appearing amidst a wedding celebration, something more imposing was required than the minimally effective trifles which Petipa had thought up. Petipa was exasperated. In general the last act was not going well for him, and he wanted to finish the production of *La Bayadère* that day no matter what. He produced something else for me in haste, still less successful. Again I calmly told him that I would not dance it. At this he lost his head completely and yelled at me in a fit of temper:

'I don' unnerstan what you need to danse? Yew can't danse one, yew can' danse other. What kin' of *talent* are yew if yew can' danse noseeng?'

Without saying a word, I took my things and left the rehearsal, which had to be cut short as a result.

The next day, as if nothing had happened, I took up with Petipa again the matter of my entrance in the last act. It was clear that his creative imagination had quite run dry. Hurrying with the completion of the production, he announced to me:

'If yew can' danse sometheeg else, then do wha' Madame Gorshenkov does.'

Gorshenkova, who danced the princess, was distinguished by her extraordinary lightness, and her entrée consisted of a series of high jumps—jetés—from the back of the stage to the footlights. Proposing that I dance her *pas*, the balletmaster wanted to 'needle' me: I was an 'earthly' ballerina, a specialist in technically complex, virtuoso dances, and in general did not possess the ability 'to fly'. But I did not back down.

'Fine,' I answered, 'but for the sake of variety I will do the same *pas* not from the last, but from the first wing.'

The latter was much more difficult because it was impossible to take advantage of the incline of the stage to increase the effect of the jumps.

'As yew weesh, as yew weesh,' Petipa answered, and began the rehearsal.

I must add that at preparatory rehearsals I never danced, limiting myself to approximations of my *pas*, even without being dressed in ballet slippers. Such was now the case. During the *pas d'action*. I simply walked about the stage among the dancers.

The day came for the first rehearsal with orchestra in the theatre. Here of course I had to dance. The balletmaster, as if wishing to relieve himself of any responsibility for his *pas d'action*, said to the artists over and over again:

'I don' know wha' Madame Vazème will danse, she never danse at rehearsal.'

The rehearsal ran its normal course. We finally came to the last act and the *pas d'action*. I stood in the first wing, waiting for my entrance. I was seething with righteous indignation—a voice within spurred me on to great deeds. I wanted to teach the conceited Frenchman a lesson and demonstrate clearly, right before his eyes, what a *talent* I was. My entrance came. At the first sounds of the music which accompanies it I strained every muscle—my nerves tripled my strength—and literally flew out on to the stage, vaulting past the heads of dancers who were kneeling there in groups. Crossing the stage in three jumps I stopped, as if rooted to the ground. The entire company, on the stage and in the hall, broke out in a storm of applause. Petipa, who was on stage, immediately satisfied himself that his treatment of me was unjust. He came up to me and said:

'Madame, forgive, I—am a fool . . .'

That day word circulated about Vazem's 'stunt'. Everyone working in the theatres tried to get into the rehearsal of *La Bayadère* to see my jump. Of the performance itself nothing needs be said. The reception given me by the public was magnificent. Besides the last act, we were all much applauded for the scene, 'The Kingdom of the Shades', which Petipa in general handled very well. Here the groupings and dances were infused with poetry. The balletmaster borrowed drawings of groupings from Gustave Doré's illustrations of 'Paradise' from Dante's *The Divine Comedy*. I had a great success, in the variation, accompanied by [Leopold] Auer's violin solo, with the veil which flies upwards at the end. The roster of principals in *La Bayadère* was in all respects successful: Lev Ivanov as Solor, Golts as the Great Brahmin, Johanson as the Rajah, Gorshenkova as his daughter. Gerdt in classical dances, Radina, Kshesinsky and Picheau in the Hindu Dance of Scene 3—all contributed much to the success of *La Bayadère*, as did the considerable efforts of the artists Wagner, Andreyev, Shishkov, Bocharov, and especially Roller, who also distinguished himself as the machinist of the masterful destruction of the palace at the end of the ballet.

For my benefit performance the next season a new production was not envisaged. Something from the old ballets had to be revived. My choice rested with *Giselle* and excerpts from *Météora*, given for the first time with my participation in mid-February 1878. *Giselle* had not been performed on our stage for a comparatively long time. I always liked this poetical ballet with a story by Théophile Gautier, based on the Rhine legend about girls who died as brides and are transformed after death into fantastic beings, wilis. Adam's charming, melodious music matched this story perfectly. A double task was set before the ballerina here—in the first act to give a sharply dramatic performance, depicting the experiences of the peasant girl Giselle, from amorous scenes with her seducer, her false fellow villager who is in fact Duke Albert, to insanity and death from grief when his betrayal was revealed, and in the second act, light, 'otherworldly' dances of the wilis at night in a moonlit cemetery. One of the critics reproached me for this second act, pointing out that the ethereality of the wilis' dance did not suit me. My colleagues told me I brought off *Giselle* quite satisfactorily. Which of these opinions is closer to the truth I shall not take it upon myself to decide. Be that as it may, *Giselle* went into my repertoire, and I danced it several times.

The next new ballet Petipa produced for me was first performed at my benefit at the beginning of 1879. It was called *The Daughter of the Snows* and as I have indicated it must be counted among Petipa's unsuccessful works. With the public, at least, it had little success. I appeared on stage only in the second and third acts, which were devoted to classical dances; the first was filled with character dances of the northern peoples. Now I positively cannot remember the ballerina's dances—they probably did not amount to much. At the end of the ballet Gerdt, as the captain of the icebound vessel, and I and others played out a scene of 'love and rebirth', but of what it consisted I cannot now say. *The Daughter of the Snows* did not hold the stage for long, and was taken out of the repertoire.

At the wish of Alexandre II the old ballet *La Fille du Danube* was revived the next season. The tsar had once seen the celebrated Marie Taglioni in it, and wanted to experience that impression of his youth again. *La Fille du Danube*, therefore, was a ballet in some sense historical and, in any event, very archaic. Taglioni was its author, not a very gifted balletmaster who produced works exclusively for his celebrated daughter whose artistic charm concealed the weaknesses of her father's compositions. Since Taglioni's time the choreographic art in general, and matters of theatrical production in particular, have made such progress that in the 1880s *La Fille du Danube* could hardly amount to anything particularly interesting. Its story is inspired by the old Austrian legend about the Donauweibchen, and concerns the girl water-spirit who lives an earthly life and captivates a young page.

After the powerfully dramatic ballets of Perrot and Petipa it was rather 'flat'. However Petipa tried to rejuvenate this old piece with new dances, his efforts proved unsuccessful. The ballet had neither scenic effects, to which the audience was accustomed from our balletmaster's latest works, nor brilliant dances which impressed the public. Adam's music seemed very antiquated and yielded significantly to his other compositions, especially *Giselle*. Moreover, in the old days *La Fille du Danube* on our stage stood out for its luxurious *mise-en-scène*, whereas in revival, because of the economy introduced by Baron Kister, there could be no thought of luxury in this respect.

The first performance of *La Fille du Danube* in this revival was given as my benefit performance in February 1880. I took the title part, and remember having success with the classical trio in the first act with Shaposhnikova and Gerdt, and with the waltz to Lanner's music in the second act, performed with Gerdt. In general, the heroine's part was not the most effective. In the 'earthly' scenes at the beginning of the ballet I had to portray some kind of caricature, all but simple-minded. To Gerdt's incomparable performance in the role of the page I have already referred. The mounting of the ballet looked quite wretched. For all the minuses of this production, however, the public, perhaps fascinated by the legend of the furore which Taglioni created in *La Fille du Danube*, came to the theatre in throngs.

14
Libretto of *La Bayadère*
Marius Petipa and Sergei Nikolaevich Khudekov

The bayadère, defined in twentieth-century lexicography as 'a Hindu dancing girl', had been the object of elaboration and fantasy in western Europe at least since Goethe. In that master's ballade, 'Der Gott und die Bajadere', she is already voluptuous and loving, yet inaccessible and even forbidden for her association with a divinity.[1] Part vestal, part Geisha, in the nineteenth century she came to be a symbol of exotic India. In his review of Petipa's *La Bayadère*, the well-travelled balletomane Konstantin Skalkovsky attempted to distinguish imagination from reality. Conceding that fantasy is more important in a ballet than realism, and that there is no place for pedantry in assessing the new work, Skalkovsky nevertheless pointed out that

> Mr Petipa borrowed from India . . . only some external features, because the dances of this scene [the Festival of Fire] are little similar to the dances of the bayadères, which consist, as is well known, of some oscillations of the body and measured movements of the arms to the most doleful music.
>
> But if the dances of the bayadères are ethnographically inaccurate, the idea of forcing the daughter of a rajah to dance is in still greater disagreement with reality. According to Indian notions only courtesans can sing and dance, and every woman who would break this sacred decree would quickly be punished by contempt of caste.[2]

One of Petipa's great ballets, *La Bayadère* is also one of two works of his which survive in any degree of authentic preservation, from the 1870s to the present day. (The other is *Don Quixote*.) *La Bayadère* illustrates the continuing vitality of the *ballet à grand spectacle* on oriental motifs, of which *The Pharaoh's Daughter* was prototype: extravagant tableaux interspersed with episodes of an active, melodramatic love story.

The new ballet was mounted under unfortunate circumstances. The friction Vazem described between herself and Petipa was one problem. Baron Karl Karlovich Kister, Director of Imperial Theatres from 1875 to 1881, was

---

[1] In opera, she was the subject of Catel's *Les Bayadères* (1810) and Auber's *Le Dieu et la Bayadère ou La Courtisane amoureuse* (1830), and figured in Spohr's *Jessonda* (1823), to name works which achieved great popularity and wide distribution. In ballet, Théophile Gautier included bayadères in *Sacountala* (1858), one of Petipa's sources for *La Bayadère*, and Petipa had made them part of the Armenian caravan in the opening scene of *The Pharaoh's Daughter*.

[2] S., 'Mme Vazem's Benefit Performance', *Novoe vremya* [The New Time], 26 Jan. 1877, No. 328, p. 2. In making these criticisms, Skalkovsky was continuing what appears to have been a running feud between himself and Sergei Nikolaevich Khudekov, who assisted Petipa in the formulation of the libretto of *La Bayadère* and who may have been responsible for the inclusion (and possible misuse) of Indian lore and Sanskrit terms.

another. He imposed stringent budget limitations wherever he could, and felt no personal sympathy for ballet. In addition, the Italian opera, whose fortunes were equal but opposite those of the ballet throughout the nineteenth century, received the lion's share of amenities at the Petersburg Bolshoy Theatre. In the winter when *La Bayadère* was first produced, performances of Italian opera outnumbered those of ballet five or six to one. As Skalkovsky put it:

> The production of such an extensive ballet is not an easy task, especially considering that the ballet company, due to six subscriptions of Italian opera, has barely any free days for rehearsal and must rehearse scenes separately, requiring at least half a year of the most concentrated work to learn, since the balletmaster must show each artist his or her *pas*. Therefore thanks only to Mr Petipa's diligence and to the régisseur's experience does a work have good ensemble from the very first performance.[3]

Tribulations notwithstanding, Petipa brought off another important success with *La Bayadère*. 'Looking at the new ballet one can only be astonished at the inexhaustible imagination that our balletmaster possesses,' another reviewer wrote. 'All the dances are distinguished by their freshness and colour; novelties in groupings, a wealth of invention, the intelligence of the story and its correlation with the locale of the action—these are the principal merits of *La Bayadère*.'[4]

Much else was praised, but perhaps two things most of all: the spectacular pageants and Vazem's technique.

> Everything necessary to render the *couleur locale* exactly has been taken from engravings appearing in the *Graphic* and the *Illustrated London News* on the occasion of the Prince of Wales's journey. As a result we see a series of scrupulously exact tableaux of the mores and costumes of the Indians, which naturally give the ballet an ethnographic interest quite exceptional and singularly fascinating.[5]

The opening scene was cited as an example:

> Extremely effective is the 'Festival of Fire' (in the first scene), where the dances of the bayadères, now smooth and voluptuous, now vivacious and fascinating, give way to the wild and frenzied movements of the fakirs, who in a state of religious ecstasy jump through the fire and taunt their bodies with daggers and knives.[6]

The press was filled with hyperbole about Vazem's dancing. 'A miracle of the choreographic art', is how one critic described her variation in the 'Shades' scene, accompanied by Leopold Auer's violin solo.[7] 'It is difficult to evaluate that perfection with which the benefit artist, Mme Vazem, performed all the new dances in her new role, both classical and character,' declared another,

---

[3] S., 'Mme Vazem's Benefit Performance', *Novoe vremya* [The New Time], 26 Jan. 1877, No. 328, p. 2.
[4] 'Theatre Echo', *Peterburgskaya gazeta* [The Petersburg Gazette], 25 Jan. 1877, No. 17, p. 2.
[5] -DE, 'Grand-Théâtre', *Journal de Saint Pétersbourg*, 28 Jan. 1877, p. 1.
[6] 'Chronicle', *Golos* [The Voice], 26 Jan. 1877, No. 26, p. 2.
[7] 'Theatre Echo', *Peterburgskaya gazeta*, 25 Jan. 1877, No. 17, p. 2.

The incomparable talent of the first ballerina of our ballet, who has no peer at the present time in all of Europe, has long since reached its full maturity and such a degree of perfection that it would seem impossible to go further. And meanwhile, into each of her newly performed roles Mme Vazem, as if deliberately, puts new choreographic difficulties, which she overcomes with imperceptible ease, confidence, and precision.[8]

Of the leading artists who performed in *La Bayadère* one whom Vazem listed, Nikolai Osipovich Golts, warrants special mention. The Great Brahmin was to be the last important role Golts was to create before his death in 1880. Sixty years prior to the production of *La Bayadère*, Golts began his training in the Petersburg theatre school, then made his début in a leading role as Rostislav in *The Captive of the Caucasus* in 1823. Of the other ballets whose libretti are included in this volume, Golts took part in the first performances of *La Fille du Danube* as Rudolf, *Esmeralda* as Claude Frollo, *The Pharaoh's Daughter* as the Pharaoh, and *The Little Humpbacked Horse* as Ivanushka's father. His repertoire as a dancer comprised seventy-three roles, and he was also a distinguished teacher and choreographer, notably of the dances in Glinka's *A Life for the Tsar*.[9]

THE BAYADÈRE
Ballet
in four acts and seven scenes
with
apotheosis

By Mr Petipa

Music by Mr Minkus

Presented for the first time at the IMPERIAL Spb.

Bolshoy Theatre on 23 January 1877

**

St Petersburg

Editions Edouard Hoppe,

Typographer of the Imperial Spb. Theatres.

1877

DRAMATIS PERSONAE:

Dugmanta, *rajah of Golconda*
Hamsatti, *his daughter*

---

[8] 'Chronicle', *Golos* [The Voice], 26 Jan. 1877, No. 26, p. 2.
[9] 'N. O. Golts, Balletmaster of the St Petersburg Imperial Theatre', *Vsemirnaya illyustratsiya* [World Illustration], No. 582 [1 Mar. 1880], p. 193.

Solor, *a wealthy and important kshatriya [a warrior of the royal caste]*
Nikia, *a bayadère*
The Great Brahmin
Madhavaya, *a fakir*
Toloragva, *a warrior*
Four fakirs
Six kshatriyas
Two attendants of Hamsatti
A slave girl
Brahmins, Brahmacarins, Sudras (servants of the rajah) warriors, bayadères, fakirs, pilgrims, Indian people, musicians and hunters.

Permitted by the censor, St Petersburg, 12 January 1877

ACT I

SCENE I

The Festival of Fire

The stage represents a consecrated forest; branches of bananas, amras, madhavis, and other Indian trees are intertwined. At the left a pond designated for ablutions. In the distance, the peaks of the Himalayas.

The wealthy kshatriya Solor (a famous warrior) enters with a bow in his hand. Hunters are pursuing a tiger. At a sign from Solor, they run across the stage and are lost in the depths of the forest.

Solor lingers for a time and orders the fakir Madhavaya not to leave this place, that he might find occasion to say a few words to the beautiful Nikia, who lives in the [nearby] pagoda.

Then Solor exits.

The doors of the pagoda open and from the temple the Great Brahmin emerges triumphantly; behind him follow munis [monastic wise men], *rsi* [seers], *bramacarins* (Indian priests), and finally gurus in long linen garments. The priests wear [pendants made of] cords on their foreheads—a sign of brahminesque rank.

From the pagoda also emerge *devadasi* (bayadères of the first rank).*

Preparations for the festival of fire are being made. At the sides of the pagoda, and on its galleries, gather fakirs, yogas, and *fadiny* (wandering holy people).

'Where is our modest bayadère Nikia?' the Great Brahmin asks. 'I do not see her here. Order her to be called. She must adorn our spiritual procession with her dances.'

---

* Bayadères are charged with looking after the pagodas; they live in the pagodas and study with the brahmins.

Several bayadères go out after Nikia. Penitents handle the iron and the fire, touching them to their bodies. Some have daggers, sabres, knives, and other sharp instruments which they brandish; others hold burning torches.

The fakir Madhavaya also takes part in the dance, but in doing so never stops looking for the beautiful Nikia. At last the bayadère appears, veiled, in the doors of the pagoda. Illuminated by the reddish light of the torches, she attracts the general attention.

The Great Brahmin walks up to her, lifts her veil, and orders her to take part in the dances.

Nikia comes down from the steps of the pagoda and begins to dance.

The dance 'Djampo'

Then the sounds of the turti (bagpipes) and the vina (a small guitar) serve to accompany the graceful and languorous movements of the bayadère. These movements become faster and more lively, the orchestra positively thunders, and the previous dance is taken up again.

During this time the Great Brahmin does not take his enamoured gaze from the beautiful bayadère. He walks up to her while she is dancing, and says:

'I love you . . . I am going mad with love for you . . . do you want me to protect you? . . . I will make you first in our temple . . . I shall force the people to worship you! . . . You will be the goddess of all India . . . Only . . . return my love!'

Nikia takes from him his brahmin's cord.

'You are forgetting who you are!' she says, 'Look at this cord! . . . It is a sign of the high rank which you hold . . . I do not love you and never will.'

She pushes him back in horror.

'Ah!', the Great Brahmin exclaims . . . 'Mark well that I shall never forget this insult! . . . this terrible offence! . . . I shall use all my powers to take revenge on you! . . . And my vengeance will be frightful! . . .'

Nikia tries to get away. She joins the other bayadères, fills her vase from the sacred pond, and gives drink to weary travellers and those who took part in the dances.

The fakir Madhavaya continues his original dance and his fanatical flagellation.

Nikia goes up to him and offers water to cool him.

The fakir makes use of the opportunity and says to the bayadère:

'Solor is nearby . . . he wants to see you.'

Nikia is delighted with the news.

'Let him approach as soon as the celebration ends,' she answers. 'I will be at the window . . . Knock three times, and I will come out.'

'Fine, I understand. Only quiet! They can hear us.'

The fakir resumes his tortured dance, and Nikia walks away as if nothing had happened.

The ceremony ends. The brahmins order the bayadères back into the temple. Everyone leaves the stage.

The moon comes up. The windows of the pagoda are dark.

Solor enters with the fakir, sits down on a pile of rocks and anxiously awaits the appearance of his beloved bayadère.

A light appears in one of the windows of the pagoda.

The pleasant sounds of a vina (guitar) are heard.

Solor slowly approaches the window and knocks three times. The window opens and Nikia appears in it, holding a guitar.

The fakir crawls along some branches and places a plank beneath the window, along which the bayadère descends, illuminated by the moonlight.

Solor falls at her feet, then embraces her. They are happy.

'I love you,' Nikia says, 'you are courageous! . . . What grief it is that we cannot see each other often! . . .'

'I cannot live without you,' Solor answers, 'you are the air I breathe . . . my life! . . .'

'Yes, but what is to be done? . . . Look at these garments, I am a bayadère! I must keep order in the pagoda. I was destined for this calling since childhood. I cannot give it up . . . You are my only consolation in life.'

The Great Brahmin appears in the doors of the pagoda. He sees the lovers embracing. In a burst of jealousy and wrath he wants to run to them but holds back, promising vengeance. He hides and listens to their conversation, then exits.

'I know a way we can find happiness,' says Solor. 'Let us flee. In a few days I shall come for you . . . I am rich . . . You have only to agree! . . .'

'I cannot refuse you . . . I agree! Only swear to me before this temple that your heart will never belong to anyone else but me, and that you will love me your whole life! . . .'

'This I swear to you, and I call on Brahma and Vishna as witness, that I shall remain true to you my whole life! . . .'

'All right then, remember your vow . . . If you forget it, all possible misfortune will pursue you.'

'Look though, it is beginning to dawn; we must part.'

At this moment the fakir runs in with the news that the hunters are returning.

The doors of the pagoda open, and the bayadères come out to the pond for water. Unnoticed, Nikia hurriedly enters the pagoda, and Solor watches as she appears at the window again. Having heard the girls approach, Solor hides among the trees.

In triumph the hunters bring in a tiger they have killed. The kshatriya Toloragva tells Solor how they brought down the wild animal, but Solor listens distracted, and looks pensively at his beloved's window. Finally he orders the hunters to return home and goes with them, planning to return soon.

Nikia throws Solor a kiss from her window, and begins to play the same melody as before on her instrument. The Great Brahmin appears in the doors of the pagoda again. He calls on the gods as witness to his future vengeance.

ACT II

SCENE 2

The Two Rivals

The stage represents a magnificent hall in the palace of the rajah Dugmanta.

The rajah is sitting on pillows on a tiger skin. He orders that bayadères be called to entertain him, and proposes a round of chess to one of the kshatriyas. During the rajah's game of chess—a *divertissement*.

After the dances, the rajah sends for his daughter Hamsatti, who enters with her girlfriends. 'Today, my child,' says Dugmanta, 'the day of your wedding to the brave warrior Solor will be set. It is time for you to marry.'

'I agree, father . . . Only I have yet to see my bridegroom . . . and I am not sure if he will love me.'

'He is my subject . . . He is obliged to fulfil my commands! . . . Call him!'

In a few moments Solor appears. When he enters, the rajah's daughter covers her face with a veil. 'It seems you have long been aware,' the rajah says, turning to him, 'that your marriage to my daughter will soon take place.'*

'But sire,' Solor answers, embarrassed, 'I am not yet prepared to do this.'

'In childhood you were proclaimed Hamsatti's bridegroom, and now you must marry her. Come here, my daughter!'

She goes to her father, and he removes her veil.

'Behold, Solor! . . . Is she not beautiful! . . . The finest pearl in the universe! . . . I am sure you will be happy with her.'

---

* Marriages among Indians are contracted when girls are no less than seven and no more than nine years old, and boys are from twelve to fourteen. After a long wedding ceremony, at which a brahmin is present, the bride usually returns to her parents' home, where she stays until maturity. At that point there is another marriage ceremony, with other formalities.

Solor looks and is struck by the girl's beauty, but recalling his beautiful bayadère, to whom he swore eternal love, he suddenly turns away.

'You are a brave kshatriya,' the rajah continues. 'I entrust to you the fate of my lovely child, and am certain that you, as no other could, will carry out your duty in relation to your future wife, marriage to whom shall be your happiness.'

Solor is deeply troubled by the impending marriage.

Perplexed, the rajah's daughter watches her betrothed, wondering what is causing his grief.

'He does not love me! . . . I do not please him,' she says. 'But he will nevertheless be my husband . . . Not for nothing am I a rajah's daughter . . . My will must be done! . . .'

'Sire,' Solor says quietly, approaching the rajah, 'the news which you have announced to me is astounding. I openly confess that I cannot fulfil your desire.'

'What!? You make bold to disobey your rajah's command!? . . . I repeat my order to you: in three days you shall marry my daughter. Do you understand?'

Solor realizes that the rajah cannot be propitiated, and feels devastated by this fateful command.

A sudra (servant) announces the arrival of the Great Brahmin.

'Let him enter,' says the rajah.

The Brahmin enters and bows down before the worldly sovereign.

'I know a great secret! . . . I must tell you about it in private,' he whispers to the rajah, looking at Solor with hatred.

'All leave!' the rajah orders, 'And you, Solor, see that you do not forget my command.'

All exit; the Brahmin and the rajah remain alone, except for the rajah's daughter, who hides behind the portière and listens to their conversation.

In a lively narrative the Great Brahmin describes what happened the night before. He declares that Solor does not love Hamsatti, but adores the bayadère with whom he is seen every night, and wants to run away with her.

Indignant at his future son-in-law's behaviour, the rajah tells the Great Brahmin of his intention to destroy the bayadère. The Brahmin, wishing only Solor's death, is frightened at the thought of the serious danger to which he has exposed his beloved bayadère, and tells the rajah that her death will anger the god Vishna and set the god against them.

The rajah, however, will not hear of this and announces to the Brahmin that tomorrow, during the celebration in honour of Badrinata, Nikia will, as usual, dance with flowers. In one of the baskets of flowers a serpent will be concealed which will crawl out, frightened at the dancer's movements, and bite her, causing her death.

At these words the Brahmin's whole body shudders.

Hamsatti, who heard everything from behind the portière, wants to see the bayadère, and sends her slave girl to fetch her.

The rajah, completely satisfied with the vengeance he has planned, exits with the Brahmin.

Hamsatti sobs and cries in sorrow. She wants to hear from the bayadère herself that Solor adores her. The slave girl runs in with word of Nikia's arrival.

Bowing, the bayadère approaches the rajah's daughter. Hamsatti looks at her and finds her beautiful. She tells Nikia of her impending wedding and invites her to dance in her presence on that day.

Nikia is flattered by such an honour.

Hamsatti wants to see the impression it will make on the bayadère if she knows the identity of her betrothed, and points to a portrait of Solor.*

Nikia all but goes mad with grief. She declares that Solor swore eternal love to her, and that his marriage to the rajah's daughter will never take place.

Hamsatti insists that Nikia renounce Solor.

'Never!' answers Nikia, 'I would sooner die!'

Hamsatti offers her diamonds and gold, and tries to persuade her to go off to another land. Nikia seizes the jewels that the rajah's daughter is offering to her and throws them on the floor.

Hamsatti beseeches the bayadère to let her have Solor, and then to leave. With these words Nikia takes a dagger which has happened into her hand, and rushes at her rival. The slave girl, who has anxiously followed the bayadère's movements and intentions, defends her mistress with her body. Nikia, meanwhile, disappears from the palace.

Hamsatti gets up and says: 'Now she must die!'

SCENE 3
The Bayadère's Death

The stage represents the façade of the rajah's palace from the side of a garden, with masses of huge flowers and broad-leaved trees. In the distance—the tower of the large pagoda of Megatshada, which reaches almost to the heavens. In the background, the light blue of the heavens themselves. The Himalayas are thinly covered with silvery snow.

---

* I know very well that Indians did not have portraits, and used this anachronism only to make the comprehension of the story easier. (Author's note.)

At the rise of the curtain the great procession of Badrinata is in progress. Brahmins pass, then four classes of bayadères (*devadasi, natche, vestiatrissi, kansenissi*), finally pagoda servants, various Indian castes, and others. Penitents enter with burning hot irons. The rajah, his daughter, Solor and other rich Indians are brought in on palanquins.

The rajah takes his place on a platform and orders that the festival begin.

At the end of the dances, the rajah commands the beautiful Nikia to come in, and orders her to entertain the public.

Nikia comes out of the crowd with her little guitar. Her face is covered by a veil. She plays the same melody she played in Act I. Solor, placed near the rajah's throne, listens attentively to this harmonious melody and recognizes his beloved. He gazes at her lovingly.

Hardly able to conceal his wrath, the Great Brahmin watches him with suppressed malice.

During the bayadère's dance the rajah's jealous daughter uses all her strength to conceal her state of mind. Smiling, she comes down from the balcony and orders a basket with flowers to be presented to the graceful Nikia. Nikia takes the basket and continues her dance, admiring the pensive Solor.

Suddenly a snake crawls out of the basket and strikes the bayadère in the heart. Its bite is deadly. Continuing her dance, the beautiful girl appeals to Solor for help, and he embraces her.

'Do not forget your vow,' she gasps. 'You are sworn to me . . . I am dying . . . Farewell!'

The Great Brahmin runs up and offers the dancer an antidote. But Nikia refuses the flagon and throws herself once again into Solor's arms.

'Farewell Solor! . . . I love you! . . . I die innocent! . . .'

These are the bayadère's last words, after which she falls and dies.

The rajah and his daughter triumph.

As through mist a shade is seen, behind which follow will-o'-the-wisps. It grows pale and vanishes among the icecaps of the Himalayas.

ACT III

SCENE 4

The Appearance of the Shade

Solor's room in the rajah's palace.

As the curtain rises, Solor is walking around the stage like a madman, now slowly, now in wild haste. He seems to be trying to remember something. Then he falls, exhausted, on to a divan.

The fakir Madhavaya watches him with a look of profound pity, then orders snake charmers brought in (a man and a woman), to drive the evil spirit from Solor's body. (*Comic Dance.*)

Solor orders the fakir to dismiss them.

There is a knock at the door. The fakir opens it. Hamsatti enters, the rajah's daughter, with a number of women retainers. She is magnificently dressed in gold and pearls. She turns to Solor with reproaches.

The fakir informs her that he requires healing, not quarrels, whereupon Hamsatti wants to divert him, and is extremely amiable. She sits down next to him, caresses him and tries in every way to attract his attention.

Solor at last revives, and takes her hand. At this moment the melancholy strains of the bayadère's song are heard. The shade of the weeping Nikia appears on the wall. Solor trembles.

'Oh! Now my misfortunes will begin,' he says, 'I forgot my vow! Remorse will pursue me my whole life.'

'Calm down! . . . What's the matter?' Hamsatti says, and tries to console him.

'I beg of you . . . leave me . . . Tomorrow we shall see each other again! . . . Tomorrow is our wedding . . . I feel unwell just now . . . I must rest! . . .'

Sorrowful, Hamsatti withdraws, bidding him farewell until the morrow.

Solor goes over to the wall, but the shade is gone; it appears only at moments when his imagination is inflamed.

'You forgot your vow, unhappy man!'—it is as if the shade were speaking to him—'You plan to marry Hamsatti, and so to disturb my peace beyond the grave! But I still love you!'

In vain Solor tries to catch Nikia's elusive shade. 'And I love you,' he answers. 'I have not forgotten you and love you as before.'

The shade at last disappears. Solor falls unconscious on the divan. A dream comes over him and he falls asleep, never ceasing to think about the shade.

Clouds descend.

SCENE 5

The Kingdom of the Shades

An enchanted place. Soft, harmonious music is heard.

Shades appear while this music sounds—Nikia first, then Solor.

DANCES
Plastic Groupings

'I died innocent,' says Nikia's shade, 'I remained true to you. Behold everything around me. Is it not splendid! . . . The gods have granted me all possible blessings. I lack only you!'

'What must I do in order to be yours?' Solor asks her.

'Remember your vow! You promised to be faithful to me! . . . The melody that you now hear will protect you . . . , and my shade will guard you . . . I shall be with you in misfortune.'

'If you do not betray me,' Nikia continues, 'your spirit shall find rest here, in this kingdom of the shades.'

A large concluding dance of the shades.

Clouds descend.

SCENE 6
Solor's Awakening

Solor's room, as before.

Solor is lying on his divan, in a troubled sleep. The fakir enters, pauses next to his master, and looks at him sadly. Solor awakens suddenly. He thought he was in Nikia's embrace.

Servants of the rajah bring in expensive gifts and tell Solor that all preparations are completed for his wedding to the rajah's daughter.

They exit.

Obsessed with his thoughts, Solor follows them.

ACT IV
SCENE 7
The gods' wrath

The stage represents a large hall with columns in the rajah's palace.

Preparations are underway for the ritual of sipmanadi (marriage) of Solor and Hamsatti. Warriors enter, together with brahmins, bayadères, and others. Hamsatti appears, followed by her father with his retinue. When the young warrior Solor appears, the rajah orders the festival to begin.

During the dances the shade pursues Solor and reminds him of his vow.

Hamsatti, meanwhile, does everything in her power to please her bridegroom, who grieves the whole time, and never stops thinking of Nikia.

Four girls present a basket to the bride exactly like the one given to the bayadère, from which crawled the snake that bit her. Hamsatti rejects the basket in horror, as it reminds her of her rival—the cause of all her unhappiness.

Recalling the basket revives the image of the poisoned bayadère in Hamsatti's mind. The shade appears before her, the spectre of the bayadère appears to Hamsatti's troubled mind.

The rajah's daughter flees from it and rushes into her father's arms, begging him to hasten the wedding. The rajah orders the ceremony to begin.

The Great Brahmin takes bride and groom by the hand.

As the ceremony begins, the sky darkens, lightning flashes, there are peals of thunder, and it begins to rain.

At the very moment when the Brahmin takes the hands of Solor and Hamsatti to join them, there is a fearful thunderclap followed by an earthquake. Lightning strikes the hall, which collapses and covers in its ruins the rajah, his daughter, the Great Brahmin and Solor.

APOTHEOSIS

Through the rain the peaks of the Himalayas are visible. Nikia's shade glides through the air; she is triumphant, and tenderly looks at her beloved Solor, who is at her feet.

THE END

Typographer of the Imperial Spb. Theatres (Edouard Hoppe), Voznesenskii prosp., No. 58.

VIII
The 1880s

INTRODUCTION

The 1880s were marked by three important events taken up in the documents which follow: the administrative reform of the imperial theatres, the return of foreign (mainly Italian) ballerinas to the Petersburg company, and the trend towards ever more elaborate ballet production, especially at the end of the decade.

When Alexandre III became emperor in 1881 he appointed Ivan Alexandrovich Vsevolozhsky (1835–1909) to replace Baron Kister as Director of Imperial Theatres. Vsevolozhsky introduced a programme of theatre reform, the major points of which were almost a reaction to Kister: the Italian opera was abolished, and severe economic constraints were relaxed. Konstantin Apollonovich Skalkovsky (1843–1905) wrote extensively about this reform in his collection of essays, *V teatral'nom mire* [In the Theatre World] (St Petersburg, 1899), a retrospective on his career as an observer of theatre affairs. A mining engineer by profession, Skalkovsky was also a world traveller and a balletomane who wrote criticism and compiled a historical monograph, *Balet ego istoriya i mesto v ryadu izyashchnykh iskusstv* [Ballet, its History and Place Among the Elegant Arts] (St Petersburg, 1882; reprinted 1886). His tone is typically spicy and irreverent if not outwardly sarcastic. Himself a professional bureaucrat, he was in a good position to analyse the foibles of bureaucratic policy, especially in the retrospect of seventeen years which separate the publication of *In The Theatre World* from the times he is describing at the beginning of 'Ballet Reforms'.

For all his sarcasm, Skalkovsky was a stalwart advocate of the Italian ballerinas who performed in the Petersburg ballet from 1887. Their coming resulted not from a change in policy towards foreign artists, but from the abolition of the state's long-standing restrictions on theatrical private enterprise: most foreign dancers invited to the imperial stage had already been seen in the newly opened pleasure gardens and summer theatres of St Petersburg and Moscow. Skalkovsky devotes a hundred pages of *In the Theatre World* to Italian ballerinas, from which the opening section, on Virginia Zucchi, is translated here.

What made the Italians so attractive? They were virtuosas, but some, including Zucchi, were not so technically accomplished as Vazem. Some were superb mimes, and projected strong, at times sensual emotions to an audience. Inspired by Enrico Cecchetti, virtuoso male dancing was taken up by Russians. In general, Italian dancers brought a fresh way of looking at dance, and the interaction of Italian virtuosity and passion with Russian elegance and school produced the stars whose names still linger in the memory: Preobrazhenskaya, Kshesinskaya, Trefilova, Pavlova, Karsavina, and many others.

Skalkovsky's 'Ballet Musicians' is typical of his humour and shows his sympathy for a Russian artist made redundant by administrative reform. But there is a hidden note of exaggeration here. The Russian conductor Papkov took his farewell benefit, which Skalkovsky is describing, in 1890, after nearly forty-five years of imperial service and nearly four years after Riccardo Drigo's transfer to the Petersburg ballet. His retirement was neither peremptory nor markedly premature.

15

From *In the Theatre World*

Konstantin Apollonovich Skalkovsky

'Ballet Reforms'

In the preceding two chapters we have described the problems which arose from the absence of ballerinas and the desire not to invite foreigners (the last was Grantzow in 1873). Our ballet management, by its *non possumus* wanted, it seems, to outdo the Roman pope. It caused the balletomanes of the beginning of the 1880s other griefs as well, and spoiled their normally good-humoured mood. One rarely goes to the ballet in a ferocious mood, and if it is true that art 'calms the savage breast', it must be the choreographic art above all others.

The opening of ballet performances at the beginning of the 1880s was accompanied by a kind of wailing and gnashing of teeth among the balletomanes—we say balletomanes because the public which attended the Bolshoy Theatre was not numerous, thanks to the doubling of ballet ticket prices, the reason for which is still not clear. Whole rows of seats remained completely empty or were occupied by ballet retainers with their progeny and household retinue. Prices for tickets to the ballet became more expensive than for the opera; there was absolutely no logical basis for this, considering the make-up of the public which attended ballet, and the management lost more than it gained from the projected rises.

The balletomanes had good reason to complain. First of all the Bolshoy Theatre was spoiled. True, expensive brocade drapes were hung in the foyer and signs of refurbishment were visible—but on stage the electric lighting was for some reason obliterated, which made it dark. The beautiful living pictures of [Petipa's] *Trilby* were completely lost, as only the head of one dancer, the leg of another, the back of a third, etc, were illuminated—the rest remained in darkness. Astonishingly beautiful! The Germans are a very economical people, yet in Berlin the ballet stage is beautifully lit, for without light the spectacle, which speaks chiefly to the eyes, or as Professor Sechenov would say, to the spectator's mind through his eyes—is absurd. It was necessary to sit very close to the stage with huge binoculars to make out anything in such darkness and to take pleasure in it.

The stalls seats crept ever further away from the stage as the result of digging out a huge cellar for the orchestra. Such a large pit was unnecessary, for the expansion of the orchestra in the Russian opera made our singers difficult to hear, and in ballet, with its generally loud music and over-accented rhythm, a large orchestra produced such noise that it gave one a headache.

The stalls were so far removed from the stage that the dancers—to express it technically—could no longer manage to 'aim' [their glances] at the 'objects' [of their interest], and this deprived our sylphides of liveliness; in addition, that happy disposition of mind was not to be seen which ought to have been there as a result of the management's generosity—which doubled and to some even tripled their salaries. Reduced were the salaries only of ballerinas and certain balletic invalids, who danced 'out of gratitude'—a special technical phrase—some twenty years beyond their normal pension time.

Despite the fact that ballet ticket prices were doubled and bureaucrats came to oversee dancers' tricots and slippers, ballet seasons did not offer any consolation. Ballerinas from abroad were not expected, and the public was left with those earlier ballerinas who had delighted our gazes for more than ten years. These were of course talented artists, but it wouldn't have been a bad idea to look at someone new, all the more so since the public was not paying money out of respect for artists' long service, but simply out of the wish to be entertained. Only the corps de ballet was supplemented by several Moscow dancers, men and women. But since the Moscow corps de ballet at that time was distinguished only by its fat legs, this little delighted the hearts of balletomanes embittered by 'reforms'.

And it didn't help that the ballet troupe was placed under the supervision of the retired Colonel Frolov. Experience probably showed that merely high-ranking officers were not up to the ballet, that it required staff officers. We propose, however, that the matter would not be handled better if they assigned an infantry general to the ballet. It ought to be possible to create some genuine attraction for the public putting aside all military aesthetics.

Besides corps de ballet dancers, from Moscow they invited the first dancer Mr Manokhin. This put the local ballet world and its satellites into a flurry. One must point out that our best male dancers had either attained a Methuselean age or, taking advantage of their indispensability, did not think it necessary to overwork at practice, limiting themselves to arm waving, which they considered 'mime'—whereas if for women dancers daily practice is essential, a man, with his characteristically coarser structure, must work twice as hard.

Why do we need male dancers, they said right and left at the time of a début, when even abroad the question is being raised of their complete abolition and substitution by women dressed as men? 'Even in Paris there are no longer any remarkable male dancers; we cannot call [the Moscow dancer] Espinosa good, but since when does Moscow set the trend?' In Moscow Mr Manokhin received a salary of 400 roubles per year. That horrifed many; what if they give him 600 roubles here?! In ballet such a figure seemed more astonishing than 120,000 made by the directors of some banks, who work with their heads even less than dancers do.

The friends of male dancers sit in the upper reaches of the theatre, lady dancers' friends in the first rows of the stalls. Each group has its sympathies; defending their favourites, the galleries sometimes boo all lady dancers in succession. Once Mme Vorobieva danced her variation when not one note of the music could be heard thanks to the roar from the gods. Such a 'revolution' in the heights resulted from a misunderstanding; they thought that Mr Litavkin could not repeat his variation because he had received a chilly reception from the parterre.

Several variations for soloists, for no apparent reason, were suddenly shortened. Why? They say the management heard that in Italy such variations are not done, being somewhat reminiscent of dance examinations. But why do we have soloists if they will have no solos (variations in balletic parlance)? Secondly, if in Italy the best dancers rarely have separate *pas*, a ballerina dances five or six scenes without tiring, and here, in *The Wilful Wife* for example, the dances of two ballerinas take up no more than five minutes. Finally, solos resemble a dance examination when they are inartistically juxtaposed, one right after another. It is the balletmaster's responsibility, using each artist's variety of means, to place her variation in its proper place and blend it into the general ensemble.

Unfortunately, Mr Petipa's works of this time were weaker than his earlier ones. Too much was demanded of our esteemed balletmaster, especially since it is not enough for the balletmaster just to compose a ballet; he must teach it and produce it and then follow all the performances—an immense labour. Not counting the new ballet, dances for [Anton Rubinstein's] opera *Nero*, which take up an entire act, and separate *pas* for various other operas, Mr Petipa had to compose in one season yet more dances, for the operas *Richard III* [of Gaston Salvayre] and [Saint-Saëns's] *Philemon et Baucis*.

There was simply no time to prepare a new ballet. Thus ballets were given which either bored everybody or for which the public never had much affection in the first place. Since these ballets had few dances in them, it was naturally much easier, for artists awaiting a pension, to dance in them. Execution was flabby and lifeless, the company undisciplined;

some danseuses became mamas, others grandmamas, while a majority put on weight. Obviously the new management threatened to ruin the ballet, just as Baron Kister had ruined the French theatre in Petersburg.

Production in old ballets was horrifying. In *The Naiad and the Fisherman* every decoration had cracks in the Neapolitan sky, which as we all know the real Italy doesn't have, and which exist only in the mind of the drunken merchant in a famous comedy by Ostrovsky. The width of the Maryinsky stage, 8 arshins [about 17 feet] greater than the stage of the Bolshoy (while its depth is less by a third), did not permit the transfer of ballet décors, so they used decorations from forgotten operas. Transformations and drops were also astonishing for their primitiveness. Nothing like it had been seen since the mystery plays of the thirteenth century! Nor was it easy dancing in the corps de ballet; the whole morning taken up at rehearsals of new ballets, and evenings in operas, and this at a salary for first soloists of five roubles per performance—a sum for which the lowliest cabbies work. Salaries, however, were quickly and significantly raised.

Despite the threefold management of régisseur, balletmaster and special functionary in charge of ballet, a lot of rubbish was given in place of the ballets announced in the repertoire for which the public had gathered. In these circumstances, without much ado, the most important *pas* and variations were deleted without substitutions. It was like giving an opera and deleting its best arias or quartets. A lack of artists might have justified this, but many artists, including the most talented, sat around with nothing to do. And why should the public care, paying twice as much money at the new rate, for various insignificant backstage considerations? If there was nobody to dance, the management ought to have forced Messieurs the régisseur, balletmaster, and special functionary to perform a *pas de trois*, and not put before the public the scraps of a work. To give only part of the goods for money is a deceptive trade practice—an act forbidden by statute in the legal code.

The following case demonstrates the theatre supervisors' indifference to the dictates of the public. Although the affiche announced that [Petipa's] ballet *A Midsummer Night's Dream* was to be given, for some backstage reason they performed only *half* the ballet, and, moreover, the worse half. The entire first part, in which Mme Vazem danced, was cut, although this ballerina calmly danced the very same day in another ballet, which means that her legs were in good order. After the half of *Dream* which was deleted, only the part where Bottom treats himself to hay was left, but the ballet managers might better have kept the hay and treated the public to dances. *La Vivandière* has two acts which they joined into one; this caused, as in the stories of Nemirovich–Danchenko, the sun to rise twice on the same day, once from either side of the stage!

In other countries a whole company takes part in ballets, since the more people dance in a ballet, the more beautiful it is. Here the theatre works rather like official service: the fewer involved the better. Therefore a fair portion of our ballet company was always sitting in the auditorium with the customers, and watching how the other part leapt about. This seating practice, however, corresponded to an optical illusion advantageous to the management. Frightful prices combined with lack of novelty kept ballets up to 1885 from being well attended. This way you look at the house and many people seem to be there. It was easy to explain. According to new regulations, twenty-five seats in the auditorium belonged to the house; they were mostly given to bureaucrats (recently created in the theatres in view of staff reductions). If you add unoccupied dancers and a cadre of balletomanes, you have a 'public'. At least it was less boring than dancing in front of empty seats.

Suddenly a new misfortune befell the ballet. Because of intrigues by the Russian opera, the ballet lost its Sunday matinees for a time. Thus the ballet—this purely festive entertainment, the favourite of children—came to be given on days when only a few inveterate balletomanes could attend, and it became unbearable even for them to go to ballet on Tuesdays, and then again on Thursdays, only a day later. By what stretch of the imagination these changes were made, in defiance of public custom, is simply incomprehensible.

Then still another misfortune struck: the planned rebuilding of the Maryinsky Theatre (to what end is unknown) was not finished on time. To give performances in the Bolshoy Theatre was nevertheless 'shameful', in the opinion of the gentlemen reformers, for it was announced to the public (no one knows by whom and when) that the Italian opera was to be abolished, since there was a danger that the Bolshoy Theatre might collapse. But it was more shameful, in the first place, to mislead the public, and much more shameful, in the second, to leave the public without the performances to which it was accustomed and without which many Petersburgers on autumn evenings would not know what to do with themselves. Most curious of all, however, was that having abandoned the Bolshoy Theatre, the reformers added *féerie* to the repertoire of the Maryinsky. You can imagine the mess when three types of spectacle began to battle with each other for days on which to perform. And Messieurs the régisseurs had no little battling to do: after losing more than a month as a result of prohibitions on performances on Saturdays, the eves of religious holidays and on Christmas, the number of free days in the season was extremely limited. Losing Russian opera and ballet for more than a month also deprived the management of at least 60,000 roubles in receipts, while almost all the expenses remained, except the expense for lighting.

In spring-time in Petersburg there are many fewer theatrical presentations and concerts [than in winter], so that ballet performances could have excellent houses, but for some sage reason they are given infrequently, at the same time the ballet company receives some twenty-thousand roubles' salary in good order. Performances could easily be continued for another two weeks or so, but the management has never utilized the material, so to speak, it already has in hand and paid for.

'Italian Ballerinas in Russia'

Towards the middle of the 1880s our poor ballet was headed for complete decline. The public patronized it little and its receipts were small, whereas the personnel of the company, the balletmaster's experience, the excellent traditions preserved in the theatre school as regards dances, and the considerable resources expended on ballet should have allowed it to occupy, as before, an outstanding place among the capital's stages. More than anything else the system of backstage favouritism hastened the decline of the ballet, once the pride of Petersburg which the city constantly flaunted before foreigners. The occupation of choreographic art was looked upon the same as government service, for which reason a table of ranks was applied to it.

Forgetting that talents, even in the art of cooking, as Brillat-Savarin bears witness, are not made but born, they began to fill vacancies among first dancers with second dancers, and among ballerinas with first dancers. Such promotion was perhaps very humane, for why shouldn't some girl, jumping around her whole life at the bottom of the salary scale, jump her way little by little to the top? But the public was indifferent to such Montionesque rewards. It made no difference to the public whether someone's diligence merited an incentive; it needed an artist who would fascinate it; to elevate somebody linearly in the ballet was exactly like promoting choristers in opera to prima donnas. The ballet also needed variety. Its circle was comparatively limited, especially among the masses who did not distinguish the subtleties of the art, and since nowhere as in ballet is interest concentrated on one person, the ballerina, albeit technically first class, will in the final analysis come to bore the public which previously followed her every *pas* and movement. For that reason even Taglioni, who enraptured all Europe and who was considered incomparable, had long since ceased to attract a public in her fourth or fifth year.

If that is true regarding first-class talents, the public is in worse shape still when offered ballerinas whom everyone has seen in secondary roles for ten years and whose talent has already been more than well scrutinized.

For that reason everywhere—even here in earlier times—if a talent turns up in school, they guard it for a time and do not show it off, and after having trained it to the utmost, they release such a person immediately to responsible parts. The first impression is frightfully important on the stage, and if the opinion is formed that an artist is second-rate, the public's bias is impossible to overcome, even if the highest stage of development has been reached by work and talent.

For one must not forget that lifting the leg high in battements and cabrioles is not enough to become a ballerina. One can have an excellent technique and still not have the right to be a prima ballerina. To entice a public one must combine grace, mime, elegance, and good looks such as occur but rarely.

These elementary truths were disregarded here, and it is no surprise that the public cooled towards the ballet. Mme Vazem was an outstanding dancer, but it wearied the public to watch her only for more than ten years running, whereas Mme Sokolova, considered to be our other ballerina, was distinguished by gracefulness alone. On the retirement of Mme Vazem, who understood with truly artistic instinct that it is better to leave the stage earlier than necessary rather than later, Mmes Gorshenkova, Nikitina, and Johanson filled the ballerina's vacancy thus produced. They were all excellent soloists; they could take first parts in small ballets successfully; but if they could do the same in large ballets, then one fact—that the public already knew them by heart—indicated that it would be impossible for such dancers to attract attention and receipts. The professors could approve one dancer or another as much as they liked, but if she didn't draw a crowd, she was little use to the theatre.

In addition, our talented male dancers were all middle-aged, and some were quite elderly. Among the younger set there was not one one excellent classical dancer or a good comic. Messrs Johanson, Stukolkin, Kshesinsky, and Gerdt had no candidates to replace them. The only superior element of our ballet was its staff of first and second women dancers.

They said that the public grew cool towards ballet because it was not in fashion. That is not correct; as an art dance will always please, for it is in human nature. They assured us that Italian opera was not in fashion when it was poorly attended, but one had only to invite Patti, Nilsson, Masini—and opera became fashionable. Even in Russian opera, members of the 'mighty handful' were declaring long ago that Italian music was opposed to the cultivated Russian ear, that it needed Chaldean music. But the success of *Traviata* and *Il Barbiere* when performed by excellent singers showed that the whole issue resided in the artists, and not in theories. It is the same with ballet; it was always fashionable in Petersburg when Taglioni, Elssler, Grisi, Grantzow, Muravieva, etc., were dancing; but when mediocrities were dancing, even in the 1840s and 1850s, the auditorium was completely empty. This can easily be proved in Wolf's *Chronicle* [*of the Petersburg Theatres* . . ., 3 vols., St Petersburg, 1877–84].

And in truth one must be an inveterate balletomane to look at *The Humpbacked Horse* for the 150th time or at *Zoraya* for the fiftieth, given in exactly the same production, without the slightest changes, and with the same available personnel. Most of the public, especially when faced with the extremely high ticket prices for ballet introduced in 1881, finds repetition like this very uninteresting. But the public would find interesting the masterful performance of a new, talented ballerina, and so the way to make matters right in the ballet would be to engage some star, a solution which we have earnestly and unsuccessfully advised.

It was impossible, they objected, for two reasons to have a ballerina who would attract the public: first there wasn't money to hire her, and second, no such ballerina was to be found in Europe. Both reasons were false. If the money was available to expend thousands on unnecessary ironwork and drapery and on the production of monstrous operas by the various Salvayres and the boring dramas of Bornier, then 10,000–15,000 roubles could be found for an excellent ballerina. Common sense said that it was better to expend 20,000 and be ahead 50,000 in receipts than to spend 10,000 and be the same 50,000 in deficit.

To assert that there were no ballerinas in Europe, whatever the balletmaster's claim on his official missions abroad, was also ridiculous. Even though the public here would be satisfied with the ones we had if only it were given some variety, several excellent Italian dancers, technically far superior to Russian ballerinas of recent times, were available, and Milan is always preparing more. Zucchi, the 'Divine Virginia' at whose very mention Italians blow a kiss into the air, would have an immense success in Petersburg, especially in a ballet mounted to suit her winning talent. But they intended to invite her after she had aged a bit, so they could tell the public, 'There you have the much-vaunted Zucchi!' No few such old ladies have visited us already. If three or four ballerinas perform at the same time in Paris, London, and Milan, it would easily be possible to find someone to engage. In any case, without some such engagement our ballet would pine away completely, and if that were the management's goal, and not to display so much backstage diplomacy, one could simply close the theatre school, disband the company and retain a couple of dozen dancers for opera divertissements.

For opera, however, especially Russian, no such effort could be made. The management, its experience to the contrary but in slavish imitation of Paris, tried most unsuccessfully to unite choreographic presentations with operatic ones. In [Anton Rubinstein's] *The Demon*, for example, much labour was expended on the dances, first dancers performed them, the most splendid costumes cost huge sums, and as a result the public booed the dancers, who were in no wise at fault except to delay for ten minutes the pleasure of hearing Mme Raab and Mr Vasiliev II sing.

Our ballet's saviour turned out wholly unexpectedly to be Mr Lentovsky, who invited Zucchi in the summer of 1885 to his [garden theatre] Sans Souci. The Muscovite magician and sorcerer got the idea of signing her from reading our *feuilleton* in *The New Time*, in which we described Milan and Zucchi's dances, and mentioned that Mme Ferni Germano pointed out the ballerina to us.

VIRGINIA ZUCCHI

The news that Zucchi—the Divine Virginia, as the Italians called her—the most celebrated contemporary ballerina, was to dance at the Sans Souci, met with disbelief. Sceptics thought it was perhaps another dancer of the same name. Only with Zucchi's arrival were all doubts dispelled; *la celebra danzatrice* was in fact to dance here, to the solace of balletomanes sitting on the shores of the Neva and weeping over ballet's decline, and to dance the entire summer at that. The choreographic art began to shine on the banks of the Large Nevka, in most unsuitable surroundings. Mr Lentovsky, however, put on quite splendid shows: he hired the well-known London–Moscow balletmaster Hansen, and a corps de ballet partly from Milan, partly from Vienna.

The public still did not believe it, and for several weeks of the Italian ballerina's sojourn here before her début, a literature about her was created in prose and verse. The ballerina's origins and age were disputed. One of our newspapers called the ballerina 'young', another attributed to her—*horribile dictu*—an age of forty-two! Zucchi was in fact about thirty-five, but she called herself twenty-eight. Born in Parma, at the age of seven she began to study dancing in Milan with Hus and Corbetta and then with Ramaccini and Blasis, who put the final touches on her training. Zucchi made her début as a ballerina in Padua in Montplaisir's ballet, *Brahma*, which has remained her favourite ballet.*

Having won celebrity, she danced in Rome, Naples, Madrid, Milan, Turin, Berlin (1876–8), Paris (in 1883), and London, everywhere with huge success. For the final consecration of her reputation Zucchi lacked only Petersburg's approval, a city considered a first authority in Europe as regards ballet.

---

* Evil tongues declare that before this she danced in the *corps de ballet* in Florence.

We never doubted that she would enchant our public. In any event, Zucchi created a curious impression in Petersburg, where, thanks to excellent traditions, there were many savants of choreography. The severity of these critics could hardly stand its ground against the eyes of the Milanese Circe, with that expression even photographs cannot capture. And Zucchi was interesting not just on the boards. With Italian liveliness of mind she joined an innate sense of merriment and naturalness. She spoke excellent French and a little German, and her likeness made all our illustrated journals popular.

For some time we hadn't seen such a mass of spectators as gathered at the Théâtre Sans Souci for Zucchi's début at the première of the *féerie*, *An Extraordinary Journey to the Moon*. All the seats were sold, and dozens of people paid a rouble to stand in the aisles. And to stand for your money meant you did a lot, since the performance, which began after 8.00 p.m., went on until 2.00 in the morning, and although most of the public stayed, towards the end one could read on every face, 'Fine by me if it were a little shorter . . .'

The Journey to the Moon was adapted from Jules Verne's famous science-fiction novel and was originally produced in Paris at the Théâtre du Châtelet, after which it was given with great magnificence in London at the Alhambra, whence Mr Lentovsky bought part of the costumes and accessories to produce the *féerie* in Moscow. There also it enjoyed great and prolonged success, which gave the indefatigable producer of Moscow entertainments the idea of familiarizing Petersburg with it as well, having invited the famous Italian ballerina to enhance the attraction.

Although the corps de ballet, contracted abroad, did not arrive at full strength, Mr Lentovsky, to avoid losing time and cooling public expectations, decided to begin his journey to the moon on 6 July (a banner day in the history of our ballet)—a decision all the more legitimate since the *féerie* required several hundred artists, paying the salary of whom without having them work was disadvantageous. In general, the hopes of the public were justified. The story of the *féerie* was carefully and cleverly set forth; the subject-matter justified magnificent fantastic decorations, numerous processions and various ballets, enlivened with the music of Offenbach.

For a majority of the spectators, in their elegance quite unlike the habitués of our suburban pleasure grounds, interest in the performance did not reside in the *féerie*. These people came to see Zucchi, who easily won public favour. Received quite warmly at her entrance—a rarity for artists making their débuts here, even famous ones—she was, if it may be expressed thus, 'heard out' with astonishing attention; the public, it seemed, did not want to miss the artist's slightest gesture. The same attention was shown in the eleventh scene of the féerie, where the ballerina appeared again.

It was impossible to judge Zucchi on the basis of two short divertissements interpolated offhand into a *féerie*. No male dancer in the company could partner her in *tours de force*. For a talent like hers a completely different field of activity would be needed: only on a large stage, in a real ballet, could she express herself fully. Nevertheless, everything Zucchi performed she performed excellently. Strict judges wanted to see her in dances where her strength or elevation, and the purity of several *temps* difficult for their speed, could better be assessed; but there were no two opinions in matters of gracefulness, plastique, originality, mime, as well as firmness of pointe and aplomb.

For those naïve souls who expected the famous Zucchi to leap about the stage and dance back and forth across the ceiling, we note that the virtue of her dancing, even abroad, was recognized not in its acrobatics or the strength of hackneyed academic pirouettes, which any excellent dancer with strong legs can do, but rather in the elusive art of dancing lightly, merrily, elegantly, with spirit and fire, without the slightest apparent effort, and also originally, boldly, gracefully, and in the highest degree girlish and charming. At a time when dancers of plastique were normally heavy and awkward, and ethereal dancers were like spiders and could rarely manage their long arms, Zucchi represented a remarkable balance of all choreographic virtues.

Her entire figure was elegant; her head, with its whimsically tousled hair, vivified by the eyes which took on, when necessary, an astonishing, passionate expression; for all that her mime was not of the eyes alone, but of the entire body, amazingly well proportioned. At a time when a majority of ballerinas had legs developed to the detriment of the other parts of their bodies, Zucchi, who possessed the legs of Diana, also had full, beautiful arms with delicate joints, rounded shoulders, a well-developed chest and back in which there was more poetry than in half the contemporary Italian poets taken together.*

If the troubadour Peire Vidal were resurrected he could repeat, 'Madame, it seems that I am seeing a deity when I look at your beautiful form.'

That Zucchi produced an impression at least commensurate with her powers was shown in the waltz on pointe to the music of the famous song, 'Do you remember?', when she compelled the public unanimously to demand an encore; contrarily, the public was cooler in the second scene of

---

* For some reason this phrase has enjoyed an extraordinary success. It has literally created a whole literature in verse and prose and is often repeated in ironic reference to the author of this book. It turns out that by accident he almost paraphrased the words of Bulgarin, who in 1840 in Pesotsky's *Repertoire*, in reference to Taglioni, wrote that in her dances there is much more poetry than in all the works of Byron. Heine also refers to the 'poetry' of Taglioni's legs.

the ballet, where the ballerina displayed better technique than in the first scene, and had a chance to show her artistry as a mime.

Zucchi's dances were produced by Mr Hansen with meagre stuff; he managed quite satisfactorily, although at times he could be criticized for a certain monotony of *pas*. He kept Offenbach's music, although the public would not have called him to account if for Zucchi something had been interpolated, as they say, quite 'from another opera'.

The ballerina's next débuts were in the *féeries The Golden Apples* and *The Forest Vagrant*. The fifth scene of *The Golden Apples* was a small, quite original ballet consisting of allegories of various games, beginning with innocent play with dolls and ending with baccarat, roulette, and that not at all innocent game called love. Zucchi could represent this type of game superbly in her dances, for her dancing was every bit as elegantly erotic as the poetry of Ovid or de Musset. Her poses, gestures, *pas*, and arabesques were, as usual, full of harmony, her mime intelligent and lively. Whatever she did turned out elegantly and girlishly charming.

Zucchi was dressed as a gypsy in a red costume with black skirt and black tricot inserts; this went very nicely with her figure, her dark hair and eyes, full of life and fire. Her dances—a Spanish *pas*, a solo on pointe and two galops—let the artist show her lightness, gracefulness, and composure, and that greatest art of all, concealing the difficulties of execution. Zucchi danced as if it were a jest, and for her own pleasure.

Her performance would have been still better if the stage had been larger and if all the scenes of the ballet had not had the character of a divertissement, for which reason the dances ended either monotonously or just seemed endless, as they did not quite fit in with the variations which followed. But to a *féerie* it is understandable that one cannot apply the standards and requirements of a ballet, where for months a ballet-master ponders the logical transitions from one dance to another.

Zucchi enjoyed a brilliant success. Mr Lentovsky erred in putting off the ballet scene until midnight; it was quite difficult for the public and the balletomanes to eat apples every day in anticipation of seeing the famous ballerina's back.

In *The Forest Vagrant* the gifted artist had to dance without advance notice and consequently without rehearsal. She selected a *grand pas caractéristique* with cabrioles. It turned out that she danced character *pas* as well as classical. Beginning with the costume, which cleverly outlined her elegant form, and ending with the execution, graceful and finished, everything bore an artistic character.

'Ballet Musicians'

On certain islands of the Pacific Ocean they have a custom: in order not to have to feed old people, they eat them. The custom is called 'kiki'. The theatre management apparently wanted in some degree to imitate this custom. This 'kiki' was committed in 1890 on Mr Papkov, and he was devoured, though figuratively.

The history of this Russian conductor is most instructive. Being first violin of the Moscow theatres, in 1862 he received the post of assistant to Lyadov, now deceased but then conductor of the ballet orchestra in St Petersburg. Soon afterward, on Lyadov's death, Papkov was made conductor. For more than twenty years he peacefully waved his baton to everyone's satisfaction, when suddenly, with the abolition of the Italian opera, it became necessary to appoint Mr Drigo. Without much ado the latter was designated first conductor in the ballet, and the worthy Mr Papkov second. The German troupe was abolished at the same time, and it too had a conductor under contract. What to do? The decision was simple: commit 'kiki' on Mr Papkov, retire him on a small pension, and make Mr Kanegisser second ballet conductor.

In all of this not the slightest attention was paid to the fact that Mr Kanegisser had hardly ever been to a ballet in his life and that the second ballet conductor supervises rehearsals, for which one must play the piano or the violin. Mr Kanegisser, alas, played only the bassoon! A magnificent spectacle—seeing our ballerinas leaping about to the sound of a bassoon. This would be more suited to the Smorgonsky Academy, where young bears are taught to dance to the sounds of a fife.*

Thus Mr Papkov's whole fault was that he was born Papkov, and not Papkini or Papkenhof. Then undoubtedly 'kiki' would not have been committed on him and the audience would not have had to hear out the public reading of addresses where the entire company declared to Mr Papkov that he was irreplaceable. But if he were so good, then why was he replaced, and at that by foreigners? One small newspaper declared, however, that Mr Drigo was invited to the ballet because his name derived from the verb *drygat'* [that is, 'to kick or jerk one's legs'].

It is even more curious that when we did not have a single music school, all the conductors and their assistants in the state theatres were Russian; now, when in Russia (not counting the Kingdom of Poland) there are two conservatoires and more than a hundred music courses, classes, and schools with all manner of 'rights and privileges' (especially the right to

---

* Some time before this Mr de-Sieni was made régisseur of the ballet troupe; prior to his régisseurship of the Italian opera he was, it is said, a salesman in a hat shop.

devour one's diploma from hunger), all our theatre conductors are foreigners.

Was not Mr Papkov retired because with his Russian name he seemed to be some kind of contradiction to the general order of things, recalling the not so distant past when a Russian name did not prevent a musician from advancing?

Still earlier than Mr Papkov they committed 'kiki' on Mr Minkus, the talented Czech composer. Mr Minkus served in the state theatres for thirty-one years. In the 1860s he was already inspector of music in the Imperial Moscow Theatres, and after Pugni's death was appointed, in 1871, composer of ballet music in St Petersburg. For the modest salary of 2,000 roubles Mr Minkus was obliged to write a new ballet every year, rework old ones, compose supplementary *pas*, and the like. In all Mr Minkus wrote sixteen new ballets; the best of them—*Fiametta, La Bayadère, Zoraya*—are distinguished by the clear and lively melody so necessary for dances. His march from *Roxana* was the favourite piece of Emperor Alexandre II, who in general did not love music. Several units of our troops stormed the Plevna to the music of this march. Mr Minkus also worked in Paris with Delibes, and in collaboration with him wrote the music for the ballets *Néméa* and *La Source*. The latter has made the rounds of Europe. With the new theatre administration Mr Minkus's pay was quickly doubled, and then, with the same dispatch, it was decided that a composer of ballet music was no longer required.

Since that time music has been commissioned from various people, sometimes successfully, as with Glazunov and Tchaikovsky (in *The Sleeping Beauty*), but for the most part unsuccessfully. Rubinstein showed us that ballet music demands a specialist. Not one balletmaster wanted to set dances to his music for the ballet *The Grapevine*, for the simple reason that not one ballerina could perform them.

16
Libretto of *The Vestal*
Sergei Nikolaevich Khudekov

The 1880s were difficult for Marius Petipa. The dynastic succession in 1881 brought a period of inactivity during the official mourning, followed by preparations for an elaborate coronation ballet performed only once. At mid-decade, his creative energies flagging, he turned to revivals: of *Giselle* and *Coppélia* in 1884, *La Fille mal gardée* and *The Wilful Wife* in 1885, and *Esmeralda* in 1887.

Although Petipa at first opposed inviting Virginia Zucchi to perform on the imperial stage, it would appear that she and her successors revived his artistic fortunes. *Esmeralda* with Zucchi was an extraordinary success, the high point of her career on the imperial stage, after which Petipa went on to produce some of the finest ballets in his repertoire for Italian ballerinas. The end of the 1880s was taken up with two works of unprecedented magnificence: *The Vestal* and *The Talisman*, given for the benefit performances of Elena Cornalba in 1888 and 1889 respectively. *The Vestal* marks the high point of Petipa's *ballet à grand spectacle* in the complexity of its story and extravagant production. Khudekov's libretto resembles a short story more than a scenario.

Press reports of the first performance suggest that the scenic demands of *The Vestal* may have overtaxed the resources of the company. Setting up and changing decorations extended the intervals to half an hour. The first performance was marred by a dangerous accident: two dancers representing the reveries and visions of Venus flew across the stage on to a property altar, knocking it over and upsetting the contents of a flaming urn on to the stage. The ensuing fire was put out, but not without disrupting the performance and frightening the audience, part of which left the auditorium. The première was poorly attended even before the fire, which reviewers attributed to high ticket prices for an untested work. Elena Cornalba was a first-rate Italian virtuosa and praised as such in reviews, but she was not an effective mime, and critics considered her at her weakest at the most dramatic points in her role as Amata.[1]

The Vestal was the product of administrative reform, the coming of Italian ballerinas, and a scenic conception so elaborate as to eclipse Petipa's earlier works along the same lines. If the new ballet did not wholly succeed, it warrants our attention as a prototype of better things to come.

---

[1] These observations are from unsigned reviews in the column 'Theatre, Music and Spectacle' on p. 3 of the *Syn otechestva* [Son of the Fatherland] issued on 14, 18, and 19 Feb. 1888.

THE VESTAL
Ballet
in 3 Acts and 4 Scenes
Programme by S. N. Khudekov
Produced on stage and dances composed
by the balletmaster M. Petipa
Music by M. M. Ivanov

Performed for the 1st time on the stage of the Maryinsky Theatre
in February 1888

St Petersburg
Publication of Edouard Hoppe,
Printer of the Imperial SPb Theatres

1888

Permitted by the censor. SPb, 1 February 1888

Publisher of the Imperial SPb. Theatres
(Edouard Hoppe), Voznesenskii pr. No. 53.

DRAMATIS PERSONNAE:

Julius Flac, *Roman senator*	Mr Kshesinsky I
Claudia, *his elder daughter*	Mlle Gorshenkova I
Amata, *his younger daughter*	Mlle Cornalba
Lelia, *their girlfriend*	Mlle Johanssen
Lucio, *a centurion*	Mr Gerdt
High priest	Mr Aistov
The first vestal	Mlle Ogoleit I
An augur	Mr Geltser
A warrior	***

The Roman emperor, his retinue, priests, priestesses, slaves, histrionics, bacchantes, maenads, satyrs, warriors, gladiators, various gods and goddesses of ancient Rome, nymphs, etc.

The action takes place during the Roman Empire.

ACT I

A richly furnished dining-room in the home of Senator Flac. To the right, triclinia; to the left, statues, amphoras, other decorations. At the back, a curtain, behind which gardens with fountains are visible.

A feast at the home of Senator Flac. On the raised part of the dining-room are richly adorned triclinia; the tables are overflowing with viands, drinks, and fruit. On luxuriously soft triple-couches, covered with rich fabrics, leaning on bright pillows, the senator's guests—well-known Romans—recline at dinner. The place of honour is occupied by Senator Flac, who has thrown off his toga and is lying in a tunic edged in a bright purple stripe. Next to him reclines his most favoured guest, the young praetorian—the centurion Lucio.

Smoking censers spread an aroma through the air.

A bacchic melody is played in honour of wine. Beautifully dressed slaves, in short tunics and with intricate headwear, move about among the guests, serving them wine from silver vessels.

Groups of histrionics [actors], slaves dressed as goat-legged satyrs, fauns, and tailed Sileneuses, mimes with masks in their hands and on their shoulders, entertain the banquet with their lively dance.

Waving wands of yew wrapped in ivy and grapevines, the bacchantes and their friends pause in picturesque poses, encouraging the guests to revelry. They jump, throwing back their heads, now wave their wands, now join the bacchanale. The comic affectations of the dancers, to the sounds of flute and timpani, reach the point of transport. The orgy is at its height.

Senator Flac gets up from his place and orders his slaves to cease their dances. Timpani sound and the dancers, heeding their master's sign, stop and form groups around the marble statues and the smoking tripods.

'I greet you again, dear guest!' the senator says to Lucio, whom he takes by the hand and leads from behind the triclinium.

'And greetings also to you, my senator and host!' the young centurion answers.

'You were chosen by the tribunes to be a centurion. Surely you will carry out honourably the charge entrusted to you. I know that you are worthy to wear this grape branch, the graphic symbol of your title!'

'I realize that I am young to hold this distinguished title, but the tribunes found me worthy by their will.'

'Behold this sword!' continues the senator, taking it from the arms of a slave who offers the weapon. 'More than once on the field of battle it was stained with the enemy's blood. This sword belonged to my ancestors. Fate did not bless me with sons, and I have no one to whom to leave this cherished weapon. I fear that its blade will rust in the scabbard; therefore accept it as a sign of my favour, and may Mars bless you, to perform wonders of courage, to the glory of the fatherland!' says Flac, handing the weapon to the young centurion.

With a feeling of profound reverence Lucio accepts the sword.

'Today we go forth to conquer those who rise against us,' he answers. 'One look at this sword in my hands will remind me, in a foreign land, of you and your hospitable roof! This sword is a symbol of success! I swear either to die in battle or to return victorious . . .'

'I trust in the words of a valiant man, chosen by the tribunes! And now serve us good Falernian wine! Let us drink to the health of centurion Lucio, who is going to the campaign!' exclaims Flac, walking towards the triclinium. Slaves again come forward with vessels of wine and fill the guests' cups.

'I have done my duty! And now, let us delight in the nectar of the vine and return to the celebration,' the senator says.

At his sign a Chalcidian poet enters carrying a lyre. He sings a tender, anacreontic song. Its charming accents flow like waves, pouring from the strings of the lyre.

From the middle of the histrionics a slave girl-*psaltria* glides out slowly, like a swan. She listens to the caressing sounds of the lyre and performs a dance full of languor and passion. Curving her arms, she slowly turns towards the guest of honour, Lucio, with her fluttering veil. The quiet and languorous music, it seems, lulls her; then the *psaltria* beckons Lucio to her embrace, now turning, now pausing in a flirtatious pose, as if in rapture from the love song.

The gaze of all present is turned on the beautiful dancer.

Under the spell of the enchanting song and the *psaltria*'s grace of movements, Lucio rises from his place and recalls the object of his passion—Senator Flac's daughter Amata.

'I love her! Today my destiny must be known!' he resolves.

He is about ask the senator to see Amata, but a group of Andalucian dancers, who have run into the room, prevents him from doing this. He rejoins the guests, sunk in reveries.

The fiery Andalucians, castanets in their hands, perform the wild 'Gaditana' dance. Long, thick, black tresses touch their shoulders; animal skins thrown across their shoulders flutter with each twirling movement. Voluptuousness shows in every bodily movement of these dark-skinned girls, cast into slavery far from their homeland. Merging into one large group with the histrionics and the bacchantes, they whirl in a frenzy, take up picturesque poses, as if to rest, just to rush into a greater frenzy still when their fast, unbridled dance resumes.

Under the spell of wine and the intoxicating, passionate dance, the guests become extremely aroused, and are in a transport of delight. Their eyes ablaze, they clap to the beat of the music, and are about to join in the dance when Senator Flac orders the Andalucians to withdraw.

Of the numerous guests, Lucio alone is gloomy; he sits with his back to the dancers, paying them no attention. Still captivated by the poet's love song, he reflects on his own love.

'You are sad! You take no part in the celebration! What is bothering you?' the senator asks him.

'I am not in the mood for celebrating!' Lucio answers.

'And your cup is still full! You have not touched it! Or do you not like my Chian wine?'

'Hardly! Your wine is truly nectar of the gods! But now is not the time for me to fill my soul with merriment. Hear me, father of the country! My mortal body is here, but my spirit is with your daughter Amata! I love her and want to ask her to become my wife! Permit me to see her!'

'I understand your impatience! I predicted it long ago, and have a surprise for you. I saw to it that our celebration would not lack goddesses come down from heaven to entertain you, my welcome guest! Let these goddesses crown my guests with wreaths of roses, and fill our wine with the scent of rose petals!'

At the senator's sign, musicians enter who play a fanfare on brass instruments. Several negroes come in bearing a canvas secured with lances, with the inscription:

DONA CAERERIS (The Gifts of Ceres)

ATELLANA*

The guests show curiosity and pleasure anticipating the spectacle. Slaves move aside a curtain at the back, and the assembled gaze meets a tableau vivant representing Ceres on a chariot pulled by dragons; her retinue consists of Flora and Zephyr, Pomona with a satyr, and other divinities and flowers.

Ceres is acted by Senator Flac's younger daughter Amata, Pomona by his elder daughter Claudia, and Flora by their girlfriend Lelia.

Preceded by the musicians with reeds and pipes, the procession of goddesses with their retinue comes down to the guests and stops before them in a picturesque grouping, while the goddess of the harvest and fertility, Ceres, quickly takes off her veil. A wreath of ripe golden grasses lightly sets off her head: her *stola* [robe], covered with crimson poppies, flutters gently.

Lucio is struck by Amata's beauty; he cannot take his eyes off the goddess Ceres, who with her sickle indicates Pomona and Flora to reward the revellers with gifts.

* 'Atellana' was the name given to pantomimes performed during Roman feasts. The subjects of these house performances were drawn from mythology, and the principal characters were the gods of Olympus. Persons from the most important and noble families of Rome took part in these spectacles, and did not disdain to act beside histrionics, slaves, dancers, and mimes.

Light as a butterfly, barely touching the floor, Ceres flutters down from her chariot, takes a wreath and throws it to Lucio; he runs after her, but like a deer, Amata evades his arms in a few jumps and then, to the sound of cymbals, rushes to his couch again, bends over him and places a fragrant wreath on the centurion's head.

The goddess of fruit, Pomona, the perpetual object of passion of satyrs and fauns, is being pursued by a satyr who seeks her attentions, but she protests the too-persistent advances of the goat-legged idol. Zephyr, at the instruction of his beloved nymph Flora, dismisses the satyr with a light puff of wind. Meanwhile Pomona-Claudia, fire in her glance, also wants to stir Lucio's heart. She beckons to him tenderly, wraps a garland of flowers around his neck, nestles up to him, walking around him on tiptoe—all awaiting a tender smile. But the centurion looks at her coolly. His only thought is for Ceres-Amata.

After several dances in which each goddess tries to captivate the spectators with her grace and plastique, Ceres—the goddess who gave the Roman people 'golden bread'—her entire retinue holding baskets of flowers in their arms, crowns the guests with roses, pausing in picturesque groupings and sculptured poses.

While the goddesses are showering the senator's guests with roses, a rain of bouquets falls on the triple-*loge* from above.

After the dances, Claudia-Pomona is distressed that her efforts to please Lucio were in vain. Offended by the centurion's indifference, she reproaches her sister.

'Well then? *You* are the chief goddess Ceres, and we—we are your servants, lowly nymphs! Command us! You are all-powerful here! And we are subordinate to you.'

'We are all equals!' answers Amata.

'No! To you alone preference is given! We count for nothing! You are our mistress!' retorts Claudia, annoyed.

'Then let us change from goddesses back into simple mortals, and we will be equal!' answers Amata.

Servants give Amata, Claudia, and Lelia *pallae* [voluminous cloths draped around the body as robes], which they throw over their heads, across their shoulders, draping the pleats flirtatiously. At this moment Lucio appears from behind the triclinium, and walks up to Amata.

'Yes! You are equal to all the spectators except me! Amata, you alone are the chosen one of my heart, to me you will be a goddess forever!'

'And I love you also!' Amata answers, lowering her glance modestly.

'You love me? My happiness is boundless! And so tell me, my lady, do you want to be the mother of my family?' Lucio asks.

Amata, in a sign of agreement, bows her head. Together they go over to the senator and ask his consent to their marriage. Julius Flac takes the young people by the hand and turns to Lucio: 'I give to you my dear daughter, and your union will serve as an eternal joy for me, for you, and for my daughter!'

Then Lucio, as a pledge of love and fidelity, offers Amata an iron ring, which he puts on a finger of her left hand.

'And your wedding,' her father decides, 'according to our forefathers' custom, will take place in a year; meanwhile I hope Lucio will return from his campaign crowned with victor's laurels. And now, hail to their happiness! Let us drink to the health and happiness of the engaged couple! Wine!'

Just as slaves are taking around cups with wine, the sound of military trumpets is heard in the distance.

'My comrades-in-arms await me to set out on the march!' Lucio says, listening to the military music.

Several warriors enter who inform the centurion that his legions have already started the march.

'Soon the tribunes will be standing at the head of their troops! My time has come!'

With this, Lucio takes the sword and bids farewell to his bride, the senator, and the guests. Then he returns to say a last farewell to his beloved; but Senator Flac reminds him of his duty to service, and Lucio retires with the soldiers.

'To the military successes of my future son-in-law! More wine!' cries the senator and orders the histrionics and slaves to resume their dance. They perform a bacchic dance, which, however, is interrupted by the hurried arrival of a slave.

'The high priest wishes to speak to my lord!' he says to the senator.

Julius Frac throws on his toga and reverently goes to meet the high priest, who enters accompanied by several other priests.

'Greetings to Senator Julius Flac' says the high priest.

'Greetings also to you who have entered my house! I beg you to call down the blessing of the gods!' Flac humbly responds.

'But I did not come to celebrate! For that I am not to blame! I came to fulfil a consecrated duty ordered by the gods!'

'Speak, high priest! I await your order.'

'I must speak to you alone!'

'In that case, let us go into the inner chamber! And you, dear guests,' the senator continues, turning to them, 'forgive my brief absence. We shall resume the celebration presently!'

'And we shall withdraw, wishing not to be in the way,' the guests answer. Taking their leave, they exit.

Senator Flac retires with the priests to an inner room; his daughters Amata and Claudia, who remain, stop at a curtain which conceals them from the priests; they listen to what is said, but their feminine curiosity cannot be satisfied because Lucio has returned.

The sisters run to meet him. He takes Amata by the hand.

'I found one last moment to see you!' Lucio says. 'I beg you, give me something to remember you by! Although your image will never be driven from my heart, a talisman from you will defend me from the enemy's lances!'

Amata takes a bracelet from her wrist and gives it to Lucio, saying, 'May it defend you in every battle!'

Trumpets sound.

'It is time! It is time to begin the march! Farewell! I must hurry!' the centurion says, running out.

Amata wants to run after him but Claudia holds her back: 'The street is no place for you, sister! Stay here . . .'

Heedless, Amata tears herself away and runs after Lucio, just to watch her beloved, if even from a distance.

The senator and the priests return. Julius Flac is gloomy; he walks with drooping head; extreme sorrow and despair show in his every movement.

'It is the gods' will!' the high priest says, consoling him. 'You must be happy, not sad, if only that it pleases the gods to receive your daughter as a servant of Vesta! Such an honour is not granted to all, but only to the elect of the great goddess!'

Flac disregards these words of solace, and ponders how to save his daughter Amata from being chosen by the priests.

'She is already betrothed! She is promised as Lucio's wife! How can it be?' After a moment's hesitation the senator resolves: 'I shall hide Amata!'

'Where are your daughters?' asks the high priest.

'I have only one daughter . . . here she is! Take her as a vestal!' answers Flac, pointing to Claudia.

'But you forgot Amata!' Claudia whispers to her father.

'Be silent, unfortunate child! Your father knows what he is doing and saying!'

Claudia walks towards the high priest, who takes her by the hand. 'You are most worthy to be a vestal. I pray to the gods that the happy choice will fall on you!' says the priest. 'And now bring in the urn and wreaths. Summon the girls of Rome and let Vesta herself decide who is worthiest! Assemble the guests as well! Let all be witness to the great deed which must be accomplished here!'

Amata returns at this point. Still sad from parting with her fiancé, she walks slowly and sorrowfully towards the peristyle, as if unaware of what is happening in her father's house.

Seeing Amata, Julius Flac signals her to leave quickly; but Claudia by stealth points out her sister to the high priest as she was about to leave. The high priest goes over to Amata, takes her firmly by the hand, and walks up to her father.

'And who is this?' he asks.

The unhappy father hesitates.

'You wanted to hide your second daughter from me! The gods will punish you for this. Look at her! She is young, beautiful; she is fully deserving to be a servant of the goddess of chastity and purity!' the high priest says.

Lictors now enter, with priests, priestesses, young Roman women, and the senator's guests. At the sight of the priests bringing in the urn, Amata realizes what is being discussed.

'But . . . but . . . I cannot be a vestal! I am engaged!' she says to the high priest, pointing at the ring her fiancé gave her.

'Yes! But you still have not worn the red *flammeum* [bridal veil], and a priest has yet to break consecrated bread between you and your bridegroom . . . Engaged, but not married! For that reason you too must approach your destiny at the urn. The goddess Fortuna will decide your fate!' says the inexorable high priest.

Priests and augurs put down the urn, containing wreaths of scarlet verbena and one wreath of white—the flower dedicated to the goddess Vesta. The high priest steps up to the urn and announces triumphantly: 'Citizens of Rome! Let it be known to you that one of the servants in the temple of Vesta has completed thirty years of service since the day she was consecrated as a vestal. By law she is now released from the vows she gave and her place must be taken by another. By virtue of the rights granted to me, I chose from the most prominent families of Rome pure and chaste Roman women, for whom the great honour is in prospect to approach the urn and to draw out their fate. One among you, who by the gods' will picks the white wreath, will be considered worthiest, chosen by Jupiter himself to be a vestal. Noblewomen of Rome, approach!'

The priests cover the urn with their togas, and the women walk up to it. The first draws a scarlet wreath, and walks to the side with unconcealed joy. Claudia is second to approach; she also takes a scarlet wreath; feeling triumphant and delighted she runs off, joyful that she had avoided the fate of becoming a vestal. Meanwhile Amata turns to the marble statue of Cupid which stands at some distance from her, and, kneeling prays to it to protect her.

'It is you I must serve . . . not Vesta. Save me . . . protect me!'

Claudia takes her sister firmly by the hand and goes up to the urn.

'Your turn!' says the high priest.

Trembling, Amata lets her hand fall into the urn, and to her horror draws the wreath of white verbena. She gazes around, not knowing what to do, stops in an immobile pose, unable to comprehend her situation. Her gaze is fixed at one point; she is frozen in expectation; her arms fall; a feeling of despair is expressed in her face. Finally, she comes to her senses:

'No . . . it is a dream! . . . It cannot be! Lucio! . . . Lucio! . . . come to me, come to my assistance! You can save me. We swore eternal love to one another! . . .' She rushes to her father's embrace: 'Father! . . . Save me! . . .'

'I am powerless,' answers Julius Flac, lowering his head sadly.

Running from side to side as if mad, Amata turns to the young Roman women, then to her sister, begging them to save her from her fate.

'But I love him! . . . I am engaged . . . I cannot, I must not be a vestal! Sister! . . . go in my place! . . . save me!'

Her pleas are in vain. The high priest takes her by the hand and proclaims:

'The will of the gods has been done! To you, the wreath of innocence! You should not be sad, but happy to be chosen for this great honour! You are now one of six in all Rome who has the right to pardon unfortunate criminals . . . Finally, to you will be rendered all but godlike honours! . . . Thus your destiny will come to pass!'

He makes a sign to the priests, who bring forward a long purple mantle. The high priest throws it on to Amata's shoulders. Hearing the holy person's words, she finally resigns herself.

'Let the gods' will be done. Take me!'

'No, let your father give you from his hand into ours. This will signify that from this moment you are free of family ties!' decides the high priest.

The priests' servants bring in a magnificent sedan chair, on which Amata is seated; then, preceded by lictors and accompanied by priests, the newly chosen vestal is triumphantly carried out.

'To the temple of Vesta!' the high priest commands.

The procession departs, and guests surround senator Julius Flac, trying to console him. He is gloomy, and looking at Amata being carried away he says: 'I have lost a daughter and a son as well, the valiant Lucio! . . . My happiest dreams are shattered! . . .'

'But she will visit you! A vestal has entry everywhere, all the more so in her family's home! Remember, friend, that she is now the equal of the gods! Few achieve this honour. We must celebrate and be thankful that Jupiter's hand chose your house to be the most worthy! Forget your sorrow; transform it into happiness!' say his guests in consolation.

'Yes . . . true! . . . Let us forget everything! . . . and let us drown momentary sadness in wine! . . . Bring wine! . . . And you, dancers! Come forward! . . .' the senator commands.

The guests move back to the couches they occupied before; they take cups of wine and continue the feast. Slaves light the censers. Fragrant smoke streams over the table; multi-coloured shadows of blue, red, and yellow flames hover over groups of guests. Timpani sound, and a wild crowd of bacchants and bacchantes leap out before the guests with affected gestures and frenzied movements. To the sound of tambourines, sistrums, lyre, and little bells they whirl, jump, and laugh, performing the *cordax*, the favourite dance of prominent Romans in their orgies.

A bacchanale of mindless abandon.

Curtain

ACT II

A consecrated forest at the foot of the Palatine hills. On the right, the dome-shaped temple of Vesta, with a marble statue of this goddess.

It is quiet in the sacred forest. Pale rays of moonlight penetrate the dense, quivering leaves of the plane trees, myrtle, and laurels. The eternal flame burns at the altar of Vesta.

In measured steps, like a sentry, a vestal walks around the sacrificial altar, guarding and maintaining the sacred fire. She performs this duty reverently, adding fragrant oil when the flame begins to flicker, taking care during her watch to ensure that this gift sent down to earth by Phoebus not be extinguished.

Flutes and trumpets are heard far off in the mysterious silence. A procession is approaching, of lictors, the high priest, before whom torches and a laurel branch are carried, augurs and people, bearing the newly chosen vestal Amata. Claudia is also in the procession; she has come to the temple to bid her sister, who has left the family home forever, a last farewell.

The high priest approaches a sacred oak from which hangs a large copper shield. He strikes the resonant copper with a hammer. At this signal the chief vestal comes out of the temple with three lesser priestesses of the goddess of purity and chastity. Their tunics, cloaks, and headwear are perfect in their snowy whiteness. With bowed head they go to meet the visitors.

Amata descends from her sedan chair. The high priest leads her by the hand to the chief vestal. 'Accept into your midst this sixth sister, chosen by the gods for Vesta's service!' he says, giving her the wreath of white verbena.

'We greet thee, chosen the most worthy of the Romans!' the other women answer, bowing.

The high priest leads Amata in a circle around the sacrificial altar with its fire, and says to her: 'Now you must swear that for thirty years you will keep your vow of chastity! During this time you are granted the right either to be free, or to live out your time in the atrium of Vesta. Be as pure as the fire which you must preserve!'

Amata bows before the altar of the goddess and swears to serve Vesta, preserving her purity and innocence.

The high priest takes the scissors handed him by the other vestals and cuts a lock of Amata's hair. He hangs this lock in the branches of the sacred oak; then he pronounces the formula established for the consecration of a vestal:

'By virtue of the powers granted to me, in the name of the Roman people and the Quirites, I was authorized to choose a priestess and vestal for the performance of sacred rituals. Pleasing and acceptable to the gods, you I choose to be a vestal, granting to you all rights enjoyed by the servants of the goddess Vesta.'

Having said this, the high priest leads Amata by the hand to the chief vestal; as he does, he notices the initiate's ring.

'What is this?' he asks.

'The ring given me by my fiancé!' answers Amata.

'You are now committed to chastity and must put aside all thoughts of marriage. If ever you break your vow, the gods will punish you, and you face the cruel punishment of being buried alive! Remember your vow! May the chief vestal initiate you into all secrets of the religious rite of Vesta, and may you be received from my hands, beloved priestess!'

Saying this the high priest takes the ring from her finger and throws it to the ground. Claudia, unnoticed by the others, picks up the ring.

The high priest leads Amata to the chief vestal, then withdraws with his retinue. The chief vestal hands Amata the sacred flame and commands: 'Hear me, new confidante of the goddess Vesta! You must now come into your own and perform your duty! Take this lamp from my hands and let not the breath of Morpheus cloud your thinking. Stay awake, and remember to add fragrant oil to the sacrificial altar, that the sacred fire not be extinguished. May the heavenly flame shine forever!'

With this, the vestals retire to the temple.

Claudia goes up to her sister: 'Farewell, farewell, sister! You belong to the world no longer! . . . I am sure you will forget Lucio as well! . . .'

'Lucio? . . . I shall never forget him!' answers Amata.

'But he cannot be your husband? . . .'

'But I cannot dictate to my heart! . . .'

'You must forget him! Remember your vow to the high priest! . . .'

Claudia bids farewell to her sister and leaves triumphantly, certain that Lucio will now be hers.

Amata remains on watch, alone. A mysterious, sacred tremor comes over her. She looks to either side, then around her—all is quiet: but from time to time the flame on the altar flares up. The moon casts its languid, silver rays down on the marble statue of Vesta! . . .

'Everything is finished!' Amata thinks. 'Finished! Now I must forget him, forget Lucio! . . . Forget? . . . No, this is more than I can do! . . . His image stands before me as if alive . . . It will always pursue me . . . I loved him and still do! . . .' She turns to the motionless statue of Vesta.

The goddess's features seem severe to her. She covers her face with her hands, weeps, and, kneeling, falls at the threshold of the temple.

'No! . . . Have mercy, divine one . . . It is you I must serve! . . . Be gone, bright dreams of happiness! . . . Away all thoughts of love's bliss! Forgive me, chaste goddess! . . . You alone I have vowed to serve! . . . and I will meet my obligation! . . .'

She begins her rounds of the sacred altar, adds oil, prays, but her beloved's image will not go away. 'He is always before me!' Amata thinks, 'The image of my dearest will not abandon me! . . . I thirsted for Hymen to join my fate with Lucio's radiant beauty. Yes! . . . I thirsted for this . . . but happiness was only a pale spectre, momentary, and I was told, "You must be happy nevertheless, you must be equal to the gods, and to do this you must refuse love!" For what, such heavy sacrifices? Does not happiness reside in love alone? . . . I do not want these sacrifices! I do not want to pass the spring-time of my life in sorrow and tears . . .'

She looks at the marble statue of Vesta again. It seems to threaten her with a gesture of reproach. In fright Amata runs to the sacred altar again, and trembling, pours oil on the fire, and kneels, praying to Vesta for forgiveness. She prays as she circles around the sacred fire. She is extremely agitated. In a fit of religious ecstasy, fatigued, she finally drops in exhaustion next to the sacred oak, and falls asleep in the protection of its branches. As in a kaleidoscope, there appears to her

REVERIES AND VISIONS.

The temple of Vesta is plunged into darkness, and the fire on the sacrificial altar is slowly going out. From beyond the cloudy expanses flies Cupid, guided by Folly. They look around, and seeing the altar fire going out, begin to sport.

'Look, at our appearance even Vesta drew a cover over herself . . . Now our reign begins! . . . Where and to whom to direct your arrows?' asks Folly, rattling the bells which adorn his clothes and his Phrygian cap.

'The hour is nigh when the spirits of this sacred forest come out of their daytime confinement! Let us play a trick or two on them!' Cupid answers. Among the branches and the tree-trunks forest maidens appear in groups, nymphs of the forests—dryads with flowing hair, and feet which resemble the roots of trees. These protectresses of the oaks reach out towards Cupid.

'No! They are incorporeal spirits! They are blessed with immortality!' Folly observes. 'That is no place for Cupid! . . . Let us find some mortal and make sport of her! . . . That is not difficult for us!'

The forest nymphs hide in the thick foilage. Continuing his search, Folly gambols through the forest and finally notices Amata lying behind a tree. 'Over here! over here! . . .' Folly calls to Cupid. 'An arrow from your quiver! . . . Strike! . . . Look! Yes, it's a vestal, committed to chastity! . . . Ha ha ha! . . . All the better for us! Someone this fair should serve Venus, not Vesta! . . . Let us drive her from this place! . . . Let us have some fun! Behold—our victim!' Folly laughs, rattling his bells.

'Let us entertain our mistress Venus! Let us call her to our aid!' says Cupid.

Folly looks to the heavens and turns towards the brightly burning red star: 'Beauty of the universe! . . . Come down to us, to our aid!'

The star rolls down from the horizon into a group of flowers, from which fly two white doves, heralds of Venus. They coo tenderly, preen themselves, and straighten each other's feathers. Then a grotto appears, in the middle of which, outstretched on a chariot in the shape of a seashell, basks the goddess of beauty—Venus. Ash-blonde locks fall like waves over her snow-white shoulders. A girdle emblazoned with coloured stones encloses the goddess's lissom waist. The beauty's blue eyes flash beneath her brow and jet-black eyelashes. She is surrounded by nymphs holding fans—the nymphs of Hope, Sighs, Desires, and Love.

Venus makes a gesture with her transparent cloak, and dark nature in the sacred forest is quickly revived. Myrtle, roses, and other flowers consecrated to Venus spring up. The branches of the trees move apart, and cupids are seen in picturesque groupings playing on lyres, harps, and panpipes. Slowly the goddess's companions come down from the branches, rise up from the ground, and come out from behind tree-trunks. Here, swinging on the branches, a swarm of Smiles may be seen, there—Games which attract golden-winged butterflies; here, Graces who lovingly embrace each other encircled with a garland of roses, there—Hymen, Cupid's brother, with a torch in his hand and a fiery red veil on his head; here, the nymphs of Delight and Confession, next to Jealousy, Vengeance, and others.

The nymphs are all grouped before their sovereign, awaiting further instructions. The mischievous little Cupid runs up to Venus, his mother. 'We called you to present to you our latest victim!' Venus lovingly pats her dear son on his red cheeks.

'Another of your pranks! . . . Very well! Show us this victim! . . . All of us, with our joint powers, order her to bow before the power of love! . . .'

Cupid points to the sleeping Amata. 'There she is! And this is no simple mortal, but a chaste vestal! . . . In her heart even to this moment burns a passion for her beloved of an earlier time!'

'So then! . . . Let us warm this passion with the bright flames of love! . . . Bring her over here! . . .'

In a few jumps Cupid, chased by Folly, is next to Vesta's sleeping priestess. They carry Amata over to Venus; the priestess appears before the goddess in the full brilliance of her beauty and youth. Only her cloak shows Amata to be dedicated to the cult of Vesta.

Cupid, encouraged by Folly, whispers into the vestal's ears, imparting thoughts about the charms of love and recollections of the handsome centurion Lucio.

Troubled, Amata looks around among the gods who have come to earth from a world unknown to her: she runs among the nymphs, who try to twirl their new victim around; but the vestal, guided by Cupid's tender whispers, runs quickly to Venus' brilliant grotto, and, kneeling, turns to the goddess of beauty:

'Hear me, blessed goddess! Behold this veil! . . . I am dedicated to the service of Vesta's pure fires! . . . But I was not intended for this—I thirst for love! Cast over me your cloak, woven by the gods! I beg you, save me! As your slave, I bow before thee!'

'I know that love is an ornament to everything in the world!' answers Venus, playing coquettishly with her unloosed hair, 'If you will serve me steadfastly, I shall enliven your gaze with the bright light of life and love! Your beauty is a pledge of bliss! Come to me, you nymphs Expectation and Hope! Inspire this mortal with your lifegiving breath.'

Hope and Expectation come forward from a group of nymphs. They gesture to Amata with their veils; dancing, they form a grouping with her as if to whisper to her the words, 'Wait and be hopeful!' Then they lead her to the grotto.

'Excellent!' continues Venus. 'But . . . first the melancholy and gloom which obscured your mind and heart must be driven out! My realm is one of delight and laughter! . . . Come forward, dear Graces, with your constant companions Smiles and Games. Amuse our melancholy Amata! And you [Amata], take pleasure in their sport, and banish the furrows of sadness from your face: they spoil your feminine beauty.'

Gelasius, the god of laughter, enters.

'You asked for me, my queen? I hastened to appear at your call,' says the 'delight of life'.

'Yes, my ever-young friend! Rejoice, circle round Amata! . . . Drive away her melancholy! Let all her sorrows disappear! . . . Let everything around her exult and flourish!'

Led by Gelasius, the eternally young Smiles and Games, preceded by multi-coloured butterflies, begin their entertainment. Weaving back and forth with the animated flowers and the cupids, in the waft of light zephyrs, they circle around in merry pairs, dancing and trying to amuse the sad Amata. Then in harmonious order pass the Delights, bowing before Venus. They whirl in a tightly-knit round dance; then, separated again, they flit from myrtle to rose, as if wishing to waft towards the vestal the flowers' aromas, which becloud her wits and caress her heart: they pause in various groupings, beckoning her into their midst.

Gelasius, the god of laughter, tries to divert Amata by his amusing jumps: he forces her to imitate his movements, and gives her a lesson in dance. Amata follows his designs reluctantly; none of these diversions entertains her. She remains sorrowful.

'Ah! I know the cause of your sadness!' Venus says to her. 'And I understand it: when the heart suffers, it is indifferent to all laughter and merriment.'

'You have guessed my mind, divine one!' says Amata. If you want to console my poor heart, let me take pleasure in the image of my beloved! . . . I know he is far away; but the world and the power of love are your servants; you can call forth his image.'

'I shall grant your wish with pleasure.'

And Venus calls the nymph Echo. Accompanied by lovely Narcissus and doomed to silence, Echo—the personification of unrequited love—runs to the goddess of beauty, ready to obey her command.

'You, divinity who answers every call, fulfil Amata's desire! Behold this lovestruck Roman girl! She does not know if her feelings are reciprocated. Let your echo serve as an echo in the depth of Amata's heart! . . . Reflect her beloved's features as in a mirror!' Venus commands.

The nymph Echo hastens to fulfil her command.

She runs to a small rocky hillock and blows on a little golden horn. In the depths of the temple of Vesta, as if a tableau vivant, Lucio's figure is seen in military armour; in one hand a sword; in the other he holds the bracelet, Amata's gift, at which he gazes lovingly. For a moment his reflection, like Echo's echo, is seen on the cliff where the nymph stands.

Amata rushes towards her beloved.

'Let me embrace you, my dearest!' she whispers. 'Let me clasp you to my bosom. . . .'

But ... Amata's outstretched arms embrace only the air: as quickly as Lucio appeared, his vision disappears.

'Now, languish in sadness no more! You have confirmed that he loves you, as before, even in the heat of battle! You saw how he looked at your bracelet, wishing to revive your charming features in his mind! ...'

'Yes, goddess, now I am happy! I know that he has not forgotten me!' Amata announces.

'I shall complete your bliss! ... I am prepared to unite your loving hearts forever: but first I must know that you deserve to be my servant! ... I shall grant you all the joy and bliss which love confers; I shall also show you the torment and hellish pangs which love bestows. Behold my miraculous waistband! It was wrapped around me by the daughters of Themis and Jupiter, and was given the power to captivate not only people, but even gods! On one side are Modesty, Pleasures, and Fidelity, which lead to delight; on the other, Jealousy, Betrayal, and Vengeance. Let these passions create trials for you, and if you withstand temptation until the end, Hymen will clothe you and Lucio in chains of fragrant roses, and I will grant you happiness! ... Cupid, my playful son, be off ... to your duties! ...'

Obeying Venus' order, Cupid and Folly circle around Amata; Folly supervises Venus' son, laughing merrily and shaking his rattle above the vestal's head; Cupid whispers a passionate song to her about the blisses of love.

At their call appear the charms begot by love: Modesty, Fidelity, and Delight, accompanied by the personifications of flowers dedicated to Cupid. Acacias, which express pure, platonic love, make up the retinue of Modesty, carnations—emblems of passionate love—that of Fidelity, and linden flowers, the expression of conjugal love, in memory of the faithful Baucis, who was transformed into a linden tree. Myrtle and honeysuckle accompany Pleasure, as the bonds which love unites.

On the other hand, Jealousy and Suspicion, Betrayal and Sighs, Vengeance—Nemesis—and Remorse, all make an appearance. Hymen is among them.

All these fruits of human passion bow before Cupid, who dons a quiver with arrows, some of gold, others of lead. Hymen explains to Amata that people struck with Cupid's golden arrows love passionately, and are blessings to each other; and those struck by the lead arrows suffer and are harassed by suspicions.

Cupid controls all the arrows, directing at Amata's heart now a gold, now a lead arrow. As this happens, now Modesty, now Jealousy, now Fidelity or Vengeance—each takes possession of Amata, showing her on one hand the delights of love, on the other its torments and sufferings.

Amata is by turns merry and happy, or, suffering Vengeance or Betrayal, wishes in her wrath to strike an invisible but imagined object of her love, deeply imprinted on her heart. Now Pleasure attracts her with a burst of passion, nestling up to her and encircling a garland of flowers around her body; now Jealousy, directed by Cupid, places a serpent on Amata's breast which gnaws at her heart, showing her the Vengeance which appears before her, and handing her a dagger as a means of satisfying her passion.

Amata is receptive to all human passions.

Finally, after a prolonged battle of passions, Amata turns to Cupid with a prayer: 'You see! I am utterly deserving to be a priestess of your mother, the goddess of love! . . . I beg you, do not torment my sore heart any further! Let go your gold arrow and awaken me to new life! . . .'

Folly laughs uncontrollably. 'Cupid! . . . Not for nothing am I your supposed leader and supervisor! We won . . . and over whom? A virginal servant of Vesta! Ha ha ha! . . .'

'True!' Cupid answers. 'Now, Amata, your heart is quite scorched by the flames of love! Let its pure breath now calm the tempest in your heart.' Saying this, Cupid draws his pliant bow and sends from its string a gold arrow directly into Amata's heart. Joy illumines the vestal's face; passion burns in her eyes; with a smile she breathes deeply, as if anticipating the delight of an imminent meeting . . . caress . . . and kiss.

'To him! To him!' she whispers.

At this moment Lucio appears on a rise; Hymen takes Amata by the hand and leads her towards her beloved. The vestal is at the point of rushing into his embrace when the anger of the insulted goddess Vesta is heard. A thunderclap rings out, which makes Amata tremble. Frightened, she stops, unable to move.

The entire sacred forest is overcome by a dense gloom, then wrapped in flames as if all nature had caught fire. Dryads illuminated by the fire are visible in the trees. The temple of Vesta takes on a bloody aspect and is highlighted in flaming lines. Will-o'-the-wisps mingle with various fiery spectres.

Venus and her retinue are seized with horror and fright; they rapidly disperse, flying off in all directions, leaving Amata alone. Standing in the temple, the marble statue of fire and chastity, Vesta, radiating a crimson light, slowly steps down from her pedestal. With a calm, measured tread she walks to Amata, who is standing motionless in sacred fright, and touches her with her staff.

'Unfortunate creature!' Vesta says, 'You wished to break your vow of chastity and have already readied for yourself the torments of hell! . . . This time I saved you only because Venus defiled my sacred forest by her presence in it. Collect yourself, and remember your vows! . . . Your place is in my temple, not in the embraces of love!'

Under Vesta's powerful influence Amata follows the goddess as she takes her place again in the temple. The fiery visions and spectres disappear; everything returns to its prior state.

Thus the vestal's reveries come to an end.

The sun rises peacefully, illuminating Rome in the distance. The sound of approaching trumpets is heard. Amata awakens. Suffering the effects of her strange dreams, she jumps up, unable to account for what happened. She looks timidly to the sides, from one point to the next, then at the silent statue of Vesta and the flames of the sacrificial altar; she runs up to it, and with trembling hands pours oil from the lamp on to the fire.

'Yes! . . . Fate has been kind to me!' she says at last, sighing. 'It was a dream . . . and what a frightening dream!'

In the distance lictors appear with the people, coming to present a sacrifice to Vesta.

Curtain

ACT III

SCENE I

Claudia's room in Senator Flac's house

The senator's elder daughter Claudia, having finished her toilette, reposes in thought on a soft couch. She commands her slaves to withdraw, and attentively watches a divination with mirrors (catoptromancy) which is taking place in her presence. A boy sits at a table in front of a mirror, his eyes doggedly focused on its brilliant surface. Near him stands an augur, who sees that the boy does not take his eyes from the mirror.

'Do you not begin to discern the future of Senator Flac's noble daughter?' the augur asks. The boy shakes his head. He is not yet weary enough for visions to enliven his gaze in the mirror's dark perspective.

Claudia rises from her place and walks over to the fortune tellers. 'Fate does not want to reveal its commands to me!' she says. 'Let it tell me where Lucio is, and how soon he will return.' Standing behind the boy, she fixes her gaze on the mirror, awaiting an answer.

At this moment the curtain by the entrance door rises to reveal the centurion Lucio. At the sight of the fortune tellers, frozen in expectation, he stops in indecision. His image is reflected in the mirror, which faces the entrance. Claudia, not realizing that Lucio is actually there, sees only his reflection in the mirror. Frightened, she steps away from the augur, covers her face with her hands and turning around, utterly unforewarned, comes face to face with the centurion.

'No! . . . It is not a spectre!' she says. 'It is actually Lucio, alive, who stands before me!'

'Yes! It is I, Lucio, after long marches returned to my homeland and to my betrothed!'

Claudia takes a purse from the table, throws it to the augur, who retires with the boy at her signal.

'I hurried back to see your sister, my dear Amata!' says Lucio turning to Claudia.

'You are destined not to see her!' Claudia answers.

'And why? She is my bride!'

'There is no point in loving her any more! . . . What will you do with your love for her? Unrequited loved is only torture!'

'It is not true! She loves me!'

'Perhaps! But she will never be your wife!'

'I do not believe you!'

'Believe it, I tell the truth! Forget her forever! . . . There is another Roman girl who loves you passionately . . . who will give herself to you unconditionally . . . and share with you both joy and sorrow . . . A word from you and your blissful moments will begin!'

'Tell me: who is this poor girl?' asks Lucio, mocking her.

Wishing to entice him with her beauty, Claudia pauses in front of the centurion in a picturesque pose, throwing back the folds of her clothes. She looks him straight in the eye and says, with directness: 'That poor girl is standing before you! . . . I have long nourished an insatiable passion for you. I have long been tormented, long suffered . . . A mysterious power draws me to you irresistibly . . . Destiny has taken pity on me . . . My sister can never be your wife . . . I will take her place . . . I will give you happiness, joy . . . love . . . I will be your friend, your slave . . . only love me in return!'

She looks at him pleadingly.

Lucio listens to her with an impatient air. 'Never . . . never . . . it cannot be! . . .'

Claudia, reaching out, rushes to him, but Lucio rebuffs her.

'Away! . . . You thought to entice me but miscalculated . . . My heart will forever belong to Amata! . . . Tell me where she is!'

'Very well then . . . but know that she is betrothed . . .'

'That is false . . . you lie!'

'I do not lie; I am telling the truth. She is dedicated . . . to Vesta, the goddess of chastity! You dare not love a vestal!'

'It is not true! It is a conspiracy! Where is your father?—I must go to him!'

Lucio makes quickly for the door, but Claudia stops by the curtain and bars his way. 'I will not let you go until you have decided!' she says, turning to Lucio again.

'What do you want from me? Speak!'

'When you are convinced that I spoke truly in telling you that Amata is a vestal, will you be ready to share a place in your heart? See for yourself—here is your ring, which is now mine. Never doubt... that I will be able to love you more passionately than my sister... I will worship you forever...'

Lucio silences her passionate confession: 'Your prayers are in vain... never!... never!...' He pushes her away from the door just as Senator Flac is coming in. Claudia runs out in despair.

'Greetings, brave warrior Lucio!' the senator says, sadly.

'And greetings to you, Senator!' answers the centurion, hardly able to restrain his emotion. 'But—is there sadness on your face? I was not expecting such a greeting from the father of my bride!...'

'Before you stands an unhappy but innocent father!'

'Where is your daughter Amata? Why did she not meet her fiancé?'

The senator remains silent, his head lowered. 'You are silent?... Then tell me the truth!... Speak... Where is Amata?'

'She is praying for you in Vesta's temple! She is a vestal!'

'I do not believe what I am hearing!'

'It is the bitter truth! It was fate's pleasure. The choice fell upon her... no request, no prayer of mine—nothing helped. At the gods' command the priests took her from my house!'

Agitated, Lucio at first controls himself, but soon can no longer contain his emotions, and reproaches the senator: 'You not to blame?!... Then who is at fault for my woe? Who took my happiness from me? For an entire year I was away... I fought... and struck down my enemies, never forgetting Amata even in the heat of battle! I lived only for her; her image never left me for a moment. Here is the wreath which I received as an award for my valour!... The wreath I wanted to place at her feet... And now,... What is it to me? There is no joy for me in life!... I need nothing... without her there is no glory... no honour... no interests. Take this prize... See how little I value it.'

In despair Lucio tears off the wreath and tramples on it.

'What are you doing?' the senator asks. 'I realize you are unhappy; but you are young. You can still find some consolation in life. Any Roman woman would be honoured to be your wife!'

Lucio, reproachful, interrupts him: 'How can you say that? Amata's better does not exist in all Rome!... Why did I return? Why did I not die on the battlefield?... Death... yes, death!... And I shall find it. There is no love . . . no life! . . . Farewell, love . . . and life . . . To me only one fate remains—death!'

He draws the sword from its scabbard and is about to plunge it into his breast when Amata runs in, runs to Lucio and tears the weapon from his hands. 'This ancient sword,' she says, 'was given to you by my father to battle the enemies of the fatherland, and you want to spill your own blood . . . Shame, Lucio! . . .'

'Amata? . . . here . . . and wearing a vestal's veil? In an instant you have saved me . . . you have resurrected me for love, for happiness . . . Is it not so? . . . Speak! . . .' He goes to her and tries to embrace her, but Amata steps back.

'Do not come close! . . . Do not defile my sacred garment with your daring hands! Have you forgotten? . . . I—am a vestal! . . .'

'But even vestals have hearts within their breasts! . . . Or is yours silent? No, I don't believe that! . . .'

'You are right . . . I love you as before! . . .'

'You love me? . . . Then you will be mine! . . .'

He tries to embrace her again, but Amata stamps her foot and with an imperious gesture commands him to withdraw.

'Leave? . . . You are driving me away! . . .' he says; 'But I ask . . . I beg you . . . cast off the robes of a chaste priestess—and be mine! . . . Behold . . . how I thirst to press you to my heart . . . One caress from you will heal my aching heart . . . I beg you on my knees . . .'

He falls at her feet. Amata raises him up.

'But . . . dearest, this cannot be! . . . As long as I serve as a priestess of Vesta! . . . no one dares touch me . . .'

'You say "dear", which makes everything possible! Let us flee . . . in an hour we will be in the open country . . . far from Rome and from our judges. Let us flee, and in each other's embrace we will know the bliss and the happiness of love.'

'Flee! . . . Where? . . . Why? . . .'

Amata is undecided for a time; undecided for a moment, she tells him:

'Never, never! . . . If I flee . . . all Rome would be aroused . . . and would rise up against my father's house . . . To flee would dishonour my family line! Romans would never forgive or forget such an insult to sacred Vesta! . . . No! . . . What you propose is awful! . . .'

'Ah! . . . then you never loved me! Your passionate avowals were only false promises! . . . Your vows—empty sound!'

'You reproach me in vain! . . . I am not at fault! Fate decided matters this way . . . it condemned me to fulfil this awesome duty.'

'If you love me, let us flee! . . .'

'No!'

'But you will be mine! . . . If not of your own free will, then I will forcibly tear away the vestal's veil!'

Lucio rushes towards her; he is about to tear away her veil when Amata falls to her knees and begs him: 'Have mercy on me! . . . Mercy! I do not fear death! I am prepared to flee with you anywhere . . . for one minute of bliss let them bury me alive . . . Punishment I do not fear . . . but shame . . . I fear the shame.'

'Then off with this veil of chastity! I will not spare it, either here or at the altar of the goddess . . .'

He seizes the veil, but Senator Flac restrains him, stands between him and his daughter, and offers a sharp rebuke: 'Fool! What are you doing? In a fit of impertinence and mindless passion you forgot that Amata is a priestess of Vesta . . . You wanted to insult a sacred object . . . Leave this place! . . .'

Trumpets sound in the distance. Lucio listens.

'You and the entire world are conspiring against me! . . . Listen! . . . I hear the sounds! . . . Of gladiators entering the circus to do battle! . . . My soul and heart were killed here, and there, in the circus, I shall end my sufferings . . . Behold my end! . . . To the arena! . . . To the arena! . . . There I can find death!'

Amata holds him by the arm, begging him to abandon his intention, but Lucio is adamant. He tears himself from her grasp and runs out.

Amata takes off her veil and puts on the mantle Claudia left behind. It has occurred to her to save her beloved at the circus. Claudia enters. 'To the circus! Follow him! . . . Follow him!' Resolved, both sisters quickly exit.

SCENE 2

The Coliseum in Rome

The Coliseum is overflowing with people impatiently awaiting the emperor's arrival. A solemn procession in honour of the gods begins beneath the triumphal arch. The following take part in the celebration: first the children of noble Romans, after them riders, chariots, athletes, warriors; after them come dancing, grimacing slave-satyrs covered with animal skins, preceded by a choir of satyr-musicians with flutes and kitharas; next, servants of various temples preceded by lictors, all carrying vessels of different kinds, augurs' rods, priests' caps, and sacred paraphernalia. They carry statues of gods on sedan chairs: Jupiter with lightning in his hands and an eagle at his feet, Mars with *Salii* priests, Bacchus surrounded by bacchants and bacchantes, maenads, and others.

The emperor with his retinue, priests, distinguished guests and vestals all ascend to the podium. They take their places, and the emperor signals to begin the games, dances, and battles.

Gladiators of different names—*mirmilles*, netters [*retiariusii*], lassoers, *secutores*, *andabatae*, and others—march around the stadium amidst joyful cries of the crowd and the sounds of loud music. With lowered weapons, they bow before the emperor's throne.

'Hail to the emperor! Those condemned to death greet thee! The dead greet thee!' echo other groups of gladiators.

The procession ends. Trumpets sound—a signal for the entertainments to begin.

Twelve priests of Mars—*Salii*—enter, led by their high priest, the *Praesul*. Copper helmets, *galeri*, glisten on their heads; metal cuirasses gird their torsos; a purple mantle sewn with gold flutters around their shoulders; they are armed with swords in one hand and shields in the other. Behind them their attendants march in a procession—twelve servants of Mars, young Roman women dressed in military armour. The *Salii* and the priestesses of Mars jump and leap, waving their swords. Various formations mark the performance of their sacred military dances.

After the dances of the *Salii*, two *andabota* gladiators enter, to the accompaniment of trumpets. As their faces are covered [they are wearing helmets with visors down], they cannot see each other. With a javelin in each hand they pursue each other, trying to inflict blows. Their movements are extremely comical, like playing Blind Man's Buff, though this is a game of grim consequences. One of the combatants finally hits the other's head so hard that his helmet and visor fly on to the arena floor, and the *andabota* lies prostrate, bare-headed, begging for mercy. The victor removes his helmet, and pointing a javelin at his opponent's chest, turns to the emperor and asks: 'Life or death to the vanquished?' The vestals, sitting in the box, and behind them the people, raise their right thumbs in a sign of mercy.

The emperor rises from his place and throws two wooden staffs into the arena: 'You both were brave and agile!' he says, 'you both fought with utter contempt for death! For this I grant you life and freedom! Slaves, you are both free!'

Hearing this kindness, the gladiators raise their staffs and bow to the emperor and the people; then, embracing each other joyously, they leave the arena.

The trumpets' invocation sounds again, and into the stadium run two warrior-women, from Thrace and Macedonia. The heads of these maenads, with tousled hair, are adorned with wreaths made of interwoven serpents; they hold knives and live snakes. They perform a wild dance; their movements show the unbridled heat of passion, reckless valour, intoxication, and a thirst for vengeance. They wave knives at each other, try to attack each other with their snakes, fall in a fit of languor, as if frozen, then jump up again, whirl in a frenzy and finally, brandishing their weapons and their snakes, run quickly out of the circus.

After this unruly dance, a magic grove representing Mount Parnassus, with the source of the Inachus in the middle, rises slowly from beneath the arena. Apollo is standing on Parnassus, in the full splendour of his beauty and magnificence, with a lyre in his hands. He invites the muses, of whom he is patron, to the festival. All nine muses gradually descend from Parnassus, bowing to their patron. Each muse appears with her attributes.

Clio—the muse of history—is crowned with a laurel wreath, and holds a trumpet, for the glorification of exploits, and a book on which a line from Virgil is inscribed: *Clio gesta canens transacti tempora reddit* [Clio, singing of past deeds, recounts the times].

Euterpe—music. The personification of joy and merriment, she is very elegantly dressed, with a wreath of flowers on her head and a flute in her hands.

Thalia—the muse of comedy and epigram; she wears a wreath of ivy; she holds a mask and a pen; a little monkey accompanies her, the symbol of imitation.

Melpomene—the muse of tragedy; she is draped in a heroic costume, and holds a dagger in one hand, a sceptre in the other; she wears buskins; a frightening mask is hanging at her side. Two companions accompany her: Terror, wearing a lionskin and holding a horn and a shield with the likeness of the Medusa's head, and Compassion, wearing an olive wreath and holding a cedar branch.

Polyhymnia—the muse of rhetoric and cheerful songs. Her head is adorned with precious stones; she is dressed in white; in her hands, a sceptre.

Erato—the muse of love; the patroness of light poetry, love songs, and elegies; she is crowned with a wreath of myrtle and roses; in her hands, a lyre; she is accompanied by winged cupids holding torches and doves.

Urania—celestial; the muse of astronomy, dressed in the colour of the heavens, studded with stars; on her head a radiant crown; she holds a sphere encircled with planets and stars.

Calliope—eloquent, the muse of epic poetry, her brow crowned with a laurel wreath; in one hand a horn, in the other a book.

Terpsichore—the muse of dance; she wears a feather head-dress; in her hands, a tambourine. The vestal Amata represents Terpsichore; she has joined the actors in order to follow Lucio's actions, for he has come to the circus seeking death.

The muses greet Apollo, their patron, who invites them to quench their thirst in the spring of Inachus. He says to them:

'Today is our festival! Therefore show us your prowess, you who are blessed with the godlike spark! Show us the sciences and arts which have been entrusted to you by the will of the gods!'

Each muse represents her callings and purposes. They rival one another in elegance of poses, plastique, and gracefulness. After a series of dances, Apollo announces the reason that he called them to Parnassus.

'I called you to this festival to proclaim in your persons, charming muses, the eternal union of science and art! As a sign of this unity, join hands and perform a round dance together!' The muses form a close circle to signify the absence of dissention in the sciences and arts, that each supports the other.

'I do not know which of you deserves preference!' says Apollo. 'You are all equally precious to me! Here then for all is one wreath! Crown with these laurels the one whom you yourselves deem worthiest!'

The muses cluster around Terpsichore-Amata and give her the wreath: 'We gathered on Parnassus to dance and celebrate! And you have no equal among us, Terpsichore, as one to captivate with grace and dances; so to you, by right, belongs this wreath!'

The muses and their companions surround Apollo, forming a picturesque collective grouping; then the enchanted forest of Parnassus disappears beneath the arena.

Amata remains at the circus, concealed behind the columns. Running from place to place searching for Lucio, she looks under the gladiators' visors, but all her searches are in vain.

After the festival of the muses on Parnassus, two gladiators have come out into the arena: a Mirmillo, and the centurion Lucio in Raetian armour. The Mirmillo wears a helmet with gold ties; he is holding a small shield and sword; the Raetian is armed with a trident and holds a net in his other hand; he is covered by a helmet decorated with a silver fish.

As brothers in their profession, the gladiators enter embracing each other. Bowing to the emperor, they separate and begin to fight. The Raetian-Lucio devises how best to catch his opponent in his net. The Mirmillo stops, bends down and leans his whole body on one knee. In one deft jump the Raetian is beside the Mirmillo and throws his net over him; the Mirmillo, bending to the ground, jumps to one side and dodges the net, which was about to enmesh him. The Raetian runs around the arena, the Mirmillo in pursuit; the Raetian reaches the place where his net lay. He seizes it and throws it deftly at his opponent again, just as he was about to deliver a mortal blow.

Shaking the net off his shoulders, the Mirmillo attacks his opponent ever more aggressively, while the Raetian, brandishing his trident, attempts to retrieve his net again. But the Mirmillo fends off the trident with his shield. The battle continues, and finally the Mirmillo wins. With his sword he knocks the trident out of his opponent's hand, and the Raetian's helmet and visor from his head.

Lucio is disarmed; laid low on the ground, he lies in the victor's power. Holding his foot against the Raetian's breast, his sword at the ready, the Mirmillo turns to the imperial box:

'Death or mercy?'

'I do not want mercy! . . . Strike! . . . I thirst for death! . . .', answers Lucio, and tearing the sword from his opponent's hand, he stabs himself.

Amata, seeing the vanquished Lucio, runs to him and wants to save him, but it is too late! In his death agony Lucio recognizes his beloved, and with the words, 'I love you,' dies.

Amata rushes to his body, embraces it and then, at the sight of approaching warriors, quickly draws the sword from Lucio's breast and kills herself.

APOTHEOSIS

The triumph of the goddess of chastity, Vesta.

Curtain

17
'Marius Ivanovich Petipa'
Mikhail Mikhailovich Ivanov

Mikhail Mikhailovich Ivanov (1849–1927) wrote the music of *The Vestal*. A professionally trained composer, Ivanov was also a music critic who for decades contributed to the Petersburg daily *Novoe vremya* [The New Time]. His commission to write *The Vestal* was the result of another of Vsevolozhsky's reforms, the abolition of the post of official ballet composer in the imperial theatres. Ludwig Minkus, the last person to fill that post, took his farewell benefit in 1886; Ivanov and Tchaikovsky were among the first non-specialists invited to compose a ballet.

Ivanov wrote 'Marius Ivanovich Petipa' at the time of the balletmaster's death, as a memorial to him and a memoir of their collaboration on *The Vestal*. It was published in *Novoe vremya* on 12 July 1910.

I became acquainted with Petipa when the late Director of Theatres, Ivan Alexandrovich Vsevolozhsky, commissioned me to compose *The Vestal* in 1888. Before that I had never met Petipa, either in the theatre or socially. In the theatre it was difficult to see him because he was always backstage, where I, on principle and perhaps because of a certain odd indifference, never tried to go, even in my student days. Petipa rarely made an appearance in the auditorium. Moreover, I did not meet him because I rarely attended the ballet, to which, despite all the philosophers and aesthetes of antiquity and the latest times, I remained indifferent.

I first had occasion to chat with him when it had been decided to produce *The Vestal*. I cannot remember whether I met him at Ivan Alexandrovich Vsevolozhsky's or at the home of Sergei Nikolaevich Khudekov, author of the libretto of *The Vestal*, but after that I saw him often. To write music without the balletmaster is impossible. Not only every number, but almost every section requires the balletmaster's approval. It is without question composition *à deux*. Perhaps in earlier times official composers of ballet music—Minkus, Pugni, and others—could do without the balletmaster's instructions as a result of practice and their familiarity with ballet. But novice composers could not take one step without the balletmaster's word; their imagination would cause them to digress and compel them to write music of dimensions impossible for dance. This uninitiated quality in most new ballet composers is so great that they confuse a balletic variation, which is a dancer's solo, with a variation in the purely musical sense. The two have nothing in common but the name, however, and composers fall into confusion when balletmasters carry on conversations with them about variations. The balletmaster, stimulating the imagination of his musical collaborator, must, however, restrain the composer's inclination to 'overflow with thought'. 'I need so many bars for the representation of such and such a mental state,' the balletmaster says, so many for this other, and not one extra. Otherwise the result will be heavy and boring, and the ballerina will not be able to dance the unneeded measures.'

I had occasion to learn about the specialist composer's requirements, of which I had but a vague understanding before, from Petipa when I set to work on the ballet. It is not a complicated science, and to understand it is not difficult. One must realize, however, that Petipa in the early stage of his acquaintance with every composer had to repeat his pedagogical course on choreography and its musical requirements. But he was patient, a true foreigner as regards courtesy, and never complained of the boredom which he probably suffered when explaining to novices the ABCs of his art. He listened through every page patiently, invariably submitting it to his censorship. He loved piquant music, lively and animated rhythms, but he never evaluated what had been written except for the arrangement of its parts—whether their length and dimensions, their general plan, corresponded with the dances he had already conceived.

As I mentioned, I rarely attended ballet performances. From conversations with Petipa I immediately sensed that he was a classic of his art. At that time there was no hint of today's tendencies, which Petipa so opposed when he was retired. Nor could there have been any talk of choreographic successes à la Duncan, because this barefoot dancer, with her pseudo-classical conceptions, had not yet appeared. Petipa also strove for stylization, to use this fashionable theatrical term which expresses nothing, but he saw in it something quite different from what our current priests of Terpsichore are seeking. The ideas of Perrot, Didelot, Vestris, and Noverre himself were sacred to him and guided his whole life. His father was himself a balletmaster in Brussels, and although little Marius protested when his father taught him dancing—for the occupation seemed demeaning—fate nevertheless played a joke on him. How he was as a dancer I don't know because I never saw him, for he had finished his performing career before I started attending the Petersburg theatres. But he became one of the most celebrated balletmasters of his time. Moreover, he occupied an absolutely exceptional position, since he became the head of the Petersburg ballet just when interest in the balletic art was declining precipitously in the rest of Europe.

Whatever the causes—economic, social, or aesthetic—that fact is indisputable. Since the 1840s in Europe, if dancing did not cease, it was rarer than before. Moralists even claim that the number of marriages in France, for example, declined noticeably after an earlier passion for dancing cooled: witty conversation is a good thing, but there is far more poetry in the turn of a waltz, which before one knows it puts one at the altar of Hymen. Not only in France did people come to dance less: a serious interest in grand choreographic ballets disappeared everywhere in Europe, and state-run ballet schools began to close or fell into decline, for governments reduced the funding set aside for their maintenance. Ballet occupied third place in the theatrical life of Vienna, Dresden, Berlin, Milan, and Paris, preserving its exceptional position only in Petersburg.

As ballet has never been a national art among us—it would be silly to say it was—and as it has always been headed exclusively by foreigners, it is natural that Petipa was an imitator of his predecessors' traditions. He of course perfected them, developed them, promoting their growth in keeping with his talent without abandoning their previous line of development. He also had to produce character dances and he did this beautifully, as his opera dances bear witness. The lezginkas in *Ruslan* and *The Demon*, or the mazurka in *A Life for the Tsar*, which he produced, called forth stormy applause on their merits, and excelled by far the danced rubbish which we see now in these operas. Strange as it seems, this chief of classical choreography began his career as a character dancer in France and Spain. Perhaps the customs and recollections of youth never completely subsided in him, but as a balletmaster he was drawn only to classicism.

The story of *The Vestal* was classical, though it had nothing in common with classical choreography. Sergei Nikolaevich Khudekov's ballet called for a large number of character dances, of which we three—the author of the libretto, the balletmaster, and I—made extensive use, but Petipa's sympathies during the composition of *The Vestal* lay exclusively with classical *pas* and *adagio*. I clearly remember the urgency and enthusiasm with which he spoke to me of the significance of the divertissement on 'Parnassus' in this ballet (it seems this is what the *grand pas d'action* in the second act was called; at this point I've already forgotten), of the effect which it should produce, referring insistently the whole time, as to an unattainable model, to the corresponding scene in [Perrot's] *Faust*, where the scene of 'seven deadly sins' is similar in character to this divertissement. I never saw *Faust*, and have no notion of this scene now; but if I am not mistaken, Marius Ivanovich himself took part in it when he was still a first dancer.

As Khudekov wanted to familiarize us in detail with his thoughts on the production of *The Vestal*, Petipa and I were invited to his estate in the Ryazan district. It was summer; the order for the ballet had been issued only at the end of May, and *The Vestal* had to be produced by December:

little time remained for the composition a big three-act, even four-act ballet (as the last act of *The Vestal* was made up of two large scenes). Along with the late Skalkovsky and Grinev (such, it seems, was the surname of Mme Vazem's husband), Khudekov is one of our few savants of choreography, and was very demanding on his collaborators. He went into every detail, judging it from all aspects: historical, archaeological, scenic-aesthetic, and even musical. To visit him in the country was therefore essential.

I must say in passing that those few days spent in Khudekov's home discussing the plan of *The Vestal* number among the most pleasant recollections of my life, thanks perhaps to the host himself and Petipa. Both of them set to work with absolutely youthful enthusiasm; this was pleasant to observe, not to mention our host's courtesy. Khudekov's understanding and directions were inexhaustible, and were of a most unexpected nature besides. I would add that his knowledge of the Roman world, a favourite subject of mine since childhood, was brilliant: I am convinced that Khudekov knew the mores of ancient Rome far better than many professors of Roman history. In this respect Petipa could not compare with Khudekov. Everything the latter said was utter novelty to Petipa, but his talent and imagination enabled him to perceive vividly everything that he was hearing literally for the first time, and he subsequently demonstrated this in the production of *The Vestal*, in which the Director of Theatres himself took a large part as well. On this occasion Petipa listened quietly through a course in ancient history and archaeology, but then rewarded himself for his enforced silence with stories from his past and his artistic life. In this he was indefatigable, and it was then that I became familiar, in lively exposition, with the content of his memoirs, published some three years ago. Let me say that this little booklet does not provide the slightest notion of how interesting his stories are. Petipa's spoken discourse flows without hesitation and delay, whereas the memoirs, dry and uninteresting, do not match their verbal presentation at all. Clearly, he did not wield a skilful pen, and the memoirs could only have been written in French. In Russian he made the most comical mistakes, despite the fact that he had lived here more than fifty years.

I was able to evaluate his patience and capacity for work when they began to try out the music of *The Vestal* in the theatre school. Every rehearsal lasted at least three to four hours. Petipa was always present, never late, and the last to leave—an example of the most zealous performance of duty. Naturally, with a balletmaster like that all personnel had to be attentive. The rehearsal hall of the theatre school had always been the cherished dream of Petersburgers in search of amusement. How many stories were played out here from the days of Katenin, Griboyedov,

Begichev, Yakubovich, and others! I thought of this when I walked into it for the first time, but remained indiffferent despite the crowd of charming faces and the attractive figures of ballerinas and students which one encounters there.

The world of ballet produced a pleasant impression on me, and all the ballerinas I met—Varvara Alexandrovna Nikitina, Marie Mariusovna Petipa, Vera Vasilievna Zhukova, Maria Nikolaevna Gorshenkova and the rest—who took part in the ballet were sincere, kind, unaffected, and unconstrained. In contrast, the Italian star who danced in *The Vestal*, Cornalba, left a poor impression on me—tall, unlovely, and awkward—yet she took the principal role of Amata, the heroine of the ballet. Cornalba was, if not hard of hearing, unable to make out rhythms, even simple ones. Strictly speaking, she should not have been given a role in *The Vestal*, but since the ballet was produced at the height of the Petersburg balletomanes' attraction to Italian dancers, which began with Zucchi and ended with Legnani, the role of Amata naturally went to Cornalba. Besides, she danced well, perhaps the best of all our ballet personnel at that time with the exception of Gorshenkova. As a person, Cornalba did not produce a favourable impression on anybody, whereas the other Italians who came to Petersburg, beginning with Zucchi and Dell'Era, were scrupulously courteous and attentive, even to persons who did not review ballet performances.

Cornalba nevertheless behaved correctly in all respects. She did not ask me to change anything, although she complained to her friends the inveterate balletomanes of the 'non-dansante' quality of the music of *The Vestal*, which later produced repercussions in the press when her friends' notices of the ballet were published. These notices affected the fate of *The Vestal*, since the ballet world always has and still does consist of a close-knit, amiable circle of artists and spectators whose thoughts and impressions, on both sides, never disagree. Only later was the importance of *The Vestal* recognized: it was the first' serious attempt in Russia to produce a symphonic genre in ballet, a fact noted by Skalkovsky and Pleshcheyev, the two confirmed historians of our ballet. Tchaikovsky's *Swan Lake*, written long before *The Vestal*, still followed an earlier path and consists only of a series of waltzes. In *The Sleeping Beauty*, Tchaikovsky, and later other composers moved on to the new path, but only after *The Vestal*: this is an indisputable fact.

I, however, intend to say nothing more about the place *The Vestal* occupies in the history of our ballet; I am remembering it because it permits me to recall my acquaintance with someone who loved his work as passionately as Marius Ivanovich Petipa did. I can imagine how he must have responded to the revolution in choreography which Fokine and

Duncan so easily set in motion. He responded to it negatively. The new direction displeased him intensely, contradicted his whole integrated view of art. It is strange, however, with what ease inveterate balletomanes change their tune. I recall Konstantin Skalkovsky's disdain for Italian ballets like *Excelsior*, *Amor*, and others; he found that choreography in Italy, though better there than in other countries, was in complete decline; that jumping, running around the stage and skipping passed for dances there; that gracefulness had disappeared, and with that the significance of choreography as an art was shaken, for its significance was reduced exclusively to *féerie*. But when a number of brilliant Italian dancers appeared in Petersburg—Zucchi, Giuri, Dell'Era, Sozzo, Bessone, Limido, Brianza, and others—he changed his mind, though what he had already penned could not be struck down, as a tree by an axe. In any event, Skalkovsky placed Italian ballerinas on a pinnacle even while maintaining those very views about the essence of ballet which Petipa held. Even in his sleep the latter never dreamed of all the theoretical knowledge about entrechats and pirouettes which Skalkovsky on occasion could show off. Take the first chapters of his book, *Ballet, its History and Place in the Elegant Arts*, which places the author in his technical knowledge on a level with balletmasters. True, he acknowledges borrowing everything essential from the Frenchman [*sic*] Blasis (*The Art of Dance*), who in turn took his wisdom from the tracts of Noverre, the celebrated balletmaster of the eighteenth century. But nevertheless: you of the present day—what can you show us? Who among you takes pleasure in rummaging through the books of Blasis or Noverre, when, having sprinkled your reviews with a few technical terms and added several recherché phrases in the decadent manner, you can already be considered savants? On the other hand you change your thoughts about ballet like old gloves, and what twenty years ago was considered the ruin of art you extol today.

Skalkovsky, who granted the choreographic art its conditions of time and place, the significance in it of fashion and fashionable tendencies, also recognized the necessity of progress in choreography, but clearly saw this progress at a different level, not in running around and skipping, or else he would not have judged the ballets of Manzotti so severely, ballets in which the same ballerinas distinguished themselves whom he later put forward as models for our own. Whatever his progressive views on choreography, he—like Zaretsky at the duel—was a classic and a pedant. He and Petipa died true to their views, although the former never had to struggle against the new tendency. Who knows, perhaps he too would have changed his views about the goal of choreography! As for Petipa, he remained unshakeable as a rock in his attitudes, and responded with scorn to new currents and to new persons flattered in excess of their talent by unceremonious publicity. Glancing back at the past and seeing the ease of heart with which balletomanes abandoned their earlier standards and bowed before new gods, the old artist could hardly reflect without sadness: had he not in fact been right when as a child he instinctively guessed that it did not pay to study choreography as an art, that there was nothing positive and serious in it if even grey-haired men, with the ease of Zephyr, turn their backs today on that which they had praised only yesterday?!

Otherwise, one must agree with Skalkovsky, that 'there is joined to the small circle of true adherents of choreography no few people who give no quarter to choreography at all, but who simply love to gape at beautiful women and to have dinner with them sometimes.' Apart from the circle of balletomanes, the late critic wrote, 'in the art of choreography, as in all others except the arts of drinking vodka and playing cards and gossiping, our public has not come very far.'

This undoubtedly explains the success of barefoot dancers and their imitators. When they begin to extol as something unattainable poses and movements that any young woman can perform who is not a hunchback or lopsided and who has never studied anything—of what art can we seriously be speaking?

IX
The 1890s

INTRODUCTION

Lack of persistence did not count among Petipa's faults. He tried, for example, to duplicate the success of *The Pharaoh's Daughter* of 1862 in *The Beauty of Lebanon* of 1863. This work failed and yet Petipa persevered, apparently convinced that the scenic and choreographic conception of the oriental extravaganza was valid. He was rewarded for his belief in the success of *Le Roi Candaule* and *La Bayadère*. The middling success of *The Vestal* and *The Talisman* tested his faith in ballets approaching *féerie*-like magnificence, and he was rewarded for his convictions again in *The Sleeping Beauty*.

In every case, Petipa adjusted the collaboration, changing principal dancers, scenarist, or composer to achieve superior results. In *The Sleeping Beauty* (compared with *The Vestal*) he changed all three: a new Italian ballerina in Carlotta Brianza, a new scenarist in Director of Theatres Vsevolozhsky himself, and Russia's finest contemporary composer, Pyotr Ilyich Tchaikovsky. And yet there are clear echoes of *The Vestal* in the new ballet. It was magnificent and demanding, though not to the extreme of *The Vestal*. Certain stage situations were used again: Désiré is reminiscent of Lucio in his indifference to the blandishments of other women, and his vision of Aurora has a distinguished pedigree which passes through Amata's vision of Lucio back to Aspicia's of Taor—all experienced by the protagonists in a supra-natural state.

The Sleeping Beauty was given its first public performance on 3 January 1890. The translations which follow the libretto consist of three articles written by persons who attended either the première or other early performances of the ballet. The first is a notice of the first performance by Konstantin Skalkovsky, taken from the *Novoe vremya* of 5 January 1890 (and republished with alterations in his *V teatral'nom mire*). The second is a music review by Tchaikovsky's friend German Laroche, which appeared in the newspaper *Moskovskie vedomosti* on 17 January 1890. The third is the recollection of an artist, Alexandre Benois, looking back as an old man on *The Sleeping Beauty* as he recalled it in his youth. It was translated from Book III of his *Moi Vospominaniya* [My Reminiscences], 2 vols. (Moscow, 1980).

These authors see the ballet in different ways not simply attributable to their specialities. Skalkovsky is typically matter-of-fact, reporting what was for him the here and now. Laroche is drawn to philosophical questions about national identity and about universal meanings in the fairy tale used as the story of the ballet. Benois adopts the most philosophical tone of all, writing eloquently of the profound meanings beneath the simple exterior of the fable. But he returns at the end to history, finding in his experience of *The Sleeping Beauty* the roots of that love for ballet which went on to inspire the Diaghilev circle.

18
Libretto of *The Sleeping Beauty*
Ivan Alexandrovich Vsevolozhsky

THE SLEEPING BEAUTY
Ballet-féerie
in 3 acts with prologue
Story taken from the tales of Perrault
Music by P. I. Tchaikovsky
Production and dances by the balletmaster M. Petipa

Decorations: Prologue—'Florestan's Palace', by Mr Levogt.
First act—'The Palace Garden', by the Academician Mr Bocharov and Mr Andreyev.
Second act—'A Forest Locale, and Panorama', by the Academician Mr Bocharov; 'Interior of the Sleeping Beauty's Castle', by Mr Ivanov.
Third act—'The Esplanade of Florestan's Castle, and Apotheosis', by Professor Mr Shishkov.

Machines by Mr BERGER; men's costumes by Mr CAFFI, women's by Mmes OFITSEROVA and IVANOVA; men's headwear by Mr BRUNEAU, women's by Mme TERMAIN; flowers and feathers by Mme SIMONOVA; wigs by Messrs MICHEL and FYODOROV; accessories by the sculptor Mr KAMENSKY; footwear by Mr LEVSTEDT; metal accoutrements by Mr INGINEN.

Permitted by the Censor. St Petersburg, 30 December 1889.
Typographer of the IMPERIAL Spb Theatres (Dept. of Crown Affairs), Mokhovaya 40

DRAMATIS PERSONAE:

King Florestan XIV	Mr Kshesinsky
The Queen	Mme Cecchetti
Princess Aurora, *their daughter*	Mlle Brianza
Prince Chéri	Mr Bekefi
Prince Charmant	Mr Oblakov
Prince Fortuné	Mr Karsavin
Prince Fleur-de-pois	Mr Gillert
Catalabutte, *Chief Master of Ceremonies*	Mr Stukolkin
Prince Désiré	Mr Gerdt
Galifron, *Prince Désiré's tutor*	Mr Lukyanov
A footman	Mr Orlov

The Sleeping Beauty   361

The Lilac Fairy	Mlle Petipa
The Fairy Canari	Mlle Johanson
The Fairy Violente	Mlle Zhukova I
The Breadcrumb Fairy	Mlle Kulichevskaya
The Fairy Candide	Mlle Nedremskaya
The Fairy Fleur de farine	Mlle Anderson
Carabosse, *the evil fairy*	Mr Cecchetti

Courtiers; ladies, cavaliers, pages, hunters and huntresses, guards, footmen, fairies' retinues, nurses and wetnurses, peasant men and women, and others.

SCENE 1

Prologue
The Baptism of Princess Aurora
The Gifts of the Fairies
Grand pas d'ensemble

Fairies: Mlles Petipa, Johanson, Zhukova I, Kulichevskaya, Nedremskaya and Anderson.
Retinue of the Lilac Fairy: Mlles Vishnevskaya I, Andreyeva, Legat II, Savitskaya, Matveyeva I, Lietz II, Tatarinova and Sheberg.
Pages of the Lilac Fairy: Mlles Egorova II and Ksheshinskaya.
Pages of the Fairy Canari: Mr Kil and Mlle Urakova.
Pages of the Fairy Violente: Messrs Karsavin and Voronkov I.
Pages of the Breadcrumb Fairy: Messrs Lukyanov and Kshesinsky II.
Pages of the Fairy Candide: Mlles Peters and Natarova II.
Pages of the Fairy Fleur de farine: Mlles Rubtsova and Vishnevskaya II.
Genies with large fans: Mlles Egorova I, Serebrovskaya, Rosh, Alexeyeva I. Students Ermolaeva, Golubeva, Schnering and Erler I.
Genies proferring fragrance: Mlles Sitnikova, Vertinskaya; students Davydova and Ilyina.
Girls bearing gifts: students Ilyina II, Kasatkina, Leonova, Levina, Noskova and Stepanova.

SCENE 2

The Four Suitors of Princess Aurora

Caquets des tricoteuses

Mlles Stepanova II, Lietz I, Svirskaya, Matveyeva II, Lezenskaya, Levenson I, Klimashevskaya II and Solyannikova.

Valse villageoise

Mlles Andreyeva, Legat II, Tatarinova, Matveyeva I, Slantsova, Savitskaya, Labunskaya, Ogoleit III, Lietz, II, Schedrina, Kuskova, Tselikhova, Egorova II, Korsak, Rubtsova, Perfilieva, Aistova, Vishnevskaya II, Radina, Stepanova

362 The 1890s

III, Tikhomirova, Kunitskaya, Ryabova and Pavlova. Messrs Fyodorov I, Stepanov, Gorsky, Gavlikovsky, Petrov, Ivanov, Volkov, Pashchenko, Andreyev II, Marzhetsky, Solyannikov, Fedulov, Legat, Rakhmanov, Fyodorov II, Ponomarev, Baltser, Fomichev, Voronkov III, Usachev, Alexandrov, Voronin, Belov, Yakovlev.

Grand pas d'action

Mlle Brianza, Mme Cecchetti; Messrs Kshesinsky, Cecchetti, Bekefi, Karsavin, Oblakov and Gillert.

Maids of Honour: Mlles Tistrova, Fyodorova II, Vorobieva and Gruzdovskaya.

Little girls: Students Ivanova, Noskova, Obukhova, Skorsyuk.

Pages: Students Dyakonova, Ilyina, Kuzmina, Leonova I, Levina, Lobanova, Niemann and Stepanova.

SCENE 3
Prince Désirés Hunt

Duchesses	Mlle Legat II
	Mlle Voronova
Baronesses	Mlle Labunskaya
	Mlle Zhukova II
Countesses	Mlle Ogoleit III
	Mlle Nedremskaya
Marchionesses	Mlle Lietz II

Huntresses: Mlles Kshesinskaya, Lezenskaya, Peshkova, Ryabova, Potaikova, Starostina, Onegina II, Peters.

Hunters: Messrs Voronkov I, Kshesinsky II, Leonov, Orlov, Voronkov II, Baltser, Belov, Voronin, Yakovlev, Alexandrov, Dorofeyev, Panteleyev, Fomichev.

Peasant women: Mlles Vishnevskaya II, Levenson II, Kunitskaya, Pavlova, Rubtsova, Perfilieva, Stepanova III, and Levenson.

Peasant men: Messrs Fyodorov I, Pashchenko, Andreyev II, Fedulov, Gorsky, Marzhetsky, Petrov and Gavlikovsky.

Blind-Man's Buff

Mlles Ogoleit I, Zhukova II, Voronova, Nedremskaya, Ogoleit III, Labunskaya, Legat II, Lietz II, and Mr Lukyanov.

Variations

Mlle Ogoleit I and Mr Leonov; Mlles Voronova, Zhukova II, Nedremskaya.

Farandole

Mlles Ogoleit I, Zhukova II, Voronova, Nedremskaya, Legat II, Ogoleit III, Labunskaya, Picheau I, Pavlova, Perfilieva, Kshesinskaya, Stepanova III, Lietz II, Vishnevskaya II, Ryabova, Levenson I, Levenson II, Kunitskaya, Potaikova,

Peters. Messrs Leonov, Kshesinsky II, Voronkov I, Voronkov II, Fomichev, Dorofeyev and others.

Appearance of the Shades of Aurora and her Retinue
Mlles Brianza, Petipa and Mr Gerdt.

Nymphs:

Mlles Vishnevskaya I, Scheberg, Nikolaeva, Oblakova, Andreyeva, Isaeva, Slantsova, Savitskaya, Korsak, Preobrazhenskaya, Kuskova, Egorova II, Matveyeva I, Rubtsova, Tatarinova and Radina.

PANORAMA

SCENE 4

The Sleeping Beauty's Castle
Sleeping Groups

SCENE 5

The Wedding of Prince Désiré and Princess Aurora
The Esplanade of Florestan's Castle
Entrance of the King, Queen, and the newly-weds with their retinue and with the Fairies of Diamonds, Gold, Silver and Sapphires

Polonaise
Procession of the Fairy-Tale Characters

1. Bluebeard — Mr Orlov
   His wife — Mlle Oblakova
2. Puss in Boots — Mr Bekefi
3. Marquis de Carabas (on a sedan chair) — Student Israilev
4. Goldilocks — Mlle Legat III
   Prince Avenant — Mr Voronin
5. Donkey-skin — Mlle Ogoleit II
   Prince Charmant — Mr Yakovlev
6. Beauty — Mlle Zasedateleva
   The Beast — Mr Bizyukin
7. Cinderella — Mlle Petipa
   Prince Fortuné — Mr Kshesinsky II
8. The Blue Bird — Mr Legat
   Princess Florina — Mlle Nikitina
9. The White Cat (carried on a pillow) — Mlle Anderson
10. Little Red Riding Hood — Mlle Zhukova I
    The Wolf — Mr Lukyanov
11. Ricky of the Tuft — Mr Navatsky
    Princess Aimée — Mlle Ivanova

12. Tom Thumb and his brothers, students: Stukolkin, Legat III, Osipov, Kristerson, Medalinsky, Legat II and Aslin
13. The Ogre   Mr Bulgakov
    The Ogress   Mr Chernikov
14. The Fairy Carabosse, in a wheelbarrow drawn by rats
15. The Fairy Candide and her genies
16. The Fairy Violente and her genies
17. The chariot of Fairy Canari and her retinue
18. The Lilac Fairy (carried by four large genies)

Divertissement

Pas de quatre

The Fairy of Diamonds	Mlle Johanson
The Fairy of Gold	Mlle Kulichevskaya
The Fairy of Silver	Mlle Krüger
The Fairy of Sapphires	Mlle Tistrova

Pas de caractère

Puss in Boots and the White Cat. Mr Bekefi and Mlle Anderson.

Pas de deux

The Blue Bird and Princess Florina. Mr Cecchetti and Mlle Nikitina.

Pas de caractère

Little Red Riding Hood and the Wolf. Mlle Zhukova I and Mr Lukyanov.

Pas de caractère

Cinderella and Prince Fortuné. Mlle Petipa and Mr Ksheskinsky II.

Pas berrichon

Tom Thumb and his brothers. Students: Stukolkin, Legat III, Medalinsky, Osipov, Legat II, Kristerson, Aslin. The Ogre: Mr Bulgakov.

Pas de quatre

Aurora	Mlle Brianza
Désiré	Mr Gerdt
The Fairy of Gold	Mlle Kulichevskaya
The Fairy of Sapphires	Mlle Tistrova

Entrée of the Ballet
Sarabande

Roman: Mlles Ogoleit II, Vishnevskaya I, Legat II, Kshesinskaya; Messrs Leonov, Oblakov, Gillert and Voronkov I.

Persian: Mlles Tatarinova, Labunskaya, Leonova and Kuskova; Messrs Dorofeyev, Usachev, Baltser, Alexandrov.

Indian: Mlles Nikolaeva, Shchedrina, Egorova II and Lietz II; Messrs Fedulov, Konstantinov, Belov and Fomichev.

American: Mlles Aistova, Tselikhova, Potaikova, and Vishnevskaya II; Messrs Voronkov II, Andreyev II, Panteleyev, and Marzhetsky.

Turkish: Mlles Korsak, Peshkova, Prokofieva, and Starostina; Messrs Voskresensky, Volkov, Fyodorov I, and Solyannikov.

Ensemble coda
APOTHEOSIS
(<i>Gloire des Fées</i>)

Instrumental solos performed by:
Soloists of the Court of His Imperial Majesty
On the Violin—Mr Auer
On the Harp—Mr Zabel

The cello solo is performed by Mr Loganovsky

Conductor: Mr Drigo

PROLOGUE
Scene 1

The Christening of Princess Aurora

A celebration in one of the halls of the royal palace. At the right a platform for the king and queen, and for the fairies—Princess Aurora's godmothers. At the back of the stage, an entrance door.

Courtiers: ladies and cavaliers form groupings in expectation of the entrance of the king and queen. Masters of Ceremonies show each to his or her place and explain the procedure—how in a given instance to offer congratulations to the king and queen, and also to the influential fairies, invited to Princess Aurora's christening as godmothers.

Catalabutte, surrounded by courtiers, verifies the list of invitations sent to the fairies. Everything has been done according to the king's command, and is ready for the celebration. The court is in full attendance; any moment they expect the king and queen, and also the arrival of the invited fairies.

Trumpets sound. The king and queen enter, preceded by pages, then Aurora's nurses and wet-nurses bring in the cradle in which the royal child sleeps. No sooner have the king and queen taken their places on the platform, on either side of the cradle, than the Masters of Ceremonies announce the arrival of the fairies.

Entrance of the fairies *Candide, Fleur de farine, Violente, Canari,* and *Breadcrumb*. The king and queen meet them and show them places on the platform.

Entrance of the *Lilac Fairy*, Princess Aurora's leading godmother. She is surrounded by loyal spirits, who bring in large fans and incense burners and who carry their mistress's mantle.

At a sign from *Catalabutte* pages and young girls bring in brocade pillows with gifts from the king to his daughter's godmothers. They explain to each fairy what has been chosen for her. In their turn the fairies come down from the platform to make a gift to their godchild.

Grand pas d'ensemble
The Fairies' Gifts

The *Lilac Fairy* is approaching the cradle to offer her gift when suddenly a loud noise is heard at the entrance; a page runs in and informs *Catalabutte* that a new fairy, whom they forgot to invite to the feast, is at the castle gates. It is Fairy *Carabosse*, the most powerful and evil in the land. *Catalabutte* is completely undone—how could he, thoroughness personified, have forgotten her? Trembling with fear, he approaches the king to tell him his mistake. The king and queen are very concerned; this error can bring in its wake much unhappiness for their dear child. The fairies also seem very disturbed by it.

Carabosse appears in a wheelbarrow drawn by six large rats; she is accompanied by ugly and comical pages. The king and queen beseech her to forgive *Catalabutte*'s error; he shall be punished according to her bidding. Terrified, *Catalabutte* falls at the evil fairy's feet, begging her forgiveness and promising to serve her faithfully to the end of his days.

Carabosse mocks him, laughing, and entertains herself by pulling out tufts of his hair and throwing them to her rats, who devour them. Soon *Catalabutte*'s head is completely bald.

'Although I am not *Aurora*'s godmother,' says *Carabosse*, 'I want to give her something all the same.'

The good fairies ask her to forgive the Master of Ceremonies' accidental forgetfulness, and not to spoil the happiness of this best of kings.

But *Carabosse* only laughs. Her merriment quickly spreads to her ugly pages and even to her rats. The good fairies turn away from their sister in disgust.

'*Aurora*, thanks to the gifts of her six godmothers,' says *Carabosse*, 'will be the most beautiful, the most charming, most intelligent princess in all the world; I do not have the power to deny her these qualities. But that her happiness never be interrupted—you see how good I am—she shall fall asleep the first time she pricks her finger or hand, and her dream will be forever.' The king, queen, and court are dumbstruck.

With her wand *Carabosse* makes signs over the cradle, pronouncing magic words, and pleased with the trick she has played on her sisters, the good fairies, begins to guffaw, her merriment spreading to her entire ugly retinue.

But the *Lilac Fairy*, who had not yet given her gift to the child and who was standing, shielded by *Aurora*'s cradle, now comes forward. *Carabosse* looks at her with suspicion and malice. The good fairy bends over the cradle: 'Yes, you shall fall asleep, my little *Aurora*, as our sister *Carabosse* has willed,' says the *Lilac Fairy*, 'but not forever. The day will come when a prince, under the spell of your beauty, will kiss your brow, and you shall waken from this long dream to become his helpmate, to live happily and contented.'

Carabosse, enraged, takes a seat in her wheelbarrow and disappears. The good fairies encircle the cradle, as if to protect their goddaughter from their evil sister.

(Scene)
End of the Prologue

ACT I

SCENE 2

The Four Suitors of Princess Aurora

A park at King Florestan XIV's castle. At the audience's right an entrance to the castle. The upper levels of the castle are lost in the tree-tops. At the back, a marble fountain.

Aurora has turned twenty. *Florestan* is happy that *Fairy Carabosse*'s prediction has not come true. *Catalabutte*, whose hair has not grown back to this day, comes out in a comical nightcap. He is fining several village girls for working with needles in front of the castle, and reads them the regulation prohibiting the use of needles and pins within a hundred mile radius of the royal residence. For this offence he is sending them to prison under guard.

The king and queen come out on the terrace with the four princes who aspire to Princess Aurora's hand. The king asks what the villagers have done to be sent to prison. *Catalabutte* explains the reason for their arrest and displays the material evidence. The king and queen are horrified. 'Let the guilty suffer punishment for this and never more see the light of day.' The princes beg mercy for the guilty. Not one tear ought to be shed in *Florestan*'s realm on the day Aurora turns twenty. The king pardons the villagers, but with the condition that their work be burned by the hangman in a public place. General delight. The dances of the villagers. 'Long live King *Florestan*, long live Princess *Aurora*!' The princes have yet to see Princess *Aurora*, though each has a medallion with her portrait. They are burning with the desire to be her favourite and express this to the king and queen, who assure them that their beloved daughter has complete freedom of choice, and the one she chooses will be their son-in-law and successor to the realm.

Aurora's entrance. She runs in with her maids of honour, who have bouquets and wreaths. The four princes, struck by her beauty, try to be pleasing to her, but *Aurora* dances with all four of them, giving preference to none.

(Pas d'action)

Rivalry among the princes; *Aurora's* coquettishness. The king and queen urge her to make a choice. 'I am still so young,' *Aurora* says, 'let me enjoy my freedom a little more.' 'Do as you think best, but remember that the interests of state require your marriage, that you give the country a successor to the throne. *Carabosse's* prediction worries us very much.' 'Don't worry, for the prediction to come true I must prick my hand or finger, and I take neither pin nor needle into my hands. I sing, dance and enjoy myself, but never labour.'

The four princes cluster around her and ask her to dance, as it is rumoured that she is the most graceful girl in the world.

Aurora agrees to their request. She dances to the accompaniment of lutes and violins played by her maids of honour and pages. The four princes are delighted, whereupon she tries to be all the lighter and more graceful, the more to please them. Not only the princes and the court, but also the assembled villagers follow her ethereal flights with curiosity. General delight. Dances for all. Suddenly *Aurora* notices an old woman beating time with a spindle; she snatches the spindle away from her and continues her dance with it, now as a sceptre, now imitating the work of spinners, trying to inspire the complete delight of the four who are paying her court. Suddenly her dances are interrupted, and in horror she looks at her hand, pierced by the spindle and bloodstained.

Like a mad person she rushes from side to side and finally falls unconscious. The king and queen run to their beloved daughter, and at the sight of the princess's wounded hand, realize the full force of the misfortune which has befallen them.

Then the old woman to whom the spindle belonged throws off her cloak. They recognize the Fairy *Carabosse*, laughing at the despair of *Florestan* and his queen. The four princes draw their swords and rush towards her, but *Carabosse*, with a diabolical laugh, disappears in a cloud of smoke and fire. The four princes and their retinues run out in fright. At this moment a fountain at the back of the stage is illuminated by a magic light and the *Lilac Fairy* appears within it. 'Don't worry,' she says to the despairing parents, 'your daughter is sleeping and will sleep a hundred years, but that her happiness shall not be affected, you will slumber with her. Her awakening will be a signal for yours; return to the castle; I shall watch over you.' The sleeping princess is placed on a sedan chair and borne away, accompanied by the king, queen, and the highest officials of the court. Gentlemen, pages, and guards bow to the procession. The fairy gestures with her wand in the direction of the castle, and the groups of people on the threshold and on the staircase suddenly fall asleep, as if struck with slumber. Everything falls asleep, including the flowers and the sprays of the fountain. Ivy and creepers grow up out of the earth and cover the castle and the sleeping people. Trees and large bowers of lilacs flourish magically as a result of the fairy's influence, and transform the royal garden into an impenetrable forest. The fairy's loyal spirits group themselves around her, and she commands them to guard the castle, so that no one is emboldened to disturb the calm of the people she is protecting.

(Scene)
End of the First Act

ACT II

SCENE 3

Prince Désiré's Hunt

A forest glade, with a broad river flowing at the back of the stage. The dense forest continues into the distance. At the audience's right—cliffs, covered with vegetation.

The landscape is filled with bright sunlight.

At the rise of the curtain the stage is empty, hunting horns are heard, then Prince *Désiré's* hunt, which is pursuing game in the neighbouring forest. Hunters and huntresses enter and settle down on the grass to lunch; soon Prince *Désiré* appears with his tutor *Galifron* and several courtiers of the king, his father. Lunch is prepared for the prince and his retinue. To divert the prince, hunters and ladies put on dances, practise archery, and devise various games. *Galifron* urges his student to join the company and especially to be charming to the ladies, as he shall soon have to choose a spouse from the nobility of his own homeland. Neighbouring kingdoms only have royal sons, not one royal princess to whom he might be married. *Galifron* takes advantage of the occasion to show him the young noblewomen of the land.

Dance of the Duchesses
Dance of the Marchionesses
Dance of the Countesses
Dance of the Baronesses

All these young ladies try to please the prince, but *Désiré*, glass in hand, smile on his lips, simply watches the fruitless efforts of this crowd of beautiful girls. His heart has not yet spoken, he has yet to meet the subject of his dreams, and will not marry before finding the woman he seeks.

Hunters come in to tell the prince that a bear has been trapped and if the prince wants to kill it, this is a certain opportunity. But the prince is tired: 'Hunt without me,' he tells his retinue, 'I want to rest here, this place pleases me very much.' The hunt and the court withdraw, and *Galifron*, who has drained more than one bottle, falls asleep near the prince.

No sooner has the hunt withdrawn than a boat of mother-of-pearl appears on the river; it is bedecked with gold and precious stones, and from it the *Lilac Fairy*, who is also Prince *Désiré's* godmother, steps ashore. The prince bows to the good fairy, who is favourably inclined towards him and asks him what he feels in his heart. 'You are still not in love with anyone?' she asks. 'No,' the prince answers, 'the noble girls of my homeland have not moved my heart, and I prefer to remain a bachelor than to marry someone only for reasons of state.' 'If that is the case,' the fairy says, 'I shall show you your future wife: she is the most beautiful, most captivating and most intelligent princess in the world.' 'Where can I see her?' 'I shall call forth her shade, and if she pleases you, you may fall in love with her.' The fairy waves her wand towards the cliffs, which open to reveal *Aurora* and her girlfriends, asleep. At another sign from the fairy, *Aurora* and her girlfriends awaken and appear on stage. The rays of the rising sun illuminate her with a rosy light.

The delighted prince pursues the shade, which evades him. Her dance, now tender, now lively, enthuses him more and more. He wants to reach her, but she still eludes his grasp, appears where he did not expect her, and finally disappears in a cleft of the rocks.

Mad with love, the prince runs to his godmother and falls at her feet. 'Where is this heavenly being whom you have shown me? Lead me to her; I want to see her and to press her to my heart.'

'Let us go,' the fairy says, and leads him to her boat, which gets underway immediately. *Galifron* continues to sleep.

The boat moves quickly and the landscape becomes ever more wild (panorama).

Evening comes, then nightfall; the moon illuminates the boat with its silvery light; the castle is seen from afar, and disappears again in a bend of the river. But there it finally is—the goal of their journey. The prince and the fairy disembark.

With a wave of her magic wand the fairy causes the gates to open. The entrance hall is visible, where guards and pages are asleep. The prince rushes in, accompanied by the fairy. The stage is obscured by dense clouds; peaceful music is heard.

Musical entr'acte

SCENE 4
The Sleeping Beauty's Castle

When the clouds disperse the room is visible; Princess *Aurora* is sleeping on a large canopied bed. The king and queen are sleeping opposite her in two armchairs; courtly ladies, cavaliers and pages are asleep standing up, leaning against one another and forming groupings of sleepers.

A layer of dust and cobwebs covers the furniture and the people. The candlelight is asleep, the fire in the hearth is asleep. To the left of the bed the doors open and the fairy enters with the prince. He runs to the bed, calls to the princess in vain, would awaken the king, queen, and *Catalabutte*, who is sleeping on a stool at the king's feet. Nothing avails, but only raises clouds of dust. The fairy remains a benign observer of the prince's despair. Finally he rushes to the sleeping beauty and kisses her on the brow.

The spell is broken; *Aurora* awakens, and with her the entire court. The dust and cobwebs disappear, candles illuminate the room, the fire flares up in the hearth. The prince begs the king's consent to marry his daughter. 'It is her destiny,' the king answers, and joins the young people's hands.

End of the Second Act

ACT III

SCENE 5
The Wedding of Aurora and the Prince
(The esplanade of Florestan's Castle)

Entrance of the king, queen, and the newly-weds with their retinue and the Fairies of Diamonds, Gold, Silver, and Sapphires.

Polonaise: Procession of the Fairy Tales

1. Bluebeard and his wife
2. Puss in Boots
3. The Marquis *de Carabas*, on a sedan chair, and his footmen
4. Goldilocks and Prince *Avenant*
5. Donkey-skin and Prince *Charmant*

6. Beauty and the Beast
7. Cinderella and Prince *Fortuné*
8. The Blue Bird and Princess *Florina*
9. The White Cat, carried in on a pillow
10. Little Red Riding Hood and the Wolf
11. *Ricky of the Tuft* and Princess Aimée
12. Tom Thumb and his brothers
13. The Ogre and the Ogress
14. The Fairy *Carabosse* in a wheelbarrow pulled by rats
15. The Fairy *Candide* and her genies
16. The Fairy *Violente* and her genies
17. The chariot of the Fairy *Canari* and her retinue
18. The *Lilac Fairy*, carried by four large genies

Divertissement

Pas de quatre
The Fairy of Diamonds
The Fairy of Gold
The Fairy of Silver
The Fairy of Sapphires

Pas de caractère
Puss in Boots and the White Cat

Pas de deux
The Blue Bird and Princess *Florina*

Pas de caractère
Little Red Riding Hood and the Wolf

Pas de caractère
Cinderella and Prince *Fortuné*

Pas berrichon
Tom Thumb, his brothers and the ogre

Pas de quatre
Aurora, Désiré, the Fairies of Gold and Sapphires

Entrée of the ballet
Sarabande: Roman, Persian, Indian, American, and Turkish

Ensemble coda

APOTHEOSIS
(*Gloire des Fées*)

19
Skalkovsky's Review of *The Sleeping Beauty*

THE new ballet *The Sleeping Beauty* cannot be called a novelty for our public because, according to newly issued directives, two dress rehearsals of it were given with the admittance of the public, and since we have, even in the highest society, an extremely large number of enthusiasts to watch for free, it is understandable that what is called 'all Petersburg' has already seen the new ballet. This, however, did not keep the public from overflowing the Maryinsky Theatre on Wednesday, 3 January, all the more since rumours about the ballet's success at rehearsals fired curiosity, especially among the adherents of the composer Mr Tchaikovsky, whose numbers were considerable.

The public's appetite was satisfied; the new ballet had success, much more than other recent ballets, though in story and dances it yielded, for example, to *The Vestal* of the selfsame Mr Petipa. At first it was thought that the subject of the new ballet was to be one of the 'sleeping beauties' of the infamous Bureau of Records [a gibe directed at the indolence of the civil service]; it turned out, however, that it was taken not from the Bureau but from Perrault's celebrated tale 'La belle au bois dormant', adapted by none other than Mr Vsevolozhsky himself. Such a collaborator compelled the entire ballet administration to respond to the business at hand in the most attentive manner, except Messrs the decorators, because of whose artistic slowness the ballet did not appear on schedule, which in turn caused a perceptible reduction in the ballet's takings for a whole month—a deficit of approximately twenty thousand roubles!

After this introduction, having poured eau-de-cologne into the ink—out of respect to such an important matter—let us embark on a review of the new ballet.

The Sleeping Beauty opens with a long prologue representing Princess Aurora's christening celebration and the appearance of the Fairy Carabosse; it is not very interesting and should be cut by half or two-thirds, since the new ballet is very long without it. The second scene reveals a beautiful decoration (Mr Bocharov) of a garden in Aurora's palace; the first tableau with the knitters could also be shortened with profit. Then come dances occasioned by the presentation of the princess's suitors: a pretty *valse villageoise* performed by the corps de ballet in beautiful costumes, which pleasantly set off Mlles Ogoleit III, Nikolaeva, Kuskova, and other 'adornments' of our corps, and then a *grand pas d'action*. Mlle Brianza appears in the latter in a very effective bright red costume which goes beautifully with the Italian ballerina's black hair and eyes. All her dances: entrée, *adagio*, and variations on pointe are extremely elegant, masterfully and freely performed. The scene ends with the princess dancing with a spindle given to her by the evil fairy, pricking her finger, and in torment... falling asleep. The entire court falls asleep with her, and they sleep so long that a century's worth of vegetation grows out of some of the courtiers' mouths.

The following scene consists of three parts. First Prince Désiré celebrates with his court in a forest on the hunt. The Louis XIV costumes are beautiful. They play Blind Man's Buff and perform very nicely produced old French dances, ending with a farandole. Mlles Nedremskaya, Zhukova II, [and] Voronova pleasantly enliven this tableau with their prettiness.

The prince is left alone; Mlle Petipa appears in the form of the good Lilac Fairy and shows the prince Aurora's shade accompanied by nymphs. A number of classical *pas* ensue, where Mlle Brianza again performs a series of most difficult pirouettes and complex groupings on her 'steel pointes'. But a more 'shade' is too little for the prince; he wants something real. The good fairy seats him in a gondola and they sail to the enchanted castle, where the prince's kiss should awaken the sleeping Aurora. The journey is represented by the movement of two quite beautifully drawn decorations (Mr Bocharov) which recall, they say, those produced in Bayreuth for Wagner's *Parsifal*. By the time the castle is revealed darkness has fallen on stage, very artfully done, such darkness as you don't even find in the heads of contributors to *The Citizen* [a conservative weekly newspaper]. After two or three minutes the light returns to reveal the interior of the sleeping princess's castle. Mr Ivanov's decoration for this tableau is magnificent.

The last scene of the ballet represents an esplanade in front of the castle in late seventeenth-century style (decoration by Mr Shishkov) on which a huge masquerade is taking place, beginning with a polonaise in which characters from Perrault's best fairy tales are taking part. Then comes a long series of dances, as if intentionally laid in *pour la bonne bouche*. Good fairies perform a *pas de quatre* of precious stones. According to the affiche the diamond is Mlle Johanson. She is in fact the best of the four.

The gracious Mlle Anderson and Mr Bekefi dance a comical *pas de caractère*, 'Puss in Boots and the White Cat'. It was encored. It also enjoyed success for the clever imitation by the orchestra of the miaowing and snorting of cats. Mlle Nikitina and Mr Cechetti demonstrated their outstanding art in the *grand pas de deux*, 'The Blue Bird and Princess Florina'. Mlle Nikitina, charmingly dressed in feathers, was light and airy, like 'fluff from the lips of Aeolus', Mr Cecchetti amazingly agile and strong. In one of the groupings he threw our ballerina up on his back like a little ball from behind with his left arm. Mr Lukyanov and Mlle Zhukova perform the character *pas* 'Little Red Riding Hood and the Wolf'. One of our best soloists, Mlle Zhukova danced with expression, and the red cap on her head looked very good on her. Mlle Petipa in Cinderella's costume spiritedly danced the mazurka with Prince Fortuné (Mr Kshesinsky). After the comic *pas berrichon* (Tom Thumb and his Brothers with the Ogre), Mlle Brianza appeared in the last *grand pas*—a *pas de quatre*. The ballerina performed a number of difficult and brilliant dances, the general complexity of which, in our opinion, the public still insufficiently appreciates. If things did not go quite smoothly in places, it was exclusively the fault of her partners, and to Mlle Brianza in the role of Aurora we give the highest marks. The scene ends with a grand mazurka performed by the corps de ballet in ballet costumes of Louis XIV's time, and the whole ballet concludes with an apotheosis—a splendid living picture representing Apollo.

Apart from two or three superfluous scenes and a certain monotony in the large *ballabiles*, Mr Petipa distinguished himself as balletmaster. It requires much intelligence, a vast amount of taste, great understanding, and a rare love of work and patience to assemble such a huge piece, to work it out down to the finest details and then to teach it to a hundred people. One cannot forget that this is already the thirtieth if not the fortieth of Mr Petipa's works.

Of Mr Tchaikovsky's music our music critic will make a more detailed and competent judgement. We can observe only that the music is melodious, easily listened to, elegantly orchestrated, and pleased the public, which called for the composer several times. In places, for example in the variations of the prima ballerina, the rhythm is insufficiently distinct, which is very disadvantageous for a performer. Of course, rhythm too marked imparts vulgarity to the music, but it is necessary in dances: the latter receive from it great picturesqueness and concentrate the public's attention in the hoped-for place. The best part of the music is during Désiré's encounter with Aurora's shade in Scene 3, where there is a wonderful solo on the violincello.

The new ballet's production is extremely luxurious, the costumes—excellently drawn, partly after Doré's illustrations to Perrault's tales—are elegant. What could be more beautiful, for example, than the beautiful Mlle Petipa's costumes? The costumes are perhaps even too luxurious in their material, for made of silk with gold and similar decorations, they seem heavy for ballet. We pointed out the same defect in the ballet *The King's Command*. Of course, ballet demands magnificent production and economy is out of place, since an effectively produced ballet will sustain a larger number of performances, bring in better receipts and sooner pay for itself, but there is a limit to maginificence imposed by the very nature of art. The new ballet might justly be called: 'The Sleeping Beauty, or the Triumph of the Art of Sewing'!

The Sleeping Beauty will of course bring in receipts, fascinate the public and interest it in the sense of the spectacle, but it represents a transition to the *féerie*, and will force true lovers of the choreographic art to sigh, though their number, alas, is everywhere diminishing.

For them greater or lesser magnificence of production is secondary. They seek aesthetic pleasure in ballet. And that is possible even when dancers are dressed in white tarlatans. But then beautiful legs are needed, strict school in all the performers, excellent ensemble, colossal virtuosity in the ballerina, serious mimed acting. A huge number of extras in expensive costumes and various transformations are nothing more than a large *balagan*; it brings in receipts, especially in the capitals where people number in the millions, but such a spectacle has no ongoing public, nor a circle of enlightened connoisseurs.

The aforesaid direction is noticeable not just in ballet. In opera also nowadays bel canto is not prized, but rather the dimensions of the orchestra and the number of acts: steam machines are introduced to shift decorations, and double-basses are counted by tens; in performances of drama no one pays the slightest attention to the lines and how they are delivered, but a play is not successful unless the actresses expend five or ten thousand roubles beforehand on their *toilettes*. Such is the influence of the democratization of art.

20
'A Musical Letter from Petersburg. Apropos *The Sleeping Beauty*, the ballet of M. Petipa, with music by P. Tchaikovsky'

German Avgustovich Laroche

When an amateur wants to speak out on some specialist subject, he normally begins with an apology; I am not a specialist, he would say, I do not understand these subtleties of yours, I am a simple man, my heart dictates to me, etc.

I must proceed in exactly the same way. Speaking of the new work of Marius Petipa and P. I. Tchaikovsky, I, from the reader's standpoint, would have to be limited to the musical aspects of the subject, to speak about Tchaikovsky and keep silent about Petipa. But musical phenomena never seem to me other than as relating to the whole environment out of which they arise and on which they act in turn; public, school, era, nationality, currents of literature, and the educational arts always seemed to me extremely important for the explication of both the separate work and the whole musician. These conditions have significance even in *absolute* instrumental music, that is, in music without programmes and poetical titles, which strives exclusively towards a beautiful architecture in sound. And thus, I am all the more drawn to speak of 'subjects outside music' when the music itself springs from the terrain of another art—for example, when the music is—dramatic.

When writing of opera I have almost always digressed into the realm of literature, though there was nothing original in this. Every music critic evaluating a libretto becomes something of a literary critic for a time. For us musicians, criticism of this kind is not difficult and does not require scholarly preparation, first because we shall still judge the music from a musical point of view, and second (and chiefly) because poetry is the most accessible and the most *general* of all the arts.

Writing about ballet music is not at all the same. You want to say something about the art close to and akin to music in many respects, which inspired the musician. Just as there is nothing more widespread, accessible, and beloved than poetry, there is nothing more exclusive unto itself, more special and less popular than ballet. In some sense it can be said that we all terribly love ballet, and I agree completely with Pisemsky when he divides the Moscow students of the 1840s into a Sankovskaya faction and an Andreanova faction. But just as the hero of the old French operetta sang, 'Ce n'est pas la danse que j'aime, c'est la fille à Nicolas,' the Moscow student could say that in ballet he loves not ballet itself but rather Sankovskaya or Andreanova.

Our attitude towards ballet is much more self-involved than towards opera, precisely as in its turn opera excites a much more self-involved attitude than concerts do. In aesthetic criticism one must be able to relate objectively to facts of this kind and evaluate a number of fluctuations in judgement or confusions in thought produced by them. I repeat: we all loved Sankovskaya in our youth, and some of us continued to love her in our maturity. But ballet as an art has nothing to do with that. If the opposite were true, if choreography in the aggregate of its technical conditions and poetical resources possessed the same power of attraction as painting, for example, there would be amateur dancers just as there are amateur painters, and they would teach the family children to hold their legs on a horizontal crossbar gradually being raised as they now teach them to pay attention with a pencil. In fact ballet is an art which, excepting a very small number of people, the great mass of public doesn't know or want to know; compare the number of opera theatres with the number of ballet theatres, compare in our own Maryinsky Theatre the number of ballet patrons with those of opera, and you will see that choreography, at least in our time, is a distant, unattractive, uninteresting domain which has therefore become little accessible and little comprehended by the non-specialist.

But this art's absence of likeableness for the masses is combined with an unspeakable enchantment for the artist. Poets sing its praises, sculptors and painters represent it, and musicians wait upon it. In the musicians of our time there is a noticeable choreographic tendency: instead of specialists in ballet composition, as in the first half of our century, symphonic and operatic composers have set to writing for ballet: Lalo, Delibes, and here at home Rubinstein and Tchaikovsky are writing ballet music, as Lully and Rameau, Gluck and Haydn did in the good old days. Choreography itself—so some authorities tell us—has fallen into decline, yielding to pantomime, while music for it has progressed from the banal motifs of earlier times, loudly and monotonously orchestrated, to the choicest, most elevated style.

The Sleeping Beauty was abused by our newspaper critics. Besides the music, with which our reviewers were not always satisfied, there were complaints about the absence of drama in the ballet, about the children's fairy-tale story, and finally that the fable was taken from Perrault's French redaction. Of these charges, the last, at first glance, is most reasonable. No one prohibits our balletmasters from representing France on stage, as they represent ancient Greece, ancient Egypt, contemporary Italy, and Montenegro. To mount a ballet from the history of medieval France or France of the last Louis would be completely legitimate and natural. But *The Sleeping Beauty* does not belong to history, nor does it have an actual basis in a real locale; it is a mythic image, belonging to the legends of many peoples; the origin of the myth is lost in the gloom of prehistoric times, and all European—at least all Aryan—peoples have equal right to it. The French redaction, I submit, was selected simply because the balletmaster of the Bolshoy Theatre [that is, Maryinsky Theatre], like many of his predecessors, is a Frenchman, and therefore our ballet, despite the purely Russian constituency of the corps de ballet and most of the soloists, lives by French traditions and is nourished by French literature.

For my part I have nothing against the fact that on this occasion we danced a bit in French. Who is to keep us in the next ballet from dancing in Scottish, and after that in Spanish, Assyrian, Japanese? I do not at all like the example of French literature, especially contemporary, which through timidity and squeamishness is shut into its own country and carefully avoids foreign topics. To force ourselves out of false patriotism to deal exclusively with the homeland's life and history means to refuse our imagination the pleasure of a journey, to limit perspective arbitrarily and to restrict materials, to forget the glorious tradition of Pushkin's era, when nothing 'human' was 'foreign' to us. How an artist responds to a chosen topic is another question: was he able to preserve the Russian *style* amidst foreign *local colour*? This task is subtle and difficult, but to this question I shall return.

Another, more general reproach directed at the new ballet is that it is not drama but a 'children's tale'. If you were to say that this dramatized tale lacked stage intrigue or what it is now acceptable to call 'a beginning and an end', it would be impossible not to agree. A theatre piece with a properly tied dramatic knot is not the only, nor perhaps even the highest, type of theatre piece, but a dramatic presentation without such a knot is no doubt more difficult to watch and demands greater aesthetic development from its audience than a piece which has a plot. Be that as it may, there is no dramatic knot in *The Sleeping Beauty*. The drama is reduced to a simple anecdote: a princess pricks her finger and falls asleep, and after a hundred years a prince arrives, awakens her, and marries her. Clearly what is at fault here is not fairy tale as a genre, but this particular one.

If, broadly speaking, fairy tales are not suitable for ballet, then some new basis for it should be sought. Up to now the large majority of ballets have been based, if not on 'children's tales' then on myths, legends, and other creations of the folk imagination: *real* subject-matter, 'the living truth' in that special sense which is attached to this word in our time, was always weaker in ballet than in opera, as in opera it was weaker than in spoken drama.

Nineteenth-century realism has its significance and its fascinations. But 'the living truth' does not reside exclusively in the representation of humdrum contemporary life, towards which realism in the nineteenth century is being driven; nor does it lie in the exceptional, stupendous events so suitable for tragic subjects, any more than in contemporary life. But there was and continues to be a truth in the barely flickering distance of deep antiquity. Representing Ninus or Semiramis, Nimrod and Assur we can be no less 'true' than in a 'scene' of contemporary boulevard mores or in a novel about the Serbian War. One of the truest works of the artistic spirit, one most faithful to life—is myth, provided it is a genuine, 'real' myth and not counterfeit, not the latest fabrication, not intentionally conceived.

The fairy tale, despite its prosaic form, quite often contains within itself the most ancient and the most genuine myths: *The Sleeping Beauty*, incidentally, in its basic story similar to that of Brünnhilde who is protected by fire, is one of the countless embodiments of earth put to rest by winter and awakened by the kiss of spring; Siegfried and Prince Désiré in this sense are the same person. In the form in which we read this and stories similar to it in Perrault, the Grimm brothers, and Afanasiev, this incarnation of myth becomes a children's tale. Generations of children, one after another, have been and continue to be nourished by these works of naïve creation, unfading in their artless beauty; their taste, feeling and imagination have educated one generation after another. This education is not comprehensive or profound, but how much healthier it is than the education of those of our gentry's offspring who in childhood devoured Dumas-père, Eugène Sue, Frédéric Soulié, and Paul Feval.

Criticize children's stories as you will, you will not refute the facts that in the continuity of generations they have managed to take deep root in our imagination, that since childhood we have got used to them and loved them, that they contain some of the most profound ideas which stir humankind, nor finally the fact that under the influence of comparative mythology and its advances, supposed 'children's' tales, in contemporary eyes, have more and more become *tales for adults*, ever more revealing of their cosmogonic significance.

You do not like a ballet based on fairy tale; you don't want a ballet to resemble a *féerie*, even such a well-proportioned, elegant *féerie*, harmonious in its richness, as the one presented to you at the Maryinsky Theatre. But at the same time, being concerned about ballet, you wish it success. You know how easy it is for this magic garden to degenerate into a neglected hothouse, rarely attended and poorly cared for. Perhaps you hope to save it by introducing an element of sanguine drama. You get exited by Zucchi in *Esmeralda*, and the masterful representation of a young woman crippled by torture strikes you as the last word in choreographic progress.

In recent years sanguine drama has much extended its dominion. It has filled the small private theatres and the summer-time pleasure haunts. Where once the music-hall *chanteuse*, in simplicity of spirit, crooned stupidities and revealed her naked contours, now villains forge their snares, the dagger and revolver reign supreme. In Moscow at the Maly Theatre, thanks to the pressure of the best talents on the staff, Russian laughter has been all but muted, replaced with foreign pathos either in its original form and translated into Russian, or in Russian imitations and counterfeits.

The impoverishment of comic talents both in creation and performance comes hand in hand with the growth of public sentimentality, the pervasive pursuit of tears. In the last few years operettas have frightfully declined both in number and quality; but even before this decline set in it had managed to turn into 'serious operetta', to reveal a pretension towards melodrama. And ballet, apparently, is influenced by this contemporary trend. Here it is not laughter that is being driven out—there was never much laughter in ballet, or else it was quite cheap—but rather the childlike dream, childlike joy, childlike belief. 'Cette puissance d'évocation, qui fait que l'enfant vit dans une espèce de perpetu miracle', as Anatole France once said. I do not know how much choreography will benefit from the replacement of magic tale by 'historical drama' à la Victor Hugo, even if refashioned on to Russian mores, nor how much more interesting pirouettes, entrechats, and other subtleties of the art of dance will become if we train ourselves to seek the most profound 'purpose' for each of them, in the manner of the Wagnerian leitmotif.

I have very much gone on in the *dilettante* part of my *feuilleton*, the part for which I found necessary a preliminary apology. As a result, in the specialist part, where I can express myself without reservations and caution, I am obliged to be brief despite the richness of the theme. If I am justifying Mr Marius Petipa for having produced a *ballet-féerie*, then I am all the happier that such a powerful talent as Tchaikovsky, following the general trend of the time, has turned to ballet and promotes thereby the ennobling of musical taste in this sphere as well.

If music in its essence lives in the sphere of the undefined, then it is all the closer to the realm of myth, tale, of ancient epos generally, for primeval creation as a whole lives in the sphere of the undefined, and Richard Wagner was correct when he pointed out the profound kinship between the essence of music and the essence of epos. Only it did not follow from this to make a pseudo-Aeschylus out of the Scandinavian Edda, accompanied by an orchestra placed beneath the floor level of the auditorium, or to vie at once with Shakespeare and Berlioz, Racine and Schumann, Schiller and Gounod. For a fantastic opera like *Ruslan and Lyudmila* or for that matter an opera-ballet in the genre of Auber's *Le Dieu et la bayadère* [taking motifs from various traditions, as Wagner did in] *The Nibelungs* makes an excellent story. It is equally right to say that *The Nibelungs* and *Ruslan and Lyudmila* and *Le Dieu et la bayadère* and *The Sleeping Beauty* are all excellent stories for ballet, and that for their diversity of origins these stories are so perfectly suited to music that they could also serve, for example, as the bases of symphonic programme music. Choreography and music, symphonic poem and ballet, symphony without programme and *féerie*—all these are spheres very close to one another, and they are all suffused in the general element of the undefined and spontaneous.

In his youth Pyotr Ilyich Tchaikovsky wrote one unpretentious ballet. In its time it was produced in Moscow, where the costumes and decorations, if I am not mistaken, were old and *ad hoc*. But one already sensed in the music of *Swan Lake* a symphonist unusually gifted for ballet as a genre, a cultured musician, alien to pedantry and aloofness, able at any moment to doff the professional mantle and sit down at the piano and improvise dances, if only for a children's party. He is one of the premier melodists of our time. But up to now he has expressed this melodic gift mostly in kinds of music officially recognized as 'serious': the symphony, quartet, piano piece, in the symphonic poem, cantata, in the song and in opera of tragic content (only one opera, *The Little Slippers*, was semi-serious). Elegiac by nature, inclined to melancholy and even to a certain disillusionment, he manifested seriousness of another kind in these genres officially dubbed 'serious'—a seriousness of meditation, part sadness and part anguish, often an anxious heartache—and this, if it may be thus expressed, *minor-mode* part of his being (similar to Chopin) was seized upon and appreciated quickest of all.

But next to this Tchaikovsky there is another: the good-natured one— festive, full of blossoming health, inclined towards humour, perhaps even towards a *charge d'atelier*—and this Tchaikovsky, similar not to Chopin but to Schumann and most of all to Glinka, has been much less noticeable and recognized up to now, if we judge by the obscurity in which the most optimistic of his outpourings abides. Nevertheless, he has an inexhaustible store of dance melodies—not, of course, the kind that the operetta and ballet of 'the good old days' instilled in us—but diverse and careful in harmony, light, graceful, full of movement and tender charm. In the accompaniment of these melodies (especially in *The Sleeping Beauty*) counterpoint plays a major role (normally the counterpointing melody in the middle voice). This device may leave him open to criticism from balletomanes raised on Adam and Pugni, who have still not managed to advance to Strauss waltzes; this device has long since been adapted to Strauss waltzes, though not so richly and artistically as in Tchaikovsky.

The Russian way in music, so powerful in Tchaikovsky in recent years, is felt time and again. The music suits the costume and the character completely; it has a French nuance, but at the same time it savours of Rus. Whatever you feel, I passionately love this *French* tale accompanied by music in the Russian manner.

The point is not in local colour, which is beautifully observed, but in a more general and profound element than colour—in the internal structure of the music, primarily in the basic element of melody. This basic element is undoubtedly Russian. One can say without lapsing into contradiction that the local colour is French, but the *style* is Russian. Just as the ancient Greeks in Mendelssohn-Bartholdy are *German* Greeks, just as the Jews in Goldmark's *Die Königin von Saba* are *Viennese* Jews, and the Hindus in Delibes's *Lakmé* and in Massenet's *Le Roi de Lahore* are Parisians of the Second Empire, so the fairy-tale images of the ancient Aryan epos, transformed into *French* images by national assimilation and transmission, were subjected to new transformation under the pen of the Russian musician, assumed a new nationality, became a Russian variant. For not one of them can fully renounce the influence of the land, and blessings to Pyotr Ilyich that his development corresponded with the time when the influence of the land grew stronger among us, when the Russian spirit took flight, when 'Russian' ceased to be a synonym for 'peasant-like', and when the 'peasant-like' itself was recognized in its proper place, as but *part* of being Russian.

If in its production *The Sleeping Beauty* is one of the pearls of our theatre, then in its music it is one of the pearls of Tchaikovsky's creation. Together with *Eugene Onegin* and the composer's symphonies, his First and Third Suites, his piano concerto and the Fantasia for piano and orchestra, several of his songs and many episodes from his operas other than *Eugene Onegin*, it represents the high point for the time being which the school of Glinka has reached, that point at which the school is already beginning to free itself from Glinka and to open new horizons before now unrevealed. In this firm attachment to Glinka, in this continuous growth, in this slow and steadfast striving upwards—lie the pledge of its vitality, and so *young* is its power that any prediction about the direction of its development and the height it will achieve is premature. I hope still to return to the music of *The Sleeping Beauty* and to point out several details in it, and also to return to the definition of Tchaikovsky's talent from a completely different angle—in connection with the production being mounted in Moscow of his most 'advanced' musical dramas—*The Enchantress*.

21

'The Sleeping Beauty', from *My Reminiscences*

Alexandre Nikolaevich Benois

It seems that more than once in these memoirs I have recalled the impression that Tchaikovsky's *The Sleeping Beauty* made on me. It is necessary now to expound this passion of mine, as it initiated in me a turn towards Russian music; from my complete ignorance of it (and even a kind of contempt), this passion brought me around to an enraptured admiration.

The première of *The Sleeping Beauty* took place in the last days of 1889 or the very first of 1890. At that time, since the days of Zucchi, I had stopped attending ballet and was not *au courant* about what was happening in it. At the same time I continued to share a prejudice, typical of our family, whereby we reacted to Russian music with scorn. To my brother Leonty, and later on to me, it seemed that Tchaikovsky was unable to create something worthy in an area where Adam and especially Delibes had done brilliant work. How could a Russian composer be so bold as to take on a Perrault tale? And too, such prejudice was in keeping with a certain claim to the effect that *The Sleeping Beauty* had met with a chilly reception at its dress rehearsal. Leonty had been present, and to him as to the majority of others gathered there, the music seemed 'without much melody', too complex and chaotic, and worst of all, *not dansante*. A rumour even circulated that the artists refused to dance to it, so incomprehensible to them did it seem.

The entire company was involved in *The Sleeping Beauty*, as well as two Italian stars of the first magnitude: Signorina Brianza and Signora [and Signor] Cecchetti. The costumes, made from drawings by Director Vsevolozhsky himself, were distinguished by their magnificence, and our best theatre artists drew brilliantly effective decorations, of which the moving panorama was especially well liked. But all this 'didn't help'. Stories were even going around that the emperor, who was sitting in the first row of the stalls (and not in his box on the side), did not favour Tchaikovsky with a single word, that he turned his back to Vsevolozhsky and immediately headed for the exit when the ballet was over. Such an inauspicious response by the tsar ought inevitably to have led Vsevolozhsky to submit his resignation, and to the withdrawal of the ballet itself. But all this was only talk . . . Things turned out quite differently.

I was not at the dress rehearsal or the first performance, and saw *Beauty* for the first time probably at the second performance. In any event, it occurs to me that it was a matinée in the New Year's holidays, and this enabled Dima Filosofov and me to attend the performance. And I must confess that this first impression of *Beauty* was for me if not some kind of revelation, enough to make me leave the theatre feeling that I had attended a very splendid feast. What I saw and heard showed me, in any event, something worthy of attention, and concerning several passages of the music I sensed a kind of premonition that they might turn out completely to my taste. I was simply not yet resigned to believe what was already then being born in the secret reaches of my heart. At the same time I very much wanted to go to *Beauty* again and soon—chiefly to listen to this music.

And then for the second time I 'couldn't believe my delight'... Possibly it helped that I immediately acquired a piano arrangement of the new ballet and my pianist friends Nouvel and Pypin played through for me everything I found of interest. The principal themes, the most important moments of the music, were impressed in my memory, and a great deal 'became clear'. It turned out that Tchaikovsky's music was not only excellent and charming, but that *this was the very thing* that I *had somehow always been waiting for*. Already at the second performance I attended, it was not the spectacle or the dances or the performance or the artists which captivated me, but the music which won me over, something infinitely close, inborn, something I would call *my* music. In a word, I *fell in love* with Tchaikovsky's music, and Pyotr Ilyich himself (in our circle it was acceptable to call him by the more familiar 'Uncle Petya') became someone very close to me, though I never had occasion to meet him personally.

By now I did not miss one performance of *Beauty*, and once contrived (in Butter Week, during which matinees were given in addition to evening performances) to see this ballet four times in seven days. Moreover, I began to listen to Tchaikovsky's music wherever possible, both in concerts and playing at home. My passion for *Beauty* led me quickly to love the whole of *Eugene Onegin*, which before then I had responded to with distrust. I had known only excerpts from the opera, and it struck me as being not far removed from Massenet and Ambroise Thomas, composers whose merits I do not dispute, but who had by then ceased to be my favourites.

Analysing now the way *Beauty* possessed me then, I see that my passion was based least of all on lyrico-melodious elements, which include the music of the celebrated 'panorama' and the leitmotif of the Lilac Fairy. On the contrary, I never tired of listening (and indeed, would not tire even now) to the entr'acte between the hunt scene and the 'awakening', or to all of Fairy Carabosse's music. It was genuine Hoffmann, music which led into that fantastic, terrifying yet sweet world so fully reflected in the stories of my favourite writer, a world of captivating nightmares, a world close at hand yet inaccessible. This mixture of peculiar truth and persuasive invention had always especially attracted and frightened me at once. I had already sensed it at the time when my brother Isha drew his strange stories for me, or when Albert improvised his little tales, which music made fully convincing. One cannot compose the 'vile' music of contemptible Carabosse without having experienced the influence of some evil spirit's power. Otherwise one could not impart to this music that malicious quality of jest which gives it a special infernal acuity. In exactly the same way Tchaikovsky could not have composed the entr'acte with its music of genius without recalling that sweet languor which one experiences as a child in a half-somnolent fever sinking ever deeper into oblivion without stopping to detect precisely the echoes of reality fading ever further into the distance . . .

How many and how generously strewn throughout the score of *Beauty* are its various splendours, from the bold march in the prologue to the dances of the fairy tales! It is impossible to count them all. And how well all this is done, how well it sounds, what a welling forth of sound everywhere! *Beauty*, possesses another trait as well (which I find in *The Queen of Spades* and *The Nutcracker* also), namely something we once referred to with the faulty term 'epochal quality' and which, finding no better expression, we then called by the no less faulty term 'passé-ism'. Pyotr Ilyich was unquestionably one of those people for whom the past had not completely and forever disappeared, but continued somehow to live, intertwined with current reality. This trait is a most treasured gift, something in the nature of grace; it broadens the perspective of life, and thanks to it the very 'sting of death' is not so threatening. Thoughts of death never forsook Tchaikovsky; concerning it he 'knew whereof he spoke' (I am reminded of the fourth movement of the Sixth Symphony). Death never ceased to 'stand over his shoulder'; its close proximity tormented him, poisoned the joy of being, but at the same time he never wavered in the utterly certain knowledge that death did not end everything, that life continues beyond the grave. And his was not some abstract 'idea' of life, something formless and incorporeal, but something fully realizable. And this real sensing of something beyond beckoned to him. It drew him into the kingdom of the shades; he felt that there, far beyond urgent cares, somehow one even breathes freer and easier; there a communion will be renewed with those dearest to us, new encounters of incomparable richness can occur there. In this kingdom of the shades, moreover, not only individuals but whole epochs, their very atmospheres, continue to live.

And Tchaikovsky could summon forth the very atmosphere of the past with the enchantment of music. His success in re-creating the atmosphere of France in the days of the young Sun-King in *The Sleeping Beauty* was something to which only a person absolutely deaf to the sounds of the past could remain indifferent. The entire hunt scene, the games and dances of the courtly people in the forest, and also every turn of musical phrase characterizing Prince Désiré possess a 'genuineness' which is not at all the same as an intelligent counterfeit of antiquity or some kind of stylization.

This applies as well to much of the music in the last act of *The Sleeping Beauty*, which ends with an apotheosis to the melancholy chords of the song 'Vive Henri IV' that once all but served as the theme of a royal hymn. Tchaikovsky achieved an astonishing power in the triumphal march at the beginning of the act (the gathering of the guests who have arrived for the wedding of Aurora and Désiré), and an astonishing *profundity* in the sarabande in the style of Lully. I myself am endowed with this gift for penetration into the past; from personal experience I know how delicious it is. But in relation to Pyotr Ilyich in those days a feeling of special *gratitude* arose in me. Thanks to him my own passé-ism was made especially sensitive; it is as if Tchaikovsky revealed before me the doors through which I penetrated ever further into the past, and this past at times was even closer and more intelligible to me than the present. The magic of his sounds gave birth to the conviction that I was somehow 'coming back to myself', that I *am remembering* with special sharpness that which once was with me, of which I was witness.

Not only by this sorcery did the music of *The Sleeping Beauty* hold me captive, but also by everything in which Tchaikovsky simply remained himself and gave full vent to his ingenious and poetical imagination. Thanks to this freedom of outpouring (*liberté d'épanchement*) he succeeded in producing such indisputable masterpieces as the dances of the several of the fairies, already cited, the beautiful *pas de deux* of Aurora and Désiré in the forest, the march of the fairy tales (Polacca), the little scenes 'Little Red Riding Hood and the Wolf', 'Tom Thumb', the *grand pas de deux* of the Blue Bird, and Prince Désiré's solo which opens the final dance. All this bears the stamp of Tchaikovsky's *personal*, so distinctive taste, yet at the same time it flows together into a complete whole. To this day, listening to the music of *Beauty*, even if at the piano, I experience those same feelings I experienced in the days of my youth, when this music had just come into the world. And when I do, it is not without melancholy that I realize how all this delight was at the time in accord with my bodily and spiritual states, and how much, alas, my present condition as an eighty-year-old man is incomparably less gratifying. For I myself am standing at the threshold of the 'kingdom of the shades'.

As regards production, *The Sleeping Beauty* did not warrant my approval in all respects, although it was indisputably the result of very great and excellently co-ordinated artistic efforts. Especially successful with the public was the decoration, or rather the series of decorations which made up the moving panorama. It was masterfully drawn by Bocharov. I however, found it lacking in a fantastic element. That was a problem which occurred too often on the stage in those days when the representation of forest locales and moonlit nights was required. Finally, in the decorations of the last act and apotheosis Professor Shishkov revealed that the style of Versailles and its festivals was foreign to him. In contrast, the artist Ivanov outdid himself both in *trompe-l'œil* and mastery of technique in his decoration for the scene of the 'awakening'. In effect there were two decorations, of which the first represented Aurora's bedchamber plunged into nocturnal darkness; the pale moonlight made its way through a window overgrown by cobwebs, and in the huge, magnificent fireplace the coals were barely smouldering; in the semi-darkness it was difficult to make out the architectural details, which preserved the spirit of the French Renaissance. Then suddenly, as soon as the prince kissed the sleeping princess's hand, the room was flooded with sunlight which penetrated the remotest corners. All the statues in their niches came alive in a complex play of light and shadow, and flames blazed up merrily in the fireplace.

Much could be criticized about the costumes, for the most part created by the Director Ivan Alexandrovich Vsevolozhsky. One was struck by motley colours, juxtaposed by someone without a true sense of their harmony. And yet it cannot be said that the first production of *Beauty* was bereft of charm as regards costume. One must point out as particularly successful Vsevolozhsky's basic concept, in accordance with which the same prince who was destined to break the hundred years' somnolent torpor was identified with the person of the young King Louis XIV, in whose reign the stories appeared which Charles Perrault collected. Thanks to the shifting of the second half of the ballet into the 1660s (to a setting closer to us), one got the especially sharp sense that the soporific dream which locked away Florestan XXIV's [sic] court for a whole century ended with the return of everything to reality, a reality, however, astonishingly changed. It would seem that this poetical thought, which was the basis of the first production and the very creation of the ballet, ought to have been preserved as an inseparable feature of *The Sleeping Beauty*, all the more so since Vsevolozhsky's intelligent and poetical concept was at one with the composer and the balletmaster. In a subsequent revival of the ballet on the stage of the same Maryinsky Theatre, however, twenty years later in the days of Vladimir Arkadievich Telyakovsky's management, the artist Korovin disregarded this principle. And Leon Bakst, who created the brilliant production of *Beauty* for Diaghilev in London in 1922, did not respond consistently enough to this charming task.

The venerable balletmaster Marius Petipa was at the top of his form in *Beauty*. Moreover, this production was the magnificent crown of all his creations. And one must hope that the scenes and dances he composed will not be forgotten, but rather will continue to serve as an example and models for all producers on whom the honour is bestowed to revive the choreography of this ballet, based on Tchaikovsky's music. The richness and variety of danced and dramatic moments here are such that a single enumeration of their characteristics would take up too much space. I nevertheless cannot but single out those 'numbers' which especially captivated me. I include the whole Carabosse scene in the prologue with the grotesque dance of the creatures who accompany the old witch, the celebrated waltz in the gardens of King Florestan, Aurora's complex *pas d'action* with the suitors, all the dances of the courtiers in the forest and especially the second *pas d'action* in the same setting, and finally the procession of the fairy tales—the dances of the White Cat, the Blue Bird, Tom Thumb and his brothers. What is more, the dances of *Beauty* enchanted me not only by their cleverness and elegance, but also by their consistent stylization—the area where the revival of long past epochs entered into Petipa's charge. He was undoubtedly assisted by his French origins and the school through which he himself passed (in the days when French ballet could still, with complete justification, take pride in its traditions going back before the seventeenth century). It is the same French style upon which our own balletic school is fundamentally based. Thanks to balletmaster-Frenchmen—Didelot, Perrot, St-Léon, and Petipa, and to the French-trained Johanson—this style has been maintained in our dancing school for an entire century right up to the present time. And this style achieved its culminating point in the production of *The Sleeping Beauty*, the success of which perhaps cannot be considered the accomplishment of Petipa alone, for it was the sum of creative strivings previously directed at the same ideal.

I shall not speak in detail of the performers. The fact is, that however elegant Gerdt in his majestic role as the prince, however brilliantly the perfectly young and very pretty Brianza performed her role, however sidesplitting Stukolkin (in the role of Catalabutte, Marshall of the Court), however nightmarish Cecchetti in the role of Carabosse, the same Cecchetti who was charming in the dance of the Blue Bird, however excellent the dozens of other dancers who played fairies, peasants, genies, courtiers, huntsmen, fairy-tale personages, and the like—their collective personal mastery and the charm of each artist flowed together in the beauty of the ensemble. Thus I had at that time the good fortune to see a genuine *Gesamtkunstwerk*. In experiencing it one is especially vexed at the ephemeral nature of all theatrical art, which does not acquire a 'permanent existence'. By some *miracle*, so many first-class artists were brought together for the creation of a genuine masterpiece, precious in all its details, but alas, these people have one after the other departed this life. The decorations faded, the costumes wore thin, and the masterpiece slowly wasted away after some ten to fifteen years. And no reconstruction avails. The conditions, the technical devices all change, are forgotten, even at a time, it would seem, of the most pious observance of traditions.

I am left to take personal solace in the fact that I nevertheless saw this masterpiece, in all its freshness and brilliance, at a time I could take full pleasure in it.

My delight in *The Sleeping Beauty* generally brought me back to ballet after I had grown cool towards it, and I passed this rekindled passion along to all my friends, gradually making them 'genuine balletomanes'. Thus was created one of the principal conditions whereby in a few years we ourselves would become active in the ballet, and this activity brought us worldwide success. I hardly err if I say that were it not for my *violent* attraction to *Beauty* (and before that to *Coppélia*, *Giselle*, and *The Pharaoh's Daughter* with Zucchi), had I not communicated this enthusiasm to my friends, there would have been no 'Ballets Russes' and the balletomania born of its success. Whether this was good or bad is another question, to be decided by each person according to its merits. I, however, must express here the profound regret that on many occasions this mania assumed undesirable forms then unforeseen. And I must further regret that it was my close friend and to a large extent my student who initiated those heresies (from my point of view) which converted ballet into a fashionable affectation and a kind of competition of the most unworthy, shocking devices. What has happened to those ideals with which we appeared in 1909 before the judgement of the weary, blasé Parisian public, already ripened for snobbism? . . .

22

Libretto of *Raymonda*

Lydia Alexandrovna Pashkova

The remainder of the 1890s was rich in accomplishment for Petipa and the Petersburg ballet. After *The Sleeping Beauty* Petipa went on to collaborate with Lev Ivanov on *The Nutcracker* and *Swan Lake*, to revive *La Sylphide* and *The Little Humpbacked Horse*, to produce another coronation ballet (*The Pearl*) and works for private gala performances at Peterhof, as well as to create many new public works, including *Bluebeard*, the ballet which marked his fiftieth year of imperial service.

That anniversary, as one might expect, was filled with sentiment and congratulation, and *Bluebeard* is an appropriate ballet to mark the beginning of the final stage of Petipa's career. While he would never have acknowledged it (or perhaps even noticed), Petipa was being revered in distinctly valedictory tones. This was especially clear in responses to the two big productions of 1898. One was a revival of *The Pharaoh's Daughter*, the affirmation by a new generation of audiences of the significance of that work. The other was a new ballet: *Raymonda*.

Raymonda was neither Petipa's last ballet nor his last successful one, but it is the last to achieve a stature and longevity comparable to his other acknowledged masterworks. It is in many respects an echo and an epigone of *The Sleeping Beauty*. The long-since stereotyped situations—courtiers anticipating the protagonist's arrival, the vision scene which forecasts the coming together of lovers after tribulation, the ritual of hospitality which permits no one, even a villain, to be excluded—are all to be found here. And the transparent borrowings and conflations of character and event from obvious predecessors speak to the sturdiness (and overuse) of certain devices: Raymonda's entrance as a blend of the *Beauty* prologue with Aurora's entrance, the enchantment scene where all present fall asleep, the White Lady as a supernatural protectress similar to the Lilac Fairy, Abderakhman as a hybrid of Carabosse and von Rothbart in *Swan Lake*.

Raymonda stresses ceremony over substance, dancing over story. The exhaustive lists of dances and performers and the fussy subdivisions of the text in the libretto call attention to individual numbers at the expense of the story-line. This could also be said of *The Sleeping Beauty*, but the effect is more noticeable in *Raymonda* because the story itself lacks its predecessor's symbolic content. Petipa emphasized dancing for its own sake in ballets subsequent to *Raymonda*, continuing his turn away from the narrative dramas on which he had built his reputation towards brief, plotless ballets of a kind later taken up by Fokine. *Raymonda* and *The Magic Mirror* are exceptional because a scanty narrative is spread over a long time span, and executed choreographically in the grand manner associated with the dramatic ballets of previous decades.

Critics of the first performance, as if sensing the new emphasis, directed their remarks at the artists to the exclusion of the story, especially at Pierina Legnani, the last and perhaps greatest Italian virtuosa to visit St Petersburg, for whom *Raymonda* was produced. 'Despite the unusual difficulty of her role (she appeared in all four scenes),' one critic wrote, 'Mlle Legnani danced superbly—with incomparable grace, plastique, and strength of movements.'[1] Petipa was also praised—warmly, but without the hyperbole of earlier times:

> Mr Petipa's creativity, his highly artistic taste and mastery, spoke in *Raymonda* in their full beauty and power. The balletmaster carefully avoided any 'stunts' striking for effect—everything in *Raymonda* is artistic, intelligent, and beautiful. For that reason *Raymonda* had a grandiose success, not seen for a long time on the ballet stage . . . The public unanimously expressed its delight and astonishment at Marius Ivanovich Petipa's phenomenal artistic activity . . .
>
> The balletmaster M. I. Petipa was given a laurel wreath, and from a small circle of balletic habitués and several artists a gold wreath in the form of a lyre with the inscription 'Au grand maître-artiste'. The composer Alexandre Glazunov was presented with a wreath and an address by the ballet company. In general last evening's performance had a particularly celebratory character. After the second act the public ovation did not cease for a long time, and when the curtain fell at last, for a long time after that the applause could be heard by which the ballet troupe *in corpore* honoured Marius Petipa and Alexandre Glazunov.[2]

RAYMONDA

Ballet in 3 Acts (4 Scenes)
(Subject taken from knightly legends)
by Mme L. Pashkova

Music by A. K. Glazunov

Dances and Production by balletmaster M. I. Petipa

The role of 'Raymonda' will be performed by Mlle Pierina Legnani

New decorations: Act I, Scene 1, Mr Allegri; Act I, Scene 2, Act III and the apotheosis, Mr Lambin; Act II, Mr Ivanov. Machinist, Mr Berger. Costumes: women's, Mme Ofitserova; men's, Mr Caffi; Headwear: women's, Mme Termain; men's, Mr Bruneau. Accessories, P. P. Kamensky. Wigs and coiffures, Mr Pedder. Footwear, Mme Levstedt. Metalwork, Mr Inginen. Tricot, Mme Dobrovolskaya. Flowers, Mme Revenskaya.

---

[1] 'Theatre and Music', *Sanktpeterburgskie vedomosti* [The Saint Petersburg Gazette], 9 Jan. 1898, No. 7, p. 4.
[2] B., 'Theatre Echo', *Peterburgskaya gazeta* [Petersburg Gazette], 8 Jan. 1898, No. 7, p. 4.

394 The 1890s

Produced for the first time on the stage of the Imperial Maryinsky Theatre 7 January 1898 (for the benefit performance of Mlle Pierina Legnani).

St-Petersburg.
Typographer of the Imperial Theatres, Mokhovaya, 40.
1898.
Permitted by the censor. St-Petersburg, 3 January 1898.

DRAMATIS PERSONAE:

Raymonda, Countess de Doris	Mlle Legnani
The Countess Sybille, *canoness, Raymonda's aunt*	Mme Cecchetti
The White Lady, *protector of the house of Doris*	Mlle Svirskaya
Clémence, *girlfriend of Raymonda*	Mlle Kulichevskaya
Henriette, *girlfriend of Raymonda*	Mlle Preobrazhenskaya
The Knight Jean de Brienne, *Raymonda's fiancé*	Mr Legat III
Andrei II, *King of Hungary*	Mr Aistov
Abderrakhman, *a Saracen knight*	Mr Gerdt
Bernard de Ventadour, *a troubadour of Provence*	Mr Kyaksht
Béranger, *a troubadour of Aquitaine*	Mr Legat I
Seneschal, *in charge of the castle of Doris*	Mr Bulgakov
Cavalier in the retinue of de Brienne	Mr Yakovlev
A Hungarian knight	Mr Gillert
Saracen knights	Mr Tatarinov
	Mr Voronin
	Mr Baltser
	Mr Bykov

Women, vassals; Hungarian and Saracen knights; heralds, moors, citizens of Provence, royal soldiers and servants.

In the First Act, First Scene:
La fête de Raymonde

1. *Jeux et danses*: Mlles Kulichevskaya, Preobrazhenskaya; Messrs Legat I, Kyaksht. Girl Students: Petipa I, Sedova, Belinskaya, Egorova, Andrianova, Grupilion. Boy Students: Obukhov, Osipov, Fokine, Ogniev, Baryshistov, Ivanov.

2. *Entrée*: Mlle Legnani.

Women of the Court: Mmes Natarova, Alexandrova II, Postolenko, Kil, Antonova, Efimova, Goryacheva, Semenova II.

Vassals: Messrs Navatsky, Kunitsky, Solyannikov II, Marzhetsky, Alexeyev, Fomichev, Panteleyev, Sosnovsky.

3. *Valse provençale*: Mlles Kasatkina, Erler II, Bakerkina, Pavlova, Chernyavskaya, Kunitskaya, Vaganova, Matveyeva III, Yakoleva II, Vasilieva, Golubeva II, Leonova II, Yakovleva I, Radina, Ilyina III, Dyuzhikova, Shtikhling, Rykhlyakova II, Przhebyletskaya, Erler I, Sazonova, Stepanova II, Stepanova III, Matyatina: Messrs Kusov, Gavlikovsky, Nikitin, Paschenko I, Fedulov, Fedorov I,

Aslin, Presnyakov, Voronkov III, Chekrygin, Fedorov II, Medalinsky, Sergeyev, Martyanov, Loboiko, Ponomarev, Maslov, Mikhailov, Smirnov, Kristerson, Ivanov I, Balashev, Levinson, Dmitriev.

4. *Pizzicato*: Mlle Legnani.
5. *La romanesque*: Mlles Kulichevskaya, Preobrazhenskaya; Messrs. Kyaksht and Legat I.
6. *Une fantaisie*: Mlle Legnani.

In the Second Scene:
Visions

Mlle Legnani, Mr Legat III; *La renommée*: Mlle Nikolaidis; *Gloire*: Mlles Rykhlyakova I, Geltser, Leonova I, Mosolova, Ofitserova, Borgkhardt, Trefilova, Chumakova, Repina, Vaganova, Chernyavskaya, Nikolaeva, Kasatkina, Pavlova, Kshesinskaya I, Kushova, Shchedrina, Kunitskaya, Lits, Egorova II, Bakerkina, Kunitskaya, Erler II, Ogoleit II, Oblakova, Vasilieva, Leonova II, Golubeva I, Dorina, Konetskaya, Tsalison, Ilyina III, Matveyeva III, Slantsova, Golubeva II, Andreyeva, Radina, Urakova, Levina, Dyuzhikova, Stepanova II, Vsevolodskaya, Stepanova III, Yakovleva II, Przhebyletskaya, Lobanova, Rosh.

Les chevaliers: Messrs Alexandrov, Titov, Romanov, Ivanov II, Voronkov II, Plessyuk, Vasiliev, Rykhlyakov, Voskresensky, Chernikov, Oblakov II, Terpilovsky.
Les amours: students of the Imperial Theatre School.
Scène dramatique: Mlle Legnani, Mr Gerdt. *The White Lady and Raymonda's double.*
Farfadets: Girl and boy students of the Imperial Theatre School.

In the Second Act:
Cour d'amour

1. *Pas d'action*: Mlles Legnani, Cecchetti, Kulichevskaya, Preobrazhenskaya; Messrs Gerdt, Legat I, Kyaksht.
2. *Pas des esclaves sarrasins*: Mlles Matveyeva III, Savitskaya, Konetskaya, Leonova III, Sheberg, Isaeva I, Legat, Ilyina III, Dyuzhikova, Nikolaidis, Yakovleva II, Erler I, Ryabova, Lobanova, Peters, Mikhailova, Yakovleva I, Golubeva II, Niman, Kuzmina, Ilyina II, Golovkina, Rosh, Pakhomova, Kusterer, Levinson II, Gorskaya, Temireva, Matyatina, Rakhmanova, Ilyina I; Messrs Nikitin, Kusov, Fedorov I, Alexandrov, Gavlikovsky, Voronkov II, Trudov, Fedulov, Usachev, Aslin, Ivanov I, Loboiko, Chekrygin, Balashev, Vasiliev, Romanov, Novikov, Paschenko II, Medalinsky, Kristerson, Rykhlyakov, Mikhailov, Levinson, Presnyakov, Pechatnikov, Fedorov II, Dmitriev, Martyanov, Smirnov, Maslov.
3. *Pas des Moriscos*: students of the Imperial Theatre School.
4. *Danse sarrasine*: Mlle Skorsyuk and Mr Gorsky.
5. *Panadéros*: Mlle Petipa I, Mr Lukyanov; Mlles Kshesinskaya I, Makhotina, Lits, Kuskova, Shchedrina, Bakerkina, Radina, Urakova, Borgkhardt, Kasatkina, Chernyavskaya, Pavlova, Vasilieva, Tsalison, Ogoleit III, Vaganova.
6. *Coda*: all participants.
Les échansons: students of the Imperial Theatre School.

7. *Entrée*: Mlle Legnani.
8. *Ensemble*: all participants.
9. *Dénouement*: Mlles Legnani, Cecchetti, Kulichevskaya, Preobrazhenskaya; Messrs Gerdt, Legat III, Aistov and others. *The White Lady.*

In the Third Act:
Le festival des noces

1. *Rapsodie*: students of the Imperial Theatre School.
2. *Palotás*: Mlle Preobrazhenskaya and Mr Bekefi; Mlles Slantsova, Pavlova, Kasatkina, Kunitskaya, Bakerkina, Chernyavskaya, Tsalison, Golubeva II, Radina, Dyuzhikova, Leonova II, Egorova II, Ilyina II, Peshkova, Starostina, Peters, Gorshenkova, Tselikhova, Shtikhling, Kil; Messrs. Fedorov I, Rakhmanov, Ivanov I, Levinson, Kristerson, Loboiko, Martyanov, Smirnov, Medalinsky, Trudov, Pashchenko I, Novikov, Balashev, Fomichev, Baltser, Voronkov III, Usachev, Paschenko II, Legat II, Panteleyev.
3. *Mazurka*: Mlle Petipa I, Mr Kshesinsky II; Mlles Tatarinova, Nikolaeva, Ogoleit II, Ogoleit III, Kshesinskaya I, Kuskova, Shchedrina, Urakova, Levina, Konetskaya, Golubeva I, Nikolaidis; Messrs Voronkov II, Fedulov, Alexandrov, Yakovlev, Vasiliev, Titov, Ivanov II, Ponomarev, Aslin, Presnyakov, Romanov, Rykhlyakov.
4. *Pas classique hongrois*: Mlles Legnani, Johanson, Rykhlyakova I, Obukhova, Geltser, Ofitserova, Borghardt, Chumaova, Vaganova; Messrs Legat III, Kyaksht, Oblakov I, Legat I, Gorsky, Kusov, Gavlikovsky, Nikitin and Sergeyev.
5. *Final*: all participants.

APOTHEOSIS
Tourney

Soloists of the Court of His Imperial Highness:
On the violin—Mr Auer
On the harp—Mr Zabel
Other soloists: on the harp—Mlle Virginia Charlione; on the celesta and pianoforte—Mr Griben
Conductor: Mr Drigo

RAYMONDA

ACT I

First Scene

A hall in the castle of Countess de Doris

Tableau 1

The seneschal issues orders concerning the imminent celebration of Raymonda's name-day. Bernard de Ventadour, Béranger, and several pages are fencing, others are playing on lutes, viols, etc.; several girls from Raymonda's retinue, attracted by the general festivity, cease their work and begin to dance with the pages.

Tableau 2

Ladies of the court enter, preceded by Countess Sybille; she reproaches the girls for their idleness, but in vain—no sooner does she manage to get some of the girls back to work than others begin to dance.

Then the countess turns to Bernard, Béranger, and the pages and commands them to take away their lutes and viols. 'Take care,' she says to the girls, 'Countess de Doris, famous by the name the White Lady, will punish you for disobedience; do you see this statue? This is our revered ancestor, she appears from the other world to warn the house of Doris every time one of its members is in danger, and punishes those who do not fulfil their responsibilities.' The girls laugh at Countess Sybille's superstition, and engage her in dancing. A horn sounds, proclaiming the arrival of guests.

Tableau 3

The seneschal runs into the hall and announces the arrival of a courier from the knight Jean de Brienne, bringing a letter to Jean's bride. Countess Sybille goes over to tell her niece.

The seneschal, hurriedly giving last instructions before Raymonda's arrival, goes to the guests, who are arriving with congratulations for Raymonda; meanwhile the girls scatter flowers along the path by which Raymonda will enter.

Tableau 4

Raymonda enters. She admires the flowers that pages bring in for her.

Tableau 5

Kneeling, the courier hands her the letter from her fiancé, Jean de Brienne. It informs Raymonda that King Andrei II of Hungary, under whose banner the Knight of Brienne does battle, is returning, covered in glory, to his native land. Not later than tomorrow de Brienne will arrive at the castle of Doris for his wedding to Raymonda. She is delighted. At this moment the seneschal announces the arrival of the Saracen knight Abderrakhman; Raymonda and her aunt are astonished, although Countess Sybille orders that the guest be invited in.

Tableau 6

Abderrakhman enters, greeting the entire company with great dignity. He explains that word has reached him of Raymonda's remarkable beauty and of the regal hospitality of the castle of Doris. He made bold to come and congratulate the charming name-day celebrant and to present several gifts to her. Puzzled by this unexpected turn of events, Raymonda, concealing her indignation with effort, declines the gifts, which brings Abderrakhman to despair. Meanwhile the seneschal requests Countess Sybille's permission to introduce the vassals. Abderrakhman, leading his first sword-bearer to the side, confides in him his plan to kidnap Raymonda.

Raymonda rereads and kisses the letter from her fiancé. Countess Sybille offers Abderrakhman a place at her side to watch the forthcoming dances. The Saracen accepts her invitation, but the proceedings little interest him, as his thoughts are lost in dreams of Raymonda, whom he admires the whole time.

Tableau 7

The vassals come on, noisily greeting and congratulating Raymonda.

Waltz

After the dances Raymonda commands the seneschal to prepare with all possible magnificence the reception of her fiancé, and to make ready a *cour d'amour* in his honour. Hearing this, Abderrakhman is determined to abduct Raymonda. He exits, accompanied, at Countess Sybille's order, by the seneschal; the vassals and other guests also withdraw.

It is getting dark, the moon illuminates the terrace.

Tableau 8

Only Raymonda's closest friends and troubadours remain with her. She plays the lute (*La romanesque*, performed by two pairs). Raymonda hands the lute to one of her friends and shows them a new dance (*Fantaisie*). Finally, tired from the day's excitement, she lies down on the carpet, the pages wave fans over her, and Clémence delights Raymonda's ear with a tender melody on the lute. Suddenly a magic torpor comes over them: the pages and everyone else fall asleep, except Raymonda, who looks at them with astonishment. The White Lady appears; illuminated by moonlight, Raymonda, frightened, looks closely at her. With an imperious gesture the White Lady beckons her to follow. Obedient to some mysterious, unknown power, Raymonda submissively follows the White Lady.

Second Scene

A shady park. In the depth of the park, the high terrace of Countess de Doris's castle.

Tableau 1

The White Lady moves silently around the terrace, Raymonda following her as if in a trance. At a sign from the White Lady the garden is enshrouded in mist for a time. The mist gradually disperses, revealing the figure of Jean de Brienne.

Tableau 2

Visions

Jean de Brienne and his knights, encircled by girls putting crowns on their heads.

A Large Grouping

The White Lady indicates to Raymonda the presence of her fiancé;— delighted, she rushes to the knight's embrace.

Dances and Groupings

Raymonda expresses her delight to the White Lady, who, however, disabuses her. 'Behold, and know what still awaits you,' the White Lady tells her. Raymonda tries to return to her fiancé, but comes face-to-face with Abderrakhman, who appears in the knight's place.

The stage darkens.

Tableau 3

Abderrakhman declares his love to Raymonda, but she rejects him indignantly.

Tableau 4

Apparitions appear from all sides. Raymonda implores the White Lady to save her. At this moment Abderrakhman is about to abduct Raymonda; she falls, unconscious. The apparitions, dancing, encircle the motionless Raymonda.

Dawn comes; the first rays of the coming day replace the moonlight. The apparitions disappear with Abderrakhman and the White Lady.

Tableau 5

A castle servant and Raymonda's friends and pages run on to the terrace, notice her lying in a faint, and try to revive her.

ACT II
Cour d'amour
An interior courtyard of Countess de Doris's castle

Tableau 1

Entrance of the knights, cavaliers, owners of neighbouring castles, noblewomen, troubadours, minstrels, and others invited to be present at the *cour d'amour*. Then follow the seneschal, Raymonda and Countess Sybille. Raymonda greets the assembled company and expresses pleasure at the festive decorations of the court and reception which has been prepared for her fiancé. But Raymonda cannot conceal her uneasiness at the delay in Jean de Brienne's arrival.

Tableau 2

Abderrakhman enters with his retinue. In confusion Raymonda orders the seneschal to send away the uninvited guest, but Countess Sybille persuades her that on this day no one should be denied hospitality.

Grand pas d'action

Abderrakhman, struck by Raymonda's beauty, declares his passion for her, but she is afraid and evades him. 'You must be mine, beautiful countess,' he says to her, 'I offer you a life of magnificence and pleasure.' Abderrakhman calls his slaves to entertain Raymonda.

Tableau 3

'See how agile and diverting my slaves are,' he says. (The slaves enter dancing.) Raymonda, as before, rejects him contemptuously. He commands his slaves to dance.

Tableau 4

Cupbearers fill the cups of the assembled company with wine.

Bacchanale

During this dance Abderrakhman and his slaves attempt to abduct Raymonda and steal off with her.

Tableau 5

Suddenly the knight de Brienne appears with King Andrei II and his retinue. Jean de Brienne frees Raymonda from the clutches of the slaves and attacks Abderrakhman, but the king commands the rivals to settle their quarrel with a duel. They agree. Sword-bearers bring in arms. Raymonda runs to embrace her beloved; the king then takes Raymonda to one side, and the opponents engage in battle. Jean de Brienne attacks first. The spectre of the White Lady appears in the depth of the stage. With a stroke of the sword to the head, he delivers the Saracen a mortal wound. The latter's attendants carry him out; meanwhile Abderrakhman's slaves try to steal away, but at a sign from the king the royal sword-bearers surround them. The king joins the hands of the young couple—Raymonda and Jean de Brienne.

ACT III

A garden in the castle of the Knight de Brienne
A wedding feast at which King Andrei II is present. In honour of the highborn guest a grand Hungarian divertissement is given.

APOTHEOSIS
Tourney

X
The Twentieth Century

INTRODUCTION

Petipa produced *Raymonda* in his eightieth year. Under the best of conditions he would be thought remarkable for creating seven more ballets (not including revivals) between *Raymonda* in 1898 and *The Tale of the Rosebud* 1904. But circumstances conspired to make his life more difficult. The infirmities of old age, which he had already been suffering for several years, were augmented by disruptions in the theatre administration.

In 1899, his patron and colleague Ivan Alexandrovich Vsevolozhsky retired as Director of Theatres, to be appointed Director of the Imperial Hermitage.[1] Vsevolozhsky's successor, Prince Sergei Mikhailovich Volkonsky, was a highly literate theatre man sympathetic to Petipa's aspirations but inexperienced in administration, coming to this first official service from private life; his tenure lasted barely two years. While Director, however, Volkonsky made two decisions important to the Petersburg ballet: he commissioned *The Magic Mirror*, Petipa's last staged work, and he hired as Assistant for Special Affairs Sergei Pavlovich Diaghilev, whose projects in time would lure away many of the company's finest dancers. When Volkonsky took these decisions, of course, the implications of neither decision were apparent, for *The Magic Mirror*, to music by Arseny Nikolaevich Koreshchenko, was to celebrate Petipa's fifty-fifth year of imperial service, and Diaghilev, who edited the most sumptuous volume of *Ezhegodnik Imperatorskikh Teatrov* [The Yearbook of the Imperial Theatres] ever published, was brought in for his knowledge of the visual arts, the principal expertise for which he was recognized at the time and for some years afterwards.

Vladimir Arkadievich Telyakovsky, who replaced Volkonsky, is one of the most criticized figures in Russian theatre. While he could not match his predecessor's claim to a broad general culture, he was Volkonsky's better in administrative experience, coming to the new post after two years as head of the Moscow theatres and a military career before that. By all accounts except his own, however, Telyakovsky was unscrupulous.

We may never know if nature or untenable circumstances made him so. Like Volkonsky, Telyakovsky wanted to bring change to the theatres, but found himself in the middle of competing factions, none of which he could satisfy. He had considerable sympathy for the avant-garde, especially the

---

[1] Petipa continued a professional relationship with Vsevolozhsky up to the last years of his activity. Five of the balletmaster's seven new ballets after *Raymonda* were first given private performances in the Hermitage Theatre (four were subsequently presented at the Maryinsky); the last, *The Tale of the Rosebud*, was, like *The Sleeping Beauty*, choreographed to a scenario by Vsevolozhsky; it was produced and rehearsed, but the performance, scheduled for January 1904, was cancelled at the outbreak of the Russo–Japanese War).

painters Alexandre Golovin and Konstantin Korovin. But modernist tendencies had to be balanced with the wishes of a Minister of Court and imperial family of traditional tastes, to say nothing of a cadre of well-entrenched, extremely conservative balletomanes. The smallest concession to one brought noisy reproach from the others.

Relations between Marius Petipa and Telyakovsky were always strained. Moreover, Telyakovsky had inherited *The Magic Mirror* from Volkonsky and his lack of enthusiasm for the ballet made inauspicious circumstances worse. Two changes of director had repeatedly kept Petipa from mounting a new ballet for his benefit performance. On 19 November 1901 Telyakovsky noted in his diary:

. . . I expressed the opinion that it wouldn't be bad to revive it [*The Sleeping Beauty*] the next year, for I little counted on Koreshchenko's ballet. At that point Petipa began to complain that they do not give him new ballets to produce for his benefits: he chose *Salammbò*, Vsevolozhsky left and *Salammbò* was cancelled; they decided on *The Magic Mirror*, Volkonsky left, and now I want to cancel.[2]

Telyakovsky did not cancel, but he did go forward in the mean time with another project impolitic under the circumstances. Claiming that he wished to introduce something new, in 1902 he ordered to be brought to St Petersburg the staging of *Don Quixote* produced in Moscow by Alexandre Gorsky. Petipa and the balletomanic faction took umbrage at this, claiming the production was a mutilation of Petipa's masterwork by of one of his former students.

Despite the collective failings of health, memory, and pride, Petipa, after some hesitation, went forward with *The Magic Mirror*, which was first performed on 9 February 1903. Marking fifty-five years of his imperial service, the occasion was one of great ceremony. Indeed, ceremony outstripped criticism in the first press reports of the new ballet, nowhere in more striking fashion than the balletomanes' most prominent voice, Khudekov's *Peterburgskaya gazeta*. The day after the performance it contained a long account divided into subsections devoted to the hall and the public (in which over 100 of the highest ranking guests were listed by name, in some instances with the location of their seats and descriptions of the women's dresses), the decorations, the ballet, and finally the honouring of the benefit artist, which included a roster of gifts and addresses presented to Petipa.

The fuss was being made over a ballet based on the tale we know as 'Snow White and the Seven Dwarfs', though in Pushkin's retelling (the producers' immediate source) the names were different and the dwarfs

---

[2] V[ladimir] A[rkadievich] Telyakovskii, *Vospominaniya* [Reminiscences], ed. D. Zolotnitskii (Moscow and Leningrad, 1965), p. [469], n. 10.

were replaced by gnomes. The scenario was adapted by Petipa and an unnamed collaborator (indicated by asterisks in the libretto), possibly Director Volkonsky himself. The highly stylized rendering of the fable, in prose made archaic and mystifying by the removal of all elaboration, was probably Volkonsky's contribution, the aim to intensify the expressive force of the tale, to stress its universal statement about vanity and avarice and innocence. In this respect *The Magic Mirror* is a logical successor to *The Sleeping Beauty*, as it is to *Raymonda* in the reduction of superfluous incident.

23
Libretto of *The Magic Mirror*
Marius Petipa

THE MAGIC MIRROR
Fantastic Ballet in 4 Acts and 7 Scenes
Story taken from the tales of Pushkin and Grimm
Libretto by M. I. Petipa and Mr ***
Music by A. N. Koreshchenko
St Petersburg
Publication of the Typographer of the Imperial Theatres
(Bureau of Crown Affairs)
1903
Passed by the censor. St. Petersburg, 7 February 1903

DRAMATIS PERSONAE:

The King	Mr Gerdt
The Queen	Mlle Petipa I
The Princess, *the king's daughter by an earlier marriage*	Mlle Kshesinskaya
The Prince, *her fiancé*	Mr Legat III
The Princess's nurse	Mlle Petipa II
Chief Steward of the Household	Mr Gillert
A Polish magnate	Mr Kshesinsky II
The Princess's retinue:	Mlle Pavlova II
	Mlle Egorova II
The Prince's retinue:	Mr Obukhov
	Mr Fokine

The Elder of the Gnomes

The retinue of the King and Queen, gentry, courtly women, heralds, pages, men and women gardeners, dryads, gnomes, rays of sunlight, a zephyr, everlasting flowers, a rain of stars, guards and servants.

ACT I
First Tableau
A garden in front of the palace

SCENE 1

Gardeners, men and women, are decorating the garden, weaving baskets and preparing garlands for the queen.

SCENE 2

Entrance of the king, queen, courtly ladies and cavaliers. The queen is young and beautiful, the king old; he tries to please the queen. Baskets and garlands are brought to the Queen.

Village Waltz

SCENE 3

The king orders lace merchants to be led in, whom he has summoned, together with the lace-makers and workmen of their countries, and also the merchants of old Bohemian crystal and precious stones. Among other objects there is a beautiful 'magic' mirror, which possesses the property of reflecting the most beautiful woman in the entire realm. The queen, seeing herself in its reflection, is delighted.

The joy of the king, who buys the mirror.

The queen, rejoicing, invites the courtiers to dance.

Mazurka

The king orders the mirror to be taken to the queen's room. The queen looks into the mirror again, but to her question, 'Am I the most beautiful of all?', the mirror reflects the image of the princess. The queen's horror.

In the distance loud trumpets are heard. The queen is upset . . .

SCENE 4

Entrance of the princess, the prince her fiancé, and their retinues.

The king's joy. The secret envy of the queen, who recognizes in the princess's face the image reflected in the mirror, that is, of the one woman whose beauty excels her own.

The Presentation:
Pas d'Action

At the end of the '*Pas d'Action*' the queen, extremely upset, asks the mirror again: 'Who is the most beautiful of all?' and once again the mirror presents the princess's image in all its beauty. This time the queen, in a fit of envy and anger, faints.

All rush to her assistance.

Curtain

ACT II

Second Tableau

SCENE 1 *A Park*

At the rise of the curtain the queen is on stage, her gaze fixed on the mirror. She says to herself: 'As long as the princess is here, I will not be the most beautiful of all . . . She must die.'

SCENE 2

The queen calls the princess's nurse and says to her:

'Listen carefully to my commands: you will go into the forest with the princess, and you will kill her.'

'Take pity, take pity on the young princess whom I love so much.'

'No,' answers the queen, 'there shall be no mercy. She must die, and you must obey me. If not . . . you will die . . .'

She hands her a dagger.

SCENE 3

They announce the princess's arrival. She has come to ask the queen to be present at her wedding. Hiding her envy, the queen receives the princess courteously and agrees to her request. Then the queen asks her to go into the nearby forest with her nurse to pick forget-me-nots, of which the queen is extremely fond. The princess, delighted that she can bring pleasure to the queen, kisses her hand and leaves with the nurse.

SCENE 4

As soon as she leaves, the queen walks up to the mirror and says to herself: 'Now I shall be the most beautiful of all.'

Changement à vue

Third Tableau
(Second Tableau of Act II)
A Dense Forest

SCENE 1

The princess enters with her nursemaid.

Compelled to obey the queen, the nurse is trembling. The princess looks around in all directions and says to her:

'This forest frightens me; there are no forget-me-nots here; why did you bring me so far from the castle?'

'The queen commanded me,' the nursemaid answers.

'Why?' asks the princess.

'She commanded me to bring you into this forest and to kill you.'

'Me? But why? What have I done?'

'You are more beautiful than she.' The nurse sobs.

'And you must kill me for that? No, no, that is not possible! You would not commit such a horrible deed!'

'It must be thus,' the nurse answers, weeping and holding the dagger in her hand.

The princess begs her on her knees:

'Take pity on me, on my youth!'

The nurse, powerfully moved, throws down the dagger, and embracing her, kisses her, saying:

'Stay here, try to get your bearings and find a way out of this horrible forest.'

They pray. Then the nurse kisses her warmly again, and saying 'May the Lord protect you,' she flees like one insane.

SCENE 2

The princess calls after her for help, but in vain. The nurse is too far away to hear the princess's calls. Trembling, the princess looks to find her way, walking through the trees, and trying to get out of the forest, she disappears among them.

SCENE 3

Little by little the stage fills with dryads, who have gathered for diversion.

SCENE 4

After the dryads' dances, a curtain rises at the back of the stage, and in the clear distance the queen's shade is seen, wrathfully threatening the princess's nurse for having spared her.

The nurse falls at the queen's feet and begs her forgiveness.

The queen rudely throws her back. Then the queen commands that she herself be dressed in peasant's clothes so that she may realize her own intentions, having conceived the idea of giving the princess a poisoned apple.

The princess's maid dresses her.

The vision disappears.

Changement à vue

Fourth Tableau
(Third Tableau of Act II)

A clearing is visible amidst rocky hills; huts, a cave. Next to the huts is a forge with anvils.

SCENE 5

Gnomes come out of the cave and down from the hills: some are carrying sheafs of brushwood; others are digging out passageways in the crags.

SCENE 6

The princess rushes in, frightened, for she cannot find her way out of the forest . . . Suddenly she finds herself surrounded by gnomes, who are fascinated with her and try to calm her down.

The elder of the gnomes leads her into his hut, where she is attired in a dress of tree leaves.

SCENE 7

A merry dance of the gnomes

SCENE 8

The princess appears in her dress of leaves, strewn with dew-drops. The gnomes ask her to dance—and she agrees. The gnomes stand in groups and beat time on their anvils, accompanying the princess during her dances. After the variations, the leader of the gnomes takes the princess into the hut and advises her not to admit anybody during his absence.

The gnomes depart for work with their implements.

SCENE 9

The queen, dressed as a peasant girl and carrying a basket of apples, appears stealthily, and knocks on the door of the hut.

SCENE 10

The princess comes out, meets her with compassion and gives her some bread.

In gratitude, the queen offers her an apple (poisoned).

The princess at first declines the apple, then accepts it, takes a bite, and stunned, falls as if dead. The queen gloats over her deed, and as she does fails to notice that she has dropped her kerchief; seeing that the gnomes are returning, she runs out.

SCENE 11

The gnomes' joyful return. Horror seizes them at the sight of the princess. The leader of the gnomes finds the kerchief and shows it to the others, evidence that an outsider had been there.

Curtain

ACT III

Fifth Tableau
(First Tableau of Act III)

A Forest

A dense forest; in the distance, in the middle of the stage, a large tree. A bright sunny day.

SCENE 1

The appearance of the prince in search of the princess. He is despondent, sorrowful and tired. He sits down next to the tree to rest, and falls asleep.

SCENE 2

The prince's dream. Sunlight. Dance of the sun's rays.

SCENE 3

The sun slowly sets, and the moon appears in its place, surrounded by stars (the rain of stars).

SCENE 4

Grand pas of the princess, escorted by zephyrs and stars.

SCENE 5

The prince (that is, his shade) wants to draw her into his embrace; both rush towards each other, but they cannot come together; at the very moment they are to embrace, darkness falls and everything disappears.

SCENE 6

Night. A full moon. (The stage is empty.)

The prince awakens, still under the influence of his dream. He tries to orient himself to the forest, looks around, asking himself: what path must I take to return to the castle?

He climbs the tree: as he does the other trees disappear, revealing a broad vista; at the back of the stage a valley appears, the sky is studded with stars. From the height of the tree the prince sees the castle; he climbs down and exits in the direction that he saw the castle.

Curtain

ACT IV

Sixth Tableau
(First Tableau of Act IV)

A rocky locale

SCENE 1

A torchlight procession.

The gnomes are carrying the princess away in a glass coffin covered with flowers; they place the coffin deep in a grotto.

The gnomes leave, praying.

One of them remains to stand guard—then falls asleep.

SCENE 2

Dances of the everlasting flowers.

SCENE 3

The leader of the gnomes comes forward; behind him follow the king, queen, the prince, the nurse and the king's grooms.

The gnome shows them the place where the princess rests.

The queen conceals her joy.

The king is sorrowful.

The nursemaid is crying.

The prince, in despair at the sight of the dead princess, approaches the grating. He breaks the grating and the glass coffin. At that instant the apple rolls out in front of the prince.

SCENE 4

The princess comes to life . . .

The queen stands in astonishment, as if turned to stone.

The others express their great joy.

The princess stands up (she is wrapped in a gold coverlet), glides along like a shade as the coverlet falls away; the princess sees the prince, who rushes to embrace her.

The queen's horror . . .

The king's excitement . . .

The prince's delight, and the pervasive joy.

The gnome hands the king the kerchief as material evidence of who gave the princess the apple.

The king recognizes the queen's kerchief and shows it to her, asking: 'Is this your kerchief?'

'Yes,' the queen answers, trembling with anger and fear, 'I don't understand how it came into the hands of this gnome.'

'Was it you who gave her the poisoned apple?'

'No . . . No . . .' In her fury she cannot pronounce another word.

The king asks his daughter who gave her the apple.

The prince also begs her to recount everything.

The princess answers: 'A poor peasant girl, who was carrying a basket of apples.'

At this moment the nursemaid, all in tears, no longer able to keep the secret, tells everything that happened, but the princess stops her and finishes the story. The king's frightful anger. He threatens the queen with the dungeon. A sudden derangement of the queen's mind, who, in an attack of insanity, herself tells what she ordered done, and then falls dead.

Curtain
A short entr'acte

Seventh Tableau
(Second Tableau of Act IV).

A hall in the castle.

The betrothal of the prince and princess.
In this last scene, character dances and classical *pas*.

THE END

24
'Letter to the Editor'
Sergei Pavlovich Diaghilev

Behind the scenes, *The Magic Mirror* was a débâcle. After the first performance it was removed from the repertoire in St Petersburg, to be presented complete there only once again, the next season, for the benefit performance of the corps de ballet. Two abridged performances followed in 1904. Ironically, the ballet found a home in Moscow in a revival by Gorsky, where it was performed thirty-six times between 1905 and 1911. Soon after the première, Petipa received word that he officially had been made balletmaster for life, but in reality his active service had been terminated. The unstated reason for this was, one presumes, the failure of *The Magic Mirror*.

The ballet's failure was attributed to the stylistic disjunction between Petipa's choreography and the advanced impressionistic idiom of Alexandre Golovin's paintings and costumes. In a broader perspective, the new ballet represented a clash of traditional and modern which was fast developing in the first decade of the twentieth century, to which Petipa fell victim. He made no attempt to conceal his bitterness when writing about *The Magic Mirror* in his memoirs:

> ... I must for the defence of my reputation as balletmaster relate in detail the conditions in which my last ballet, *The Magic Mirror*, was produced.
>
> Already during rehearsals of this ballet I was convinced that something was planned against me and this work, which was commissioned of me and Mr Koreshchenko by Prince Volkonsky, not Mr Telyakovsky. Mlle Kshesinskaya took an active part in all this intrigue, avenging herself on me for the fact that I had not addressed her father, now deceased, with a speech at his last benefit performance ...
>
> I came to recognize the worth and gratitude and devotion of this artist when rehearsals of *The Magic Mirror* began. The artist-decadent Golovin joined the conspiracy to please the director. When at the rise of the curtain the decoration of the gnomes drawn by this decadent was first seen, unanimous laughter broke out.
>
> At the end of the dress rehearsal I went over to the director to implore him to let me produce some old ballet for my benefit and not to show the public a new work in such an ugly form. Suddenly someone steals up to me from behind, covers my eyes with his hands and cries out:
>
> 'Bravo, bravo, Mr Petipa! It is the height of perfection! This ballet will have a huge success!'
>
> 'No, Your Excellency, this ballet will fail triumphantly.'
>
> 'Now, what are you saying? You are mistaken, Mr Petipa.'

> On the day of the first performance and of my benefit the theatre was filled to the brim, and His Majesty and the whole imperial family were present in the tsar's box.
>
> Whoever attended this performance will recall, of course, how maliciously the public laughed at everything it was shown—and indeed, that was fully understandable. As I already pointed out, the costumes and accessories, prepared by one of the director's closest colleagues, were simply caricatures, the dancers who represented *immortals* were dressed as *nymphs*, etc.
>
> In a word, it was all unspeakably bad, and even the wretched costumes and decorations which were being prepared were not completely ready. On the eve of the benefit they came to me to ask me to put it off, but I unfortunately could not consent, taking into consideration that all the tickets were sold and treating the public so unceremoniously was out of the question. I can still thank God that the court, the public and the press found it possible to exclude me from the disgraceful production of the ballet and to acknowledge the dances successfully put together by me. Nevertheless, the blow the director delivered to me behind my back was cruel, and in truth deliberately Machiavellian.[1]

Telyakovsky's explanation of the ballet's failure differs considerably from this. He too adopts a righteous tone:

> Several times Petipa turned down the production of *The Magic Mirror*, but then, in view of the wish to celebrate his benefit, agreed to mount it, all the time, however, being at odds with the author of the music and still more with the artist Golovin. He inveighed against them verbally and in the press and especially aroused the balletomanes, who, seeing here an unswerving aspiration to bring into the ballet something new, and convinced that the experiment with *Don Quixote* was only the beginning, resolved to ruin the ballet *The Magic Mirror* with a scandal, having underscored the point that even the genius Petipa could not overcome the ugliness of Golovin's decorations and Koreshchenko's new symphonic music.
>
> Long before the first performance all this was well known, but the question of what form the balletomanes' protest would take was left open. Before the performance itself everything, it seemed, was relatively peaceful.
>
> The ballet was even successful at the dress rehearsal. Petipa, met with an ovation from the artists, praised the ballet as his creation and said several cordial words to me—that he was glad to be working with me, satisfied with the result of the production, glad that such an original production was made during my time, etc.
>
> But soon after I returned home following the rehearsal, the supervisor of the production department, Baron Kusov, came to inform me that after my departure he was accidental witness to Petipa's conversation with the chief

---

[1] *Memuary Mariusa Petipa Solista* EGO IMPERATORSKOGO VELICHESTVA *i baletmeistera Imperatorskikh teatrov* [Memoirs of Marius Petipa, Soloist of HIS IMPERIAL MAJESTY and Balletmaster of the Imperial Theatres] (St Petersburg, 1906), pp. 89-92.

representatives of the balletomanes and the press. Petipa railed at length against both mounting and music of the production, and said that despite all the care he had put into it the production was impossible to save. The direction is mocking him, forcing him to work with such inept artists as Golovin and musicians like Koreshchenko. He had already warned me about this several times, but I had paid him so little attention that he, Petipa, was washing his hands of the situation—let the public itself speak out, and the highborn of this world will be convinced of how I am leading the theatre and our model Petersburg ballet.

Such an opinion from Petipa, spoken only half an hour after the reverse which he had told me, did not much surprise me, however, for I knew his duplicity and cunning quite well. Besides, he had already passed his ninth decade—what else could be expected from such an old man who was, however, outstanding in his time? . . .

On the day of the first performance everyone was very nervous. . . . When the curtain went up and the first scene was revealed to the gaze of the audience—it was so unusual and beautiful that despite the ill wishes of many, it was impossible not to take an interest in the investigation of the details of the decoration itself and the unusually picturesque groupings in the most diverse costumes. This scene no doubt produced an impression, the melodious waltz also pleased, which, to tell the truth, the artists performed excellently; Kshesinskaya and the other dancers were much applauded.

With the second scene the impression began to change. Individual outcries and chuckles were heard on account of the decoration of the park, in which the clipped high wall of vegetation did nòt please. In this scene the role of the stepmother princess, a mimed role, was performed at Petipa's insistence by his ungifted daughter Nadezhda, who was in no wise a mime artist.

The glass of the mirror snapped, and quicksilver poured over the stage, causing laughter in the hall and disturbing the dancing artists.

Taken together, all this had an influence on the mood of the audience. Unabashed by the emperor's presence, the balletomanes began to make their observations very loudly and converse with each other. Gradually the entire parterre began to buzz. In the interval in the auditorium, and especially in the corridors and in the foyer, there was a loud exchange of opinions, and criticism of the performance—apparently some kind of demonstration of balletomanes was being organized.

When in the next act the curtain flew up and a grotto filled with gnomes was presented to the spectators' view—loud concerted laughter broke out from the balletomanes in the parterre, catcalls and whistling, which momentarily grew into a veritable roar with exclamations of individual balletomanes:

'Enough! What ugliness! It's time for this decadence to end!'

One of the eldest balletomanes, General V[intul]ov, completely bald, cried out: 'Get rid of Gurlya (my wife's name) and Telyakovsky! They will destroy the theatre with their novelties!'

420 *The Twentieth Century*

I personally did not make this out in the great hubbub, but friends told me about it in an inverval.[2]

It is difficult to know who was conspiring against whom to ensure the failure of *The Magic Mirror*. But Telyakovsky's explanation, even giving account for exaggeration, demonstrates the power of the ironclad balletic conservatives who were not shy about letting their voices be heard. Advocacy of modernism clearly had its perils, but fortunately modernism was not without its advocates, and perhaps the most intrepid among them came to Golovin's defence in connection with *The Magic Mirror*.

Sergei Diaghilev's letter to the editor of the *Peterburgskaya gazeta*[3] is striking not so much for its defence of Koreshchenko and Golovin as for its ironic relationship to Telyakovsky, whose long time antagonist Diaghilev would become, partially as a result of genuine differences in outlook but possibly also because he coveted Telyakovsky's post. By defending modern artists at the expense of Petipa, Diaghilev was giving aid and comfort to his enemy. The letter is also of interest as a sample of a device to which Diaghilev would have recourse again in polemical situations: manifestos in the press. In this case he was saying what has since become obvious: that *The Magic Mirror* was a nineteenth-century ballet resisting attack from the twentieth.

From the *Peterburgskaya gazeta*, 1 January 1904

Esteemed Sir, Mr Editor!

In No. 357 of your newspaper, in a notice about the music of the ballet *The Magic Mirror* it was said that 'only in the future will this music be understood', and that it recalls the poetry of 'those who gambol about the intellectual hothouse and wink knowingly at Messrs Bely and Balmont'. Last year, at the time of the ballet's grandiose failure, the 'decadent' decorator Golovin was made to shoulder all the blame, and now the cause of failure is being sought in Mr Koreshchenko's 'decadent' music. Clearly, the cause of the ballet's indubitable failure remains unexplained.

Permit me, as a representative of that movement in art which is called 'decadent', to say a few words in this regard.

Blame for the ballet's failure does not rest with the decorations or even with the unsuccessful, heavy music. One must look much deeper for it; it

---

[2] V[ladimir] A[rkad'evich] Telyakovskii, *Vosmponimaniya* [Recollections], ed. D. Zolotnitskii (Leningrad and Moscow: 'Iskusstvo', 1965), pp. 438–41.

[3] The letter is translated from *Sergei Dyagilev i russkoe iskusstvo. Stat'i, otkrytye pis'ma, interv'yu. Perepiska. Sovremnniki o Dyagileve v 2-kh tomakh* [Sergei Diaghilev and Russian Art. Articles, Open Letters, Interviews. Correspondence. Contemporaries on Dighilev. In Two Volumes], comp. I. S. Zil'bershtein and V. A. Samkov (Moscow, 1982), i. 184–5. It was originally published in the *Peterburgskaya gazeta* on 1 Jan. 1904, in response to a review of the performance of *The Magic Mirror* as revised for the benefit performance of the corps de ballet, published in the same newspaper on 29 Dec. 1903.

lies in the very enterprise of producing this ballet—unnecessary, boring, long, complicated, and pretentious. Let not the people who devised this production think that in this case they stand above the public, 'which did not understand and appreciate their enterprise.' The music, story, plan of action and all else in this ballet were not in the least created for 'comprehension only in the future'—the whole thing is perfectly comprehensible now, and appreciated according to its merits, as an utterly inartistic, unballetic, and chiefly, an infinitely boring spectacle.

The only things that save the ballet from total devastation, at least at times, are some decorations and part of the costumes of the artist Golovin. Looking most severely at this unpopular artist's work it must be said that the only things that entertained and satisfied the eye were the beautiful first scene of the ballet, the decorations of the spring forest and garden, and, perhaps, the hall in the final scene, rather crude but nevertheless successful. The same must be said about many of the costumes. But here too the supervisors were far from being competent, having permitted next to the artistic oases I have mentioned such dry and unsuccessful scenes as both of the grottos of the importunate gnomes, the boring forest with the most unfantastic owls, and many costumes, like the contemporary dresses and hats of Mlles Pavlova and Trefilova, thanks to which these lovely dancers more resembled *café-chanteuses* than classical dancers, or like Mr Kozlov's gymnast's costume and the last garb worn by Mr Legat III, reminiscent of those little dressed-up circus horses seen in triumphal apotheoses.

The principal decadence, I repeat, lies not in the individual participants of this unfortunate presentation, but in the very idea of expending so much effort and money on such an undistinguished affair. And one feels this especially acutely at a time when not one copeck is being spent on the production of the enchanting *Sylvia* by Delibes, and the classic Russian ballet, *The Sleeping Beauty*, rests peacefully in the storerooms of the Maryinsky Theatre.

With greetings, etc.,\
Sergei Diaghilev\
Publisher-Editor of the Journal, *The World of Art*\
1904.

25
Libretto of *Le Pavillon d'Armide*
Alexandre Benois

The failure of *The Magic Mirror* solved nothing; if anything it compounded Telyakovsky's problems. Petipa did not accept retirement with grace, and when repeated attempts to hire other balletmasters to replace him proved unfruitful, his retirement seemed all the more unjust. Meanwhile two of Telyakovsky's assistants, Alexandre Krupensky and Nikolai Sergeyev, who plagued artists with spying and intrigues, were goading the company towards the strike called in 1905, although Petipa's dismissal, the default of artistic leadership, and the general political climate of the time also contributed to this unprecedented event. The company emerged from the strike divided into 'progressive' and 'conservative' factions, perpetuating the frictions which caused it in the first place.

Nor were Telyakovsky's artistic competitors looking on indifferently. Diaghilev had attacked him in the press again, and soon would bleed the imperial ballet of its finest talents with offers of money and fame abroad, which in turn made successful artists more intractable in subsequent contractual negotiations with Telyakovsky.

Le Pavillon d'Armide was produced in 1907, at a time of comparative calm. Alexandre Benois has described the early history of the ballet—his developing the scenario in the years 1900–1, his collaboration with Nikolai Tcherepnin on the music (part of which was performed at concerts as early as 1903), Telyakovsky's acceptance of the ballet for production and Benois's subsequent falling-out with the director.[1]

The three-act scenario in which Benois originally cast the work would have suited Petipa almost perfectly; only the Hoffmannesque *frisson* marked a dramatic advance over the earlier *ballets à grand spectacle*. In Act I preparations are underway for the wedding of Berthe, the Marquis Fleurbois's daughter, to Count Tracy. This involves in part camouflaging an old pavilion which spoils the view but which the marquis refuses to have torn down. It is inhabited by the shade of beautiful Suzanne de Fleurbois. An abbot jealous of Berthe's fiancé urges her to have him pass a night in the pavilion as a test of his fidelity, knowing that when he himself had done this, he had barely survived with his life.

Act II represents the night Count Tracy spends in the pavilion. An elaborate Gobelins tapestry representing Suzanne in the guise of Armida is hanging on one wall, and there is a huge clock representing Saturn as the god of time. The clock strikes midnight and the tapestry comes alive. Armida pines for René, into whom

---

[1] See Alexandre Benois, *Reminiscences of the Russian Ballet*, tr. Mary Britnieva (London, 1941), pp. 225–7.

Tracy is transformed. There are dances. Armida gives Tracy a rose and demands his eternal fidelity. Morning comes, and a shepherd's pipe is heard.

Act III takes place at a ball after the wedding of Berthe and Count Tracy. Berthe notices the rose Tracy received from Armida and asks him to give it to her. Confused, Tracy declines, a disagreement follows, but Tracy finally yields to his bride's entreaty. The portrait of Suzanne as Armida comes alive but is visible only to Tracy. While the abbot gloats and Berthe looks on in perplexity, Tracy is drawn out on to the balcony; he rushes towards Armida, oversteps her, and falls off the balcony into the precipice below. Berthe swoons; the abbot is troubled by this turn of events.[2]

The production of *Le Pavillon* presented Telyakovsky with an opportunity to solve a number of the company's problems. By 1907 Benois was a veteran of 'The World of Art' group and could serve as mediator between the management and the avant-gardists, including Diaghilev, with whom the artist maintained close professional associations. Fokine, at the outset of his creative work at this time, was the most promising choreographer to work in the imperial theatres since Petipa. In connection with *Le Pavillon*, moreover (after part of it had been produced as a graduation piece at the theatre school), he had recognized the potential of the young Nijinsky, for whom a new role was especially created in the ballet. Thus, two years before the first 'Russian Season' of ballet which Diaghilev organized in Paris, leaders critical to the restoration of the Petersburg company after Petipa's retirement and the strike were willing to undertake that task and were in fact working towards it.

A team such as this would have satisfied the conservative tastes of the court, for the collaborators on *Le Pavillon*, whatever their associations with the avant-garde and their anti-academic protests, were themselves conservative in outlook. Benois was a self-avowed 'passé-ist'; Fokine, for all his manifestos, would quickly pass his zenith as a modernist; and Tcherepnin, student of Rimsky-Korsakov, was an academic who would go on to take a post at the St Petersburg Conservatoire.

The potential for *rapprochement* implicit in these circumstances would have required both artists and management to forgive and forget. Benois had written off Telyakovsky as a boor several years earlier, and had published a merciless attack on *The Magic Mirror*. Fokine had been an outspoken supporter of the ballet strike and had directly faced down Nikolai Sergeyev, the company régisseur and Telyakovsky's deputy.

Distrust on both sides was great, and for his part Telyakovsky, who could have promoted harmonious relations with these artists, chose instead to exacerbate the distrust, or at least to look aside while his minions did. Bureaucratic interference played a part—limiting the number of rehearsals, withholding materials for costumes, placing the first performance of *Le Pavillon* on a double bill after a complete *Swan Lake*. These obstacles were aggravated by personal intrigues: the apparently capricious withdrawal of Mathilde Kshesinskaya from the title role at

---

[2] The scenario has been summarized from N[ikolai] A[leksandrovich] Solyannikov, 'Recollections', ed. N. A. Shuvalov, Leningrad, Library of the Leningrad Branch of the All-Russian Theatre Society, pp. 163–5. Solyannikov created the role of the Marquis in *Le Pavillon d'Armide*.

the last moment—Anna Pavlova stepped in to replace her—and the abusive criticism of conservative balletomanes, who came out of their anthill again, ablaze with malice aforethought, as they had at the première of *The Magic Mirror*.[3]

While there are no reports of disruption at the first performance on 25 November 1907, critics of the *Peterburgskaya gazeta* launched a one-sided assault on *Le Pavillon* in a series of articles which ran for three days after the première. One of them claimed that the ballet was 'a musical sauce made of costly ingredients by a very poor cook,' though the reviewer for *Rech'* [Discourse] found 'nothing whatever banal in the music. It is beautiful throughout and if it is not distinguished by its originality, it is always interesting in its lively melodic contour, harmony and the richness of its orchestral colours.'[4] Yet another colleague described the ballet as 'enchantment in three scenes' and thought it should have been longer, at the same time that 'Not a Balletomane' in the *Petersburgskaya gazeta* accused Benois of being a leftist and a Social Revolutionary, declared the decorations tasteless, variously described the dances as scrum in mortar, motley and blatant, anti-aesthetic leg-jerking painful to the eye, and characterized the whole as 'contact with a pure classic art by unwashed hands.' Only Nijinsky survived this diatribe unscathed.[5] The whiff of apostasy accompanied Pavel Gerdt's condemnation of *Le Pavillon* in a short interview, where he agreed that the virulent criticism was justified, complained that the variegated colour and activity must weary the spectator's eye, lamented the fact that the ballet had 'no dances whatever', and declared that his general response to the new work was 'unconditionally negative'.[6] After creating the roles of Lucio, Désiré, Abderrakhman, and many others in his forty-seven years of service, Gerdt, 'eternally young' at sixty-three, was neither grateful nor tactful after creating, at Benois's insistence, the leading male role of René de Beaugency.

Benois himself recalled:

> The ballet was performed to continuous applause, many numbers were encored, and at the end the audience simply roared. They called for the authors and the artists, and Pavlova, Fokine, Gerdt, Tcherepnin, and myself went out on stage several times. But to me the best reward for all my labour

---

[3] In an interview published in the *Peterburgskaya gazeta* (18 Nov. 1907, No. 317, p. 11), Benois struck back at the bureaucrats. Describing the one rehearsal in costume which had been permitted, he continued: 'Were I a responsible director, I would under no circumstances have permitted the presentation of a ballet in such a condition even to a special rehearsal public. An unimaginable mess occurred on stage, because the novelty of the costumes was such that nobody recognized one another and all were out of position and not with their partners.' He goes on to cite the sole rehearsal as an example of 'civil service mentality', which prevails in the theatre: the rehearsals were restricted to save money; it is better to save money, even if it means consigning a poorly rehearsed work to the warehouse early, than to rehearse it properly and let it pay for itself through repeated showings.

Benois claimed to have drafted this article himself, which caused the management to delay the première, permitting two more rehearsals (*Reminiscences of the Russian Ballet*, pp. 262–4).

[4] Not a Balletomane, 'Theatre Echo', *Peterburgskaya gazeta*, 26 Nov. 1907, No. 325, p. 4; 'Theatre and Music', *Rech'*, 27 Nov. 1907, No. 280, p. 4.

[5] 'Theatre and Music', *Novoe vremya*, No. 11391, 27 Nov. 1907, p. 5; *Peterburgskaya gazeta*, 26 Nov. 1907, p. 4.

[6] 'Downfall of the Ballet; With P. A. Gerdt', *Peterburgskaya gazeta*, 27 Nov. 1907, No. 326, p. 4.

and sufferings was that Seriozha Diaghilev, having made his way through the crowd in the lobby exiting the theatre after the performance, stopped to smother me in hugs, and, extremely excited, cried out: 'This must be taken abroad.'[7]

LE PAVILLON D'ARMIDE
Ballet-pantomime in 3 scenes
By Alexandre Benois
Music by N. N. Tcherepnin
Dances and scenes produced by M. M. Fokine
St Petersburg
Publication of the Typographer of the Imperial St Petersburg Theatres
(Chief Authority of Crown Affairs).
1907

ROLES PERFORMED

'Armida'—Mlle Pavlova II.
'Viscount de Beaugency'—Mr Gerdt, Soloist of
HIS IMPERIAL MAJESTY
Decorations by Alexandre Benois
Costumes after drawings by Alexandre Benois

DRAMATIS PERSONAE:

Viscount René de Beaugency	Mr Gerdt
The Marquis, *owner of the pavilion*	Mr Solyannikov
Battiste, *the Viscount's servant*	Mr Grigoriev
Postman	Mr Goncharov I
A Shepherd	A boy student
A Shepherdess	A girl student
Servants of the Marquis	Mr Marzhetsky
	Mr Alexeyev II
	Mr Rykhlyakov
	Mr Rakhmanov I

DRAMATIS PERSONAE IN THE SCENE
'The Dream':

The Marquis, *in the guise of King Hydraot*	Mr Solyannikov
Madeleine, *in the guise of Queen Armida*	Mlle Pavlova
Viscount de Beaugency, *in the guise of Rinaldo*	Mr Gerdt

---

[7] Benois, *Moi vospominaniya* [My Recollections], 2 vols. (Moscow, 1980), ii. 472.

Armida's confidantes:
Fenisa	Mlle Kyaksht
Sidonia	Mlle Egorova
Zorada	Mlle Karsavina
Miranda	Mlle Will
Armida's principal slave	Mr Nijinsky

Courtly gentlemen and ladies, the retinue of King Hydraot, heralds, slaves, sorcerers, shades, genies of the hours, demons, witches, odalisques, etc.

TO BE DANCED:

In the 1st scene:

L'Amour vainqueur du temps.

In the 2nd scene:

Scène de l'animation du Gobelins.
Armide pleurant l'absence de Renaud.
Scène et grand pas d'action.
Courtly ladies.
Courtly gentlemen.
Captive knights.
Almahs.
Little Blackamoors.
Adagio.
Valse Noble.
Variations.
Coda finale.

DIVERTISSEMENT

1. L'enlèvement du sérail.
2. Danse des confidantes d'Armide.
3. Sorcières et démons.
4. Evocation des ombres.
5. Danse des bouffons.
6. Danse de l'écharpe.
7. Grande valse finale.

LE PAVILLON D'ARMIDE

René de Beaugency is the fiancé of his cousin Agnes R. He presently finds himself journeying to her estate, where the marriage should take place. A frightful storm catches him unawares along the road and forces him to seek refuge in a nearby castle, which belongs to the Marquis of S. The strangest rumours circulate about this old, ruined nobleman; the peasants are convinced that he has sold his soul to the devil. And in fact, there is much that is mysterious about the Marquis's behaviour: he never receives anyone, never goes anywhere, and spends entire days in his laboratory with experiments in alchemy. He receives René cordially nevertheless, but apologizes that he cannot accommodate him in the castle itself, which is in a state of utter neglect, but must give him lodging for the night in his garden pavilion. In sharp contrast to the poverty of the main building, the pavilion turns out to be a magnificent hall.

The Marquis explains to René that the hall was built by the celebrated beauty Madeleine de S., who shone at the time of the Regent's court and was nicknamed Armida for her exploits. 'There is her portrait,' he whispers, pointing to the Gobelins hanging above the clock, the face of which represents the symbolic victory of Time over Love. 'For love of her, many have paid with their lives, and,' he adds with irony, 'her charms are so powerful that the same fate could befall you after spending the night where everything is steeped in her memory.'

René declares that he has nothing to fear in a beautiful woman who no longer exists, as he is in love with his fiancée. The Marquis, having bid him good night, retires with his servants. Feeling an irresistible fatigue, René lies down, while his faithful servant Battiste settles down in an armchair. The storm quietens.

Suddenly in the nocturnal silence the clock rings out the strokes of midnight. The bronze Cupid comes to life, commands Saturn to disappear and releases the genies of the hours from his box. Time, which draws all behind it, ceases to rule, and events from the distant past again become real. Cupid draws back the curtains of the bed and awakens René. Sounds beckon from afar, as if some celebration were underway. René tries to awaken Battiste but he is in a deep sleep. The sounds get louder.

Tableau: a trellis is illuminated from within. Little by little the entire pavilion is transformed into a magnificent garden filled with courtiers dressed in fantastic costumes. Among them, on an eminence, sits Armida, as if still wrapping around Rinaldo's shoulder a scarf she has embroidered. But in fact there is no Rinaldo. The queen has come to life and is searching in bewilderment for the hero she has captured. In vain, anguished, she questions the courtiers: he has disappeared without a trace.

But her father King Hydraot appears and shows René to Armida. With a touch of his staff René is transformed into a knight and Armida recognizes him as her beloved. At this point it seems to René that he is not dreaming, and that he has already loved this enchanting woman for a long time; he is filled with delight when she rushes into his embrace. To commemorate the joyful event Hydraot puts on a magnificent celebration in which comic, sentimental and fantastic spectacles succeed one another. The celebration ends with a dance by Armida, during which she wraps the scarf around Rinaldo to signify that he shall be her captive forever. Intoxicated by her beauty, René swears that he will love only her.

But lo, the vision wanes and darkens, and as if half asleep René sees Saturn driving Cupid before him and vanquishing him. Darkness beclouds everything, and René loses consciousness. Morning dawns, fresh and bright. Battiste goes to awaken his lord and to prepare him some chocolate. A small herd passes by the window of the pavilion, grazing on those very meadows where formerly the cultivated flower beds of Madeleine de S. were planted.

René emerges from behind the screens. He is trying to recall his experiences of the night; an incomprehensible anguish torments him. Battiste suggests breakfast but in vain, for René is completely immersed in meditations and dreams. Little by little his memory awakens, and before him, as if alive, stands the vision of the night. On the Gobelins he sees the beautiful woman he was just embracing, and sees himself in the guise of a genuflecting knight. In an outburst of delight he now tells Battiste everything that happened. The latter is horrified and begs his master to leave this enchanted place and go to his bride. At first René objects, and even avers that he has forgotten Agnes, as proof of which he gives Battiste the medallion with her portrait. But finally good sense prevails, he remembers his vow, and sends Battiste for the carriage.

Just then the Marquis approaches to inquire about his guest's health. René recognizes him as King Hydraot. Despite the Marquis's attempt to persuade him to stay for a while, René makes ready to continue his journey, feeling the need to escape the nets of this frightful person as soon as possible. Battiste feverishly makes ready and is about to lock the trunks when the Marquis hands René a scarf which he has found on the bronze clock. 'You are forgetting this valuable item—probably a remembrance of the lady of your heart?' René recognizes the scarf which Armida had wrapped around him. He rushes to the Gobelins: the scarf is no longer there! What is this? A hallucination? He is holding in his hands something which proves that what happened during the night was not a dream. René's thoughts grow confused, and he falls unconscious into Battiste's arms, who curses the maliciously laughing Marquis.

Postscript

The critical response *Le Pavillon d'Armide* generated at its first performance seems extraordinary today, the ire it aroused in balletomanes exaggerated. Such stylistic advance as it made over Petipa's ballets is modest, and resides perhaps in the choreography, more likely in the anti-academic visual evocation of seventeenth-century France, and most clearly in that merging of dream and reality which Fokine and Benois conveyed more forcefully in *Petrushka*, a device with which they gently anticipated the psychological probings of expressionist art. *Le Pavillon* may nevertheless signal that a critical point in balletic history had been passed: if, as Diaghilev seemed to be saying, *The Magic Mirror* was a nineteenth-century ballet staving off the twentieth, then *Le Pavillon*, but four years later, was a twentieth-century ballet trying to free itself of the nineteenth.

In retrospect, the production and reception of *Le Pavillon* are exceptional only in their details. The interaction of creators, producers, and audience in this case may be reduced to principles which apply to countless other first productions. Many voices which have spoken in this book—of balletmasters, dancers, musicians, critics, bureaucrats, and spectators—affirm these principles, formed out of the shared experiences of theatre people across generations.

They remind us, for example, of the constant pressures which attend any theatrical enterprise, between artists of different ranks, between artists and management, between company and audience. Glushkovsky and Didelot, Vazem and Petipa, Kister and Telyakovsky, the balletomanic factions of the 1860s and their counterparts fifty years later—all faced similar situations and reacted to them in similar ways. This likeness of stimulus and response is a sign of continuity in Russian ballet, together with the longevity of repertoire, the traditions of pedagogy, and the prolonged careers of individual dancers. Indeed, two artists—Nikolai Golts and Pavel Gerdt—had they maintained records of their lives from the beginning of school to retirement, could have given a personal account of an entire century of Russian ballet, between 1815 and 1915, including fifteen years when both were serving in the Petersburg company at the same time.

Continuity, in turn, affects change and how it comes about. Our authors seem to be saying that once publicity is discounted, once novelty is placed in context, change in ballet is gradual, carefully prepared, and largely unresponsive to administrative initiative. Historical perspective bears this out. Knowing Diaghilev's artistic directions after the First World War, for example, it is striking that within a year of his first season of ballet in Paris he was planning a repertoire of *Le Pavillon*, Act III of *The Sleeping Beauty*, and *Raymonda*, for which he implored the sixty-four year old Pavel Gerdt to come to Paris and take the role of Abderrakhman. These were not innovations, but rather a tribute to the past, and a wager on what would appeal in Paris.

Other evidence of this continuity, and of gradual change in general, lies just beneath the surface of the Diaghilev repertoire before the First World War, which owes much to the past. Disguised in *réclame*, this indebtedness takes the form of borrowed ideas, refined and transformed, to which Diaghilev's collaborators added *chic*. Fokine's harlequinades are the offspring of Petipa's *Les Millions d'Arlequin* of 1900, in which Fokine himself performed, just as his *Sylphides* are the artistic progeny of Taglioni's. The oriental extravaganzas of *Shéhérazade* and *Josephslegende* are modifications of prototypes perfected by Petipa, and yield nothing to the earlier master in scenic opulence or sensual effect. The ethnic scenes of *Petrushka* draw on the same folk ethos as those in *The Little Humpbacked Horse*. And while there is no evidence of a causal relationship between the two works, *The Rite of Spring* is seen in a different light when we realize that Charles Didelot, ninety-five years earlier, was showing Petersburgers in *Cora and Alonso* a ballet which centred on the sacrifice of a virgin to the god of spring. Didelot's ballet, moreover, was almost certainly based on another ballet of similar title he had seen in London in 1798.

Similar connections link other repertoires. Marie Taglioni arrived in St Petersburg on a wave of fashion, though her father's works preceded her in productions by Titus, and Didelot, as we have seen, adumbrated certain principles of the white ballet in *The Captive of the Caucasus*. For all that was new in them, the repertoires of Perrot and Petipa represent a synthesis of the devices and images of earlier dramatic and romantic ballets.

Whatever the social status of the balletomanes with whom Telyakovsky had to contend, their response to *Le Pavillon d'Armide*, eccentric and mirthful to us, may be more a reflection of the discomfiture which attends a break in continuity than evidence of modishness run to seed. A deep-rooted stability, which had helped the Petersburg company survive contentious issues of the moment for a hundred years, was being disrupted. Their testimony, like that of other voices which have spoken in these pages, may suffer from inaccurate data, imperfect memory, and bias. But all were witness to what they describe, and at the very least, they have left us something on which to base a judgement.

INDEX

(Cast lists and character names not included)

Ablets, Isaac (1778–1828), dancer and balletmaster 21
Abrantès (Duchess) [Laure Permon Junot] (1784–1838), memoirist 44
Actors Among Themselves (vaudeville) 112
Adam, Adolphe Charles (1802–56), composer 90, 124, 127, 209, 289, 383, 385
Adams, Polish dancer who performed in St Petersburg 167
Afanasiev, Alexandre Nikolaevich (1826–71), folklorist 380
Aimée (pension/gymnasium in St Petersburg) 123
Aix-les-Bains 199
Albert, François Decombe (1789–1865), dancer and choreographer 15, 44
Albrecht, [Eduard?], player of the cornet-à-piston 130
Albrecht, Pyotr Ivanovich (1760–1830), Bursar of the Imperial Theatres 30
Alexandra Feodorovna (1798–1860), Empress of Russia [wife of Nikolai I] 121, 123, 132, 166–9
Alexandre Alexandrovich [Alexandre III] (1845–94), Tsar of Russia 307
Alexandre Nikolaevich [Alexandre II] (1818–81), Tsar of Russia 87, 126, 132–33, 289, 322
Alexandre Pavlovich [Alexandre I] (1777–1825), Tsar of Russia 26
Alexandrinsky Theatre [Alexandrinka] (St Petersburg) 113–15, 125, 128, 153, 255, 264, 283
Alexandrov, dancer 127
Alexandrova (Schnell), Maria Mikhailovna (1833–?), dramatic artist 136
Allegri, Oreste Karlovich (1866–1954), painter of theatre decorations 393
Amosova, [Anastasia Nikolaevna (1832–88) or Nadezhda Nikolaevna (1833–1903)], dancer 164, 265–6, 274
Anderson, Maria Karlovna (1870–1944), dancer 374
André [André Schmidt], comic actor and dancer specializing in grotesque roles 22
Andreanova, Elena Ivanovna (1819–57), dancer 124, 144, 173–4, 185, 378
Andreyev, Ivan Petrovich (1847–96), painter of theatre decorations 288, 360
Anna Ioannovna (1693–1740), Empress of Russia 34

Antonin, French dancer in St Petersburg 1817–27: 22
Antonolini, Ferdinand (d. 1824), specialist composer of ballet music 24
Ariosto, Lodovico (1474–1533), poet 14
Arnaud, d' [François Thomas Marie de Baculard] (1718–1805), writer 50
Artemiev, Pyotr Nikolaevich, dancer 124
Artist (journal) 107
Aubel, Leonty Filippovich, assistant supervisor of the Petersburg Theatre School 111, 152, 154, 161, 165
Auber, Daniel François Esprit (1782–1871), composer and Director of the Paris Conservatoire 201, 203, 286, 382
Auber, Fyodor Nikolaevich (1796–1856), Inspector of the Petersburg Theatre School 137, 145, 158, 165
Auer, Leopold (1845–1930), violinist 285, 288, 292
Auguste [Poireau] (c.1780–1844), dancer and choreographer 16, 22, 31, 41, 44, 51–2, 68, 71, 73, 175
Aumer, Jean (1774–1833), dancer and choreographer 15
Azarevicheva, Nadezhda Apollonovna (1806–73), dancer 68

Babini, costume designer for Didelot's ballets 52, 70
Bach, Yu., stage decorator in the Petersburg Imperial Theatres 133
Bakst, Leon [Lev Samoilovich Rozenberg] (1866–1924), artist 390
Balet, ego istoriya i mesto v ryadu izyashchnykh iskusstv [Ballet, its History and Place Among the Elegant Arts] (Skalkovsky) 307, 355
ballet music 23–4
Ballets, féeries, and divertissements:
 Acis and Galatea (Didelot) 15, 26–8, 36
 Adventures on the Hunt (Didelot) 36
 Aelia et Mysis (Mazilier) 209
 Aeneas and Lavinia (Didelot) 36
 Alceste, or The Descent of Hercules into Hell (Didelot) 10, 15, 36
 Alina, Queen of Golconda (Didelot) 35
 Amor (Manzotti) 355
 Amour et Psyché (Didelot) 6, 33, 35, 41, 50
 Amour vengé, L' (Didelot) 35
 Apollo and Daphne (Didelot) 6, 35, 50
 Armida (Perrot) 127, 260

Ballets, féeries, and divertissements (cont.):
  Bandits, The (Petipa) 284
  Battle of the Women, The, or The Amazons of the Ninth Century (Perrot) 124
  Bayadère, La (Petipa) 234, 262, 286–8, 291–303, 322, 359
  Beauty of Lebanon, The, or The Mountain Spirit (Petipa) 130, 239, 262–3, 359
  Blue Dahlia, The (Petipa) 129, 133, 217
  Bluebeard (Petipa) 392
  Brahma (Montplaisir) 317
  Bride of Wallachia (St-Léon) 274
  Butterfly, The (Petipa) 284
  Caliph of Baghdad, The (Didelot) 8–10, 36
  Captive of the Caucasus, The (Didelot) 13–14, 16, 36, 69–78, 107, 112, 176, 239, 293, 430
  Carl and Lisbetta (Didelot) 15, 36
  Carlos and Rosalba, or the Automaton-Lover (Didelot) 36
  Caterina, Daughter of the Robbers (Perrot) 117, 120, 122, 124, 164, 260
  Celebration in the Camp, The (Ablets and Glushkovsky) 21
  Coppélia (St-Léon) 107, 323, 391
  Cora and Alonso (Didelot) 11–13, 36, 430
  Corsaire, Le (Mazilier) 127, 284
  Cossack in London, The (Auguste and Valberkh) 21
  Courageous Alonso and the Beautiful Imogene (Didelot) 35
  Daughter of the Snows, The (Petipa) 289
  Descent of Hercules into Hell, The, see Alceste
  Diable à quatre, Le (Mazilier) 124, 311, 323
  Diable amoureux, Le, see Satanilla
  Diana and Endymion (Didelot) 37
  Dido, or The Destruction of Carthage (Didelot) 36
  Don Quixote (Petipa, revived Gorsky) 291, 406, 418
  Echo and Narcissus (Didelot) 37
  Elève d'Amour, L' (Taglioni) 175
  Enchanted Forest, The (Didelot) 37
  Eoline, or The Dryad (Perrot) 127
  Esmeralda (Perrot) 118–20, 124, 161, 163–4, 174–82, 260–1, 293, 323, 381
  Euthyme et Eucharis, or The Vanquished Shade of Lybas (Didelot) 15, 36–41
  Excelsior (Manzotti) 355
  Extraordinary Journey to the Moon, An (Hansen) 318
  Fair at Makarievsk (Ablets and Glushkovsky) 21
  Father's Curse, The (Didelot) 36
  Faun, The (Didelot) 36
  Faust (Perrot) 125, 127, 260–1, 352
  Festival in the Camp of the Allied Troops (Auguste and Valberkh) 21
  Fiametta (St-Léon) 131, 269, 322
  Filatka and Fedora, or The Fair at the Village of Novinsk (Ablets and Glushkovsky) 21
  Fille du Danube, La (Taglioni) 87–8, 90–103, 175, 192, 289–90, 292
  Fille mal gardée, La (Dauberval) 31, 117, 120, 123–4, 126, 132, 323
  First of May, The, or The Fair in the Sokolniki (Ablets and Glushkovsky) 21
  Forest Vagrant, The (Hansen) 320
  Fortunate Shipwreck, The, or The Scottish Witches (Didelot) 35
  Gerta (Taglioni) 82, 86
  Giselle (Coralli and Perrot) 117, 126, 150, 173, 175, 177, 185, 210, 261, 288–90, 323, 391
  Gitana, La (Taglioni) 86, 112, 193
  Golden Apples, The (Hansen) 320
  Golden Braid, The, or The Youth of Medea (Didelot) 37
  The Golden Fish (St-Léon) 239, 274
  Grapevine, The (music by Rubinstein) 322
  Graziella (St-Léon) 129, 133, 264
  Hamlet (Didelot) 37
  Hercules and Omphale (Didelot) 15
  Hungarian Hut, The, or The Famous Exiles (Didelot) 7, 16, 36, 107, 124
  Josephslegende (Fokine) 430
  Jovita (Mazilier) 129
  Julius Caesar in Egypt (Titus) 81
  Kamul and Zabara (Didelot) 37
  Kensi and Tao (Didelot, London) 35
  Ken-si and Tao (Didelot, St Petersburg) 15, 36
  Kia-King (Titus) 15, 81
  King's Command, The (Petipa) 375
  Laura and Henry, or The Routing of the Moors (Didelot) 15, 35–6
  Lise and Colin, see La Fille mal gardée
  Little Humpbacked Horse, The (St-Léon) 69, 130–1, 218, 234–5, 238–50, 254, 260, 265–6, 269–70, 272–3, 293, 316, 392, 430
  Love for the Fatherland (Auguste and Valberkh) 21
  Macbeth (Didelot) 37
  Madcap Mind and Good Heart (Didelot) 36
  Magic Mirror, The (Petipa) 392, 405–18, 420, 422–4, 429
  Marcobomba (Perrot) 125
  Markitanka, see Vivandière, La
  Mask, The, or The Spanish Evening (Didelot) 36
  Météora (St-Léon) 129, 288
  Midsummer Night's Dream, A (Petipa) 312
  Millers, The (Blache?) 113, 116–17, 175
  Millions d'Arlequin, Les (Petipa) 430
  Misfortune in the Dress Rehearsal (St-Léon) 130
  Naiad and the Fisherman, The (Perrot) 122, 166, 168–9, 260, 312
  Néméa (St-Léon) 322
  Noble Trait, The (Didelot) 37
  Nutcracker, The (Petipa and Ivanov) 107, 387, 392
  Offering of Love, The (Didelot) 36

Index 433

Orfa (Mazilier) 209
Orphan Théolinda, The (St-Léon) 130, 264
Origin of Painting, The (Didelot) 36
Page's Pranks, The (Didelot) 37
Pâquerette (St-Léon) 129
Paquita (Mazilier) 124, 151, 174–5, 196, 217
Pardon, The (Didelot) 37
Parisian Market, The (Petipa) 128, 217
Pavillon d'Armide, Le (Fokine) 422–30
Pearl, The (Petipa) 392
Pearl of Seville, The (St-Léon) 129, 269
Peasant Wedding, The (Steffani) 121–2, 124, 126, 131, 167
Péri, La (Coralli) 175, 191, 194
Petrushka (Fokine) 429–30
Phaedra (Didelot) 15, 36
Pharaoh's Daughter, The (Petipa) 129–30, 133, 218–39, 254, 259, 262–3, 265, 280, 291, 293, 359, 391–2
Poverty and Misfortune (Didelot) 36
Prosperity and Happiness (Didelot) 37
Psyché et l'Amour (Didelot), see *Amour et Psyché*
Raoul Bluebeard, or The Mysterious Chamber (Auguste and Valberkh) 31–2
Raoul de Créqui, or The Return from the Crusades (Didelot) 8, 16, 36, 50–67, 69
Raymonda (Petipa) 217, 392–401, 405, 407, 430
Regency Marriage, A (Petipa) 127, 217
Revolt in the Seraglio, The (Taglioni) 83, 175
Richard the Lion-Hearted (Didelot) 35
Rite of Spring, The (Nijinsky) 430
Robert and Bertram (Augué) 127–8, 131, 133
Roi Candaule, Le (Petipa) 262, 280, 285–6, 359
Roland and Morgana, or The Destruction of the Enchanted Island (Didelot) 14, 35–6
Romance of the Rosebud (Petipa) 405
Roxana (Petipa) 262, 322
Salammbò (Petipa) 406
Saltarello (St-Léon) 128
Sappho (Didelot) 35
Satanilla (Mazilier) 151, 174–5, 217
Semik, or The Fair at the Maria Woods (Ablets and Glushkovsky) 21–2
Shattered Idol, The (Didelot) 36
Shéhérazade (Fokine) 430
Shepherd and the Hamadryad, The (Didelot) 35–6
Simpletons, The (Auguste) 114, 175
Sleeping Beauty, The (Petipa) 217, 322, 354, 359–73, 376–80, 382–92, 406–7, 421, 430
Sonnambule, La (Aumer) 199
Source, La (St-Léon) 322
Sparrow Hills, The (Ablets and Glushkovsky) 21
Swan Lake (Reisinger, revived Petipa and Ivanov) 354, 382, 392, 423
Sylphide, La (Taglioni) 82, 84–5, 87–8, 116, 154–5, 192–3, 201, 210, 392
Sylphides, Les (Fokine) 430
Sylvia (Delibes) 421
Talisman, The (Petipa) 323, 359
Talisman, The (Titus) 173
Telemachus on the Island of Calypso (Didelot) 25
Theseus and Ariadne (Didelot) 15, 36, 81
Trilby (Petipa) 309
Two Sorceresses, The (Titus) 116, 153
Two Thieves, The, see *Robert and Bertram*
Unexpected Return, The (Didelot) 36
Vert-vert (Mazilier) 208
Vestal, The (Petipa) 234, 324–50, 352–4, 359, 373
Vivandière, La (St-Léon) 206–8, 312
Wilful Wife, The, see *Le Diable à quatre*
Wooden Leg, The (Didelot) 36
Young Island Girl, The, or Leon and Tamaida (Didelot) 36
Young Milkmaid, The (Didelot) 36
Zéphire et Flore (Didelot) 6–7, 30, 35–6, 41, 50, 82, 175
Zoraya (Petipa) 262, 316, 322

Balmont, Konstantin Dmitrievich (1867–1943), poet 420
Baltier, costumier for Taglioni's *La Fille du Danube* in St Petersburg 90
Baranov, Vasily Venediktovich (1793–c.1836), painter 34
Barch, Ivan Martynovich, physician 132
Bassin, Laura, circus artist 114, 141
Bassin, Victorina, student of the Petersburg Theatre School 141–2
Battiste [Jean-Baptiste-Mari Pti], dancer and choreographer 15, 22, 32
Bauer, [Mikhail Grigorievich (1841–?)?], dancer 127
Baveri, teacher of singing in the Petersburg Theatre School 143
Bayreuth 374
Becker, Mikhail (1810–72), dancer and teacher 136–7, 152–3
Begichev, Vladimir Petrovich (1828–91), writer and theatre official 354
Bakefi, Alfred Fyodorovich (1843–1925), dancer 374
Beketov, V., theatre censor 233
Bely, Andrei [Boris Nikolaevich Bugaev] (1880–1934), writer 420
Benois, Alexandre Nikolaevich (1870–1960), painter, writer, ballet scenarist 359, 385, 422–25, 429
Benois, Leonty Nikolaevich, brother of Alexandre Benois 385
Benoist, François (1794–1878), composer 124, 129
Berger, Nikolai Alexandrovich, stage machinist 360, 393
Berlin 127, 134, 188, 212, 309, 317, 352
Berlioz, Louis Hector (1803–69), composer 382
Bers, physician in St Petersburg 135

434 Index

Bertin-Boieldieu, Jenny Philis (d. 1853), singer in the French opera company in St Petersburg in the early nineteenth century, wife of composer Adrien Boieldieu 32
Bessone, Emma, Italian dancer who toured in Russia 1887, 1890–91: 355
Blache, Alexis (1791–1850), choreographer 81–2, 86
Blache, Jean-Baptiste (1765–1834), choreographer 175
Blasis, Carlo (1797–1878), dancer, choreographer and teacher 317, 355
Bocharov, Mikhail Ilyich (1831–95), stage designer 127, 288, 360, 373–4, 389
Bogdanov, [Alexandre Fedotovich (1837–1888)?], dancer 129
Bogdanov, Alexandre Konstantinovich, dancer and violinist 190, 193, 201, 203
Bogdanov, Alexei Nikolaevich (1830–1907), dancer and balletmaster 131, 273–4, 284
Bogdanov, Konstantin Fyodorovich (c. 1809–77), dancer and régisseur 189, 192–5, 197, 199
Bogdanov, Nikolai Konstantinovich, dancer and pianist 189–90, 193, 201, 203, 210–11
Bogdanova (Madame), mother of Nadezhda Bogdanova et al. 191–2, 201
Bogdanova, Nadezhda Konstantinovna (1836–97), dancer 126, 185–6, 188–213
Bogdanova, Tatyana Konstantinovna, dancer 190, 199–200
Bogolyubov, Mikhail (Father), priest of the Petersburg Theatre School 141–2, 158
Bolshoy Theatre (Moscow) 185
Bolshoy Theatre (St Petersburg) 35, 50, 71, 81–2, 119, 131, 251–3, 258, 260, 269–70, 280, 282, 292–3, 309, 312–13, 379
Bornier, dramatist 316
Brianza, Carlotta (1867–1930), ballerina 355, 359, 373–5, 385, 390
Bruneau, maker of theatre headwear 360, 393
Bryullov, Karl Pavlovich (1799–1852), painter 11
Brussels 351
Bubnov, Pyotr, transfer student from Moscow to St Petersburg 151
Budapest 186
Bulakhov, [Pavel Petrovich (1824–75)?], singer 131, 143
Bulgarin, Faddei Venediktovich (1789–1859), journalist and publisher 88, 173
Bursey, stage machinist for Didelot's *Raoul de Créqui* 52
Butter Week 156, 173, 235, 254–5, 386
Byron, George Gordon Noel (1788–1824), poet 127
Bystrov, Alexandre Dmitrievich (1851–98), comic dancer 267
Bystrov, Dmitry, teacher of singing in the Petersburg Theatre School 142

Caffi [Kaffi], Ivan Iosifovich, theatre costumier 360, 393

Cannelle, teacher of French in the Petersburg Theatre School 142
Canoppi, Antonio, (1774–1832), painter of theatre decorations 52
Cavos, Catterino (1776–1840), composer and conductor 21, 24, 31, 50, 52, 70, 113
Cecchetti, Enrico (1850–1928), dancer, choreographer and teacher 109, 307, 374, 385, 390
Cecchetti, Josephina Maria (1857–1927), dancer 385
Cerrito, Fanny (1817–1909), dancer 201, 206–7, 251
Chambéry 198–9
Chameroy, Louise (1779–1803?), dancer 44
Charlemagne [Sharleman], Adolf Iosifovich (1826–1901), painter 238
Charukhina, Nadezhda Ivanovna, classroom lady of the Petersburg Theatre School 143–4
Chernoyarova, Nadezhda (1838–?), dancer 129
Chevalier (Madame), actress, sister of Auguste Poireau 32
Chistyakov, Alexandre Dmitrievich, dancer 126, 128, 151
Chitau, Alexandra Matveyevna (1832–1912), dramatic artist 114
Chopin, Fryderyk Franciszek (1810–49), composer 382
Ciardi, Cesare (1817–77), flautist 130
Cirque Ciniselli (St Petersburg) 128
Citizen, The [Grazhdanin] (newspaper) 374
Coralli, Jean (1779–1854), dancer and choreographer 175, 261
Corbetta, Pasquale, Milanese teacher of dance 317
Cornalba, Elena (1860–?), dancer 323, 354
Corps of Pages (St Petersburg) 123–4
Cossack Poet, The (play) 113
Coulon, Jean-François (1764–1836), teacher of dance 15, 44
Crimean War 185
Chronicle of the Petersburg Theatres (Wolf) 316
Cuzent, Leonard, co-proprietor of a circus in St Petersburg 114
Cuzent, Paul, co-proprietor of a circus in St Petersburg 114
Czerny, pianist 130

Dalayrac, Nicolas-Marie (1752–1809), composer 51
Dams, dancer from Warsaw performing in St Petersburg 121
Danilova, Maria Ivanovna (1793–1810), dancer 33, 68
Dante Alighieri (1265–1321), poet 288
Dargomyzhsky, Alexandre Sergeyevich (1813–69), composer 126
Dauberval, Jean (1742–1806), dancer and balletmaster 5, 31, 44, 68, 117

Daumier, teacher of German in the Petersburg Theatre School 142
Davydov, Stepan Ivanovich (1777–1825), composer 21, 23
de la Motte-Fouqué, Friedrich Heinrich Karl (1777–1843), poet and novelist 88
Deldevez, Edouard (1817–97), composer 124, 208
Delibes, Léo (1836–91), composer 322, 378, 383, 385
Dell'Era, Antonietta (1865–?), dancer 354–5
Dépréaux, Jean-Etienne (1748–1820), dancer and teacher of dance 44
Descourt, teacher of French in the Petersburg Theater School 142
Diaghilev, Sergei Pavlovich (1872–1929), impresario 185, 359, 390, 405, 417, 420–4, 430
Didelot, Charles (1767–1837), dancer and choreographer 3–20, 22–41, 50–2, 68–70, 72–3, 81–2, 85, 107, 112, 124, 173–6, 185, 189, 239, 351, 390, 429–30; character and habits 29–30, 33–4; collaboration with composers 23–5; style of choreography 6–7, 11, 14, 16, 41, 51
Didelot, Karl Karlovich (1801–55), dancer, son of Charles Didelot 35–6
Didelot, Rose (d. 1803), dancer, wife of Charles Didelot 6
Didier, Pyotr Ivanovich (1799–1852), dancer and régisseur 113, 118, 164
Dineau [pseudonym for Zh. F. Bedin and P. P. Goubeau, authors of popular plays in nineteenth-century Russia?] 125
Dmitrievsky, Ivan Afanasievich (1734–1821), actor, writer, and pedagogue 191
Dobrovolskaya, costumier for the St Petersburg Imperial Ballet 393
Don Quixote (drama) 194
Donizetti, Gaetano (1797–1848), composer 117, 284
Dor, Henriette (1844–86), dancer 236, 251, 284–5
Doré, Gustave (1832–83), illustrator 288, 375
Dranchet, Jean, painter of theatre decorations in St Petersburg 1795–1820: 52
Drigo, Riccardo (1846–1930), ballet conductor 308, 321
Dubelt, Leonty Vasilievich (1792–1862), supervisor of the Third Section of the Imperial Chancellery 135, 137, 153, 160–2, 173
Ducange, V., author of light plays and melodramas 125
Dudkin, Dmitry, transfer student from Moscow to St Petersburg 152
Dumas, Alexandre (1802–70), novelist and dramatist 380
Duncan, Isadora (1878–1927), dancer 351, 355
Duport, Louis Antoine (c.1781–1853), dancer and choreographer 15, 20, 22, 33, 44, 68

Dutaque, Jean, dancer and teacher of dance (d. c.1865) 41, 44
Dyadichkin, Vladimir Trofimovich (1833–82), assistant régisseur of the Petersburg ballet 267
Dyurova, dancer? 124

Eberhard, Ivan Ivanovich, dancer and teacher of dance 41
Efimov, Konstantin Panteleyevich (1829–1903), assistant régisseur of the Petersburg ballet 267
Efremova, Maria Alexandrovna (1835–?), dancer 264
Ekaterininsky Institute (St Petersburg) 123–4
Elizavetinsky Institute (St Petersburg) 123
Elssler, Fanny (1810–84), ballerina 31, 107, 117–18, 121, 164, 173–7, 185, 198, 201, 207, 210, 251, 271, 315
Ershov, Pyotr Pavlovich (1815–69), writer 130, 269–70, 273
Espinosa, Leon (1825–1903), dancer 311
Esther, French artist performing in St Petersburg 121
Excursion to Tsarskoe Selo, The (vaudeville) 112
Ezhegodnik Imperatorskikh Teatrov [The Yearbook of the Imperial Theatres] 405

False Fanny Elssler, The, or The Ball and the Concert (vaudeville) 118
Fèbre, teacher of French in the Petersburg Theatre School 142
Ferraris, Amalia (1830–1904), dancer 251, 271, 284
Feval, Paul (1817–87), writer and dramatist 380
Figareda, Alexandra Vasilievna, senior supervisor of the Petersburg Theatre School 138–9, 143–7, 154, 156, 159, 166–7
Fillet (Madame), teacher of French in the Petersburg Theatre School 142
Filosofov, Dmitry Vladimirovich [Dima] (1872–1940), writer 386
Fitinhof-Schell (Baron) Boris Alexandrovich (1829–1901) 128
Fitzjames, Nathalie (1819–?), dancer 266
Fleury [Bernard Novier?], comic dancer 113, 136–7
Florence 317
Fokine, Mikhail Mikhailovich (1880–1942), dancer and choreographer 85, 354, 392, 423–5, 429–30
France, Anatole [Jacques Anatole Thibault] (1844–1924), writer 381
Frédéric, see Malaverne
Friedrich Wilhelm III, King of Prussia (1770–1840) 21
Frog Princess, The (ballet scenario) 269
Frolov, Alexandre Petrovich (1819–94), Supervisor of the Petersburg Theatre School 310

436 Index

Fyodor Ivanovich (1557–98), Tsar of Russia 72

Fyodorov, decorator of Taglioni's *La Fille du Danube* in St Petersburg 90

Fyodorov, Pavel Stepanovich (1803–79), supervisor of the repertoire division of the Petersburg Theatres 119, 265

Fyodorov, wigmaker for the St Petersburg Theatres 360

Fyodorova, [Anna (1820–92)?], circus artist 115

Gaevsky, Vadim Moiseyevich (b. 1928), Soviet ballet historian 85

Gagarin, (Prince) Sergei Sergeyevich (1795–1852), Director of the Russian Imperial Theatres 1829–33: 81

Galakhov, A. A., officer of the Life-Guards Mounted Regiment 123

Gardel, Pierre (1758–1840), choreographer and teacher 15, 44

Garkush, Sofia Petrovna, classroom lady of the Petersburg Theatre School 143

Gautier, Théophile (1811–72), poet, novelist, and critic 288

Gavrilov, maker of accessories in the Petersburg theatres 238

Gedeonov, Alexandre Mikhailovich (1791–1867), Director of the Russian Imperial Theatres 1833–58: 113, 115–16, 121, 124, 135–7, 139, 144–7, 151–3, 155, 157, 166–7, 169, 185

Gelau, assistant régisseur of the Petersburg ballet 267

Geltser, Vasily Fyodorovich (1840–1908), dancer 273

Gerdt, Pavel Andreyevich (1844–1917), dancer 107, 109, 267, 274, 284, 286, 288–90, 315, 390, 424, 429–30

Germano, Ferni (Madame), soprano 317

Gillert, [Arnold Kazimirovich (1823–?)?], dancer from Warsaw performing in St Petersburg 121

Giuri, Maria (d. 1912), Italian dancer who performed at the Maryinsky Theatre in 1899 355

Giustiniani, teacher of Italian in the Petersburg Theatre School 142

Glazunov, Alexandre Konstantinovich (1865–1936), composer 322, 393

Glazunov, Ilya Ivanovich (1786–1849), publisher 90, 178

Glinka, Mikhail Ivanovich (1804–57), composer 293, 382–3

Gluck, Christoph Willibald von (1714–87), composer 378

Glushkovsky, Adam Pavlovich (1793–c.1870), dancer and choreographer 3–5, 51, 65, 68, 429

Goethe, Johann Wolfgang von (1749–1832), poet, dramatist, and scientist 125, 291

Gogol, Nikolai Vasilievich (1809–52) 260

Goldmark, Karl (1830–1915), composer 383

Golovin, Alexandre Yakovlevich (1863–1930), painter 406, 417–21

Golts, Nikolai Osipovich (1800–80), dancer 69, 107, 122, 131, 164, 237, 267, 273, 284, 288, 293, 429

Gomburov, producer of stage battles for Didelot's *Raoul de Créqui* 52

Gorshenkova, Maria Nikolaevna (1857–1938), dancer 287–8, 315, 354

Gorsky, Alexandre Alexeyevich (1871–1924), dancer and choreographer 282, 406, 417

Gounod, Charles François (1818–93), composer 382

Gozenpud, Abram Akimovich (b. 1908), Soviet musicologist 50

Graff, artiste 121

Grahn, Lucile (1819–1907), dancer 174

Grantzow, Adèle (1843–77), dancer 251, 279, 284, 309, 315

Graphic, The (journal) 292

Gredlu, Emile, French dancer and teacher who worked in St Petersburg 1837–47: 111–13, 118

Griboyedov, Alexandre Sergeyevich (1795–1829), playwright 283, 353

Grigorieva, Maria, transfer student from Moscow 153

Grigorevich, translator of *The Monkey and the Suitor* from German into Russian 115

Grimm, Jakob Ludwig Karl (1785–1863), and Wilhelm Karl (1786–1859), philologists and folklorists 380, 408

Grinev, Apollon Afanasiev (1831–83), balletomane 255–6, 259–60, 353

Grisi, Carlotta (1819–99), ballerina 124, 177, 207, 210, 251, 315

Grisi, Giulia (1811–69), soprano 142

Guy-Stéphan, Marie (1818–73), dancer 210

Hansen, Joseph (1842–1907), choreographer 317, 320

Haydn, Franz Josepf (1732–1809), composer 16, 378

Hermitage Theatre (St Petersburg) 35

Hoffmann, Ernst Theodor Amadeus (1776–1822), novelist and composer 387, 422

Hoffmann, Mina Petrovna, clsssroom lady and senior supervisor of the Petersburg Theatre School 143, 166–7

Hoffmann, Nanetta Petrovna, classroom lady of the Petersburg Theatre School 143

Homan (Miss), classroom lady of the Petersburg Theatre School 144

Hoppe, Edouard, typographer of the Russian Imperial Theatres 293, 324

Huard, teacher of dancing 41

Hugué [Hoquet?], Evgenii [Eugène] (1821–75) teacher of dance in the Petersburg Theatre School 1848–69: 262

Hugo, Victor Marie (1802–85), poet, novelist, dramatist 176, 381

Hullin-Sor, Félicité (1804-c.1860), dancer and choreographer 34
Hus, Augusto, director of the ballet school attached to the La Scala Theatre, Milan 317

Illustrated London News 292
Imperial Alexandrovsky Lyceum (St Petersburg) 123
Inginen, maker of metalwork theatre accessories for the Petersburg Imperial Theatres 360, 393
Isakov, student of the Petersburg Theatre School 111
Istomina, Avdotia Ilyinichna (1799–1848), dancer 28, 68
Istoricheskii vestnik [Historical Messenger] 135
Istoriya tantsev [The History of Dances (Khudekov)] 235
Ivan III ('The Great'), Tsar of Russia (1440–1505) 34
Ivan IV ('The Terrible'), Tsar of Russia (1530–84) 34
Ivanov, Konstantin Matveyevich (1859–1916), painter of stage decorations 360, 374, 389, 393
Ivanov, Lev Ivanovich (1834–1901), dancer and choreographer 85, 107, 131, 133, 237, 267, 273–4, 284, 286, 288, 392
Ivanov, Mikhail Mikhailovich (1848–1927), composer and writer on music 324, 350
Ivanova, Evdokia Trofimovna, theatre costumier in the Petersburg Imperial Theatres 360

Janin, Jules-Gabriel (1804–74), journalist and ballet critic 206, 208–10
Jerusalem 8
Jogel, P. A., Moscow dancing teacher 44, 46–7
Johanson, Anna Christianovna (1860–1917), dancer 315, 374
Johanson, Christian Petrovich (1817–1903), dancer 119, 166, 177, 267, 288, 315, 390
Journal de la Savoie 199

Kachenovsky, Mikhail Trofimovich (1775–1842), journalist, historian and publisher of *The Messenger of Europe* 33
Kaluga 193
Kamennyi Ostrov Theatre (St Petersburg) 130
Kamensky, Pavel Pavlovich (1858–?), sculptor and maker of theatre accessories for the Petersburg Imperial Theatres 360, 393
Kanegisser, Leonty Leontievich, bassoonist and ballet conductor in St Petersburg 321
Kantsyreva, Klavdia Ivanovna (1847–?), dancer 267
Karatygin, Pyotr Andreyevich (1805–79), actor, translator, memoirist 113
Karsavina, Tamara Platonovna (1885–1978), ballerina 308
Kashin, Daniil Nikitich (1769–1841) folksong collector and composer 21

Katenin, [Pavel Alexandrovich (1792–1853)?], critic and writer for the theatre 353
Kemmerer, Alexandra Nikolaevna (1842–1931), dancer 129, 236, 251, 263, 266–7
Kharkov 194, 197
Khudekov, Sergei Nikolaevich (1837–1927), journalist, historian and ballet scenarist 234–5, 250, 255, 260, 291, 323–4, 350, 352–3
Khvorostinin [Khvorostin-Yaroslavlsky, (Prince) Ivan Dmitrievich (d. 1612), voevoda of Astrakhan?] 72
Kiev 34–5, 197
Kister [Küster], (Baron) Karl Karlovich (1821–93), Director of the Russian Imperial Theatres 1875–81: 290–1, 307, 312, 429
Klimov, teacher of calligraphy in the Petersburg Theatre School 139
Knapich, teacher in the Petersburg Theatre School 110
Kolosova, Evgenia Ivanovna (1780–1869), dancer 19–20, 22, 31–2, 41
Komorov, steward of the Petersburg Theatre School 154
Kondratiev, Alexei Efimovich (1784–?), painter of theatre decorations 52, 71
Koni, Fyodor Alexeyevich (1809–79), writer and dramatist 82, 85–6
Koreshchenko, Arseny Nikolaevich (1870–1921), composer 405–6, 408, 417–20
Korovin, Konstantin Alexeyevich (1861–1939), painter 390
Kosheleva (Kosheva), Anna Dmitrievna (1840–?), dancer 237, 265, 266–7, 274
Kostroma 192
Kotlyarevsky, Ivan Petrovich (1769–1838), writer and dramatist 131
Kotlyarovskaya, dancer from Warsaw performing in St Petersburg 121
Kovaleva, Fedosya Grigorievna, inspectress of the choir of the Petersburg Theatre School 142
Kozlov, Fyodor Mikhailovich (1883–1956), dancer 421
Krasovskaya, Vera Mikhailovna (b. 1915), Soviet ballet historian 68, 84
Krechinsky's Wedding (comedy) 126
Krupensky, Alexandre Dmitrievich (1875–1939), Supervisor of the Bureau of the Petersburg Theatres 1907–17: 422
Kshesinskaya, Matilde Felixovna (1872–1971), ballerina 177, 308, 417, 419, 423
Kshesinsky, Felix Ivanovich (1823–1905), dancer 128, 131, 237, 273, 284, 286, 288, 315
Kshesinsky, Iosif Felixovich (1868–1942), dancer 375
Kulikov, Nikolai Ivanovich (1815–91), actor and dramatist 110, 113
Kursk 194

438 Index

Kuskova, Elizaveta Apollonovna (1869–?), dancer 373
Kusov, (Baron) Vladimir Alexeyevich, supervisor of the production department of the Petersburg Imperial Theatres 418
Kuznetsov (pension/gymnasium in St Petersburg) 123
Kvitovsky, dancer from Warsaw performing St Petersburg 121

Lachouc, Charles (1801–41), dancer and teacher 111
Lafitte, [Jacques (1767–1844), banker and political figure?] 44
Lalo, Edouard Victor Antoine (1823–92), composer 378
Lambin, Pyotr Borisovich, stage decorator of the Petersburg Imperial Theatres 393
Lanner, Joseph (1801–43), composer 290
Larionova, Maria Osipovna, classroom lady of the Petersburg Theatre School 143
Laroche, German Avgustovich (1845–1904), writer on music 359, 377
Latysheva, Alexandra Alexandrovna (1830–72), singer 143
Lausanne 199
Lavrov, student of the Petersburg Theatre School 111
Lavrova, Ekaterina Nikolaevna, singer 143, 153
Lazarev, [(Admiral) Mikhail Petrovich (1788–1851), Commander of the Black Sea Fleet?] 197
Lebedev, [Pavel Semenovich (1787–1842), court musician and?] singer 21
Lebedeva, Praskovia Prokhorovna (1838–1917), dancer 251
Lefebvre, [Dominique (d. 1812)?], dancer and choreographer 44
Legat, Gustav Ivanovich (1837–95), dancer 174
Legat, Sergei Gustavovich (1875–1905), dancer 421
Legnani, Pierina (1863–1923), ballerina 354, 393–4
Leningrad 279
Lentovsky, Mikhail Valentinovich (1843–1906), actor and proprietor of summer theatres in Moscow and St Petersburg 317–18, 320
Léon, [Antoine?], dancer 44
Leonova, Daria Mikhailovna (1829–96), singer 142
Lermontov, Mikhail Yurevich (1814–41), poet and novelist 142
Leuven, Adolphe de, ballet scenarist 208
Levogt, Heinrich, painter of theatre decorations 360
Levstedt (Madame), maker of theatre footwear 360, 393
Limido, Giovannina (1851–90), dancer 355
Linskaya, Julia Nikolaevna (1821–71), actress 125

Litavkin, Sergei Spiridonovich (1858–98), dancer 311
Livry, Emma (1842–63), dancer 284
London 3, 7, 16, 35, 82, 133, 212, 175, 266, 316–18, 430; Alhambra Theatre 318
Lopukhin, M., balletomane 269, 274
Louis XI (1423–83), King of France 177
Louis XIV (1638–1715), King of France 46, 374–5, 388–9
Louis XVIII (1755–1824), King of France 7
Lukyanov, Sergei Ivanovich (1859–1911), dancer 375
Lully, Jean Baptiste (1632–87), composer 378, 388
Lyadov, Alexandre Nikolaevich (1808–71), violinist, composer, conductor of ballet and of a ballroom orchestra 122, 163, 169, 321
Lyadova, Vera Alexandrovna (1839–70), dancer 133, 264, 274
Lyalin (pension/gymnasium in St Petersburg) 123
Lyceum (St Petersburg) 124
Lyubskaya, transfer student from Moscow to St Petersburg 153

Machiavelli, Niccolò (1469–1527), writer and political theorist 418
Madaeva, Matilda Nikolaevna (1842–89), dancer 237, 264, 266–7, 272–3, 284
Madrid 317
Maevsky, dancer from Warsaw performing in St Petersburg 121
Makarova, Alexandra Petrovna (1828–89), dancer 164
Malaverne [Maloverne, Malovergne, Malverne], Pierre Frédéric (1810–72), dancer and teacher 111, 136, 149–51, 153
Malinovsky, teacher of arithmetic in the Petersburg Theatre School 140
Maly Theatre (Moscow) 381
Malysheva (Grimardeau), classroom lady of the Petersburg Theatre School 143
Mann, physician to Princess Murat 200
Manokhin, [Nikolai Fyodorovich (1856–1915?], dancer 310–11
Manzotti, Luigi (1835–1905), dancer and choreographer 355
Marcel, Ivan Frantsevich (1801–73), régisseur of the Petersburg ballet 120, 163, 267
Maria Ivanovna, woman who recognized talent in Anna Natarova 135–6
Mario, Giovanni Matteo (1810–83), tenor 142
Marmontel, Jean-François (1723–99), poet and dramatist 11
Marochetti, physician of the Petersburg Theatre School 110, 136
Martynov, Alexandre Astafievich [Evstafievich] (1816–60), comic actor 110
Martynov, Andrei Astafievich, student in the Petersburg Theatre School 110–11

Maryinsky Theatre (St Petersburg) 250–1, 264, 267, 312–13, 324, 373, 378–80, 390, 394, 421
Masini, Angelo (1844–1926), tenor 315
Massenet, Jules (1842–1912), composer 396
Matier, costumier for Taglioni's *La Fille du Danube* in St Petersburg 90
Matveyevsky, teacher of literature in the Petersburg Theatre School 140
Maximov, Alexei Mikhailovich (1813–61), actor 115–17
Mazilier, Joseph (1801–68), choreographer 124, 127, 174–5, 206, 209, 261
Mekhelin, A., theatre censor 178
Mendelssohn-Bartholdy, Jakob Ludwig Felix (1809–47), composer 383
Menier, dancer from Warsaw performing in St Petersburg 121
Messenger of Europe [Vestnik Evropy] (journal) 33
Meyer (pension/gymnasium in St Petersburg) 123
Meyerbeer, Giacomo (1791–1864), composer 128
Michel, wigmaker in the Petersburg Imperial Theatres 360
Mikhailovsky Theatre (St Petersburg) 112–13, 128
Mikhelson, tutor in the Petersburg Theatre School 110–11
Milan/Milanese 69, 316–18, 352
Milon, Louis-Jacques (1766–1845), dancer and choreographer 15
Milonov, Mikhail Vasilievich (1792–1821), poet 33
Miloradovich, (Count) Mikhail Andreyevich (1771–1825), Governor-General of St Petersburg 30
Minkus, Alois Ludwig [Louis] (1826–1917), composer 131, 284, 293, 322, 350
Monkey and the Suitor, The (play) 115–16
Montague, dancer 116
Montplaisir, Hyppolyte Georges (1821–77), composer 317
Monvel, Jacques-Marie Boutet de (1745–1814), actor and dramatist 51–2
Morchinsky, dancing teacher in Moscow 44
Moscow 20–2, 30–1, 34–5, 45–7, 49, 116, 122, 126–7, 151–3, 175, 185–6, 189, 192–3, 195, 198, 217, 236, 269, 273, 279, 307, 310–11, 317–18, 321–2, 359, 378, 381–2, 406, 417
Moskovskie vedomosti [Moscow News] 359
Moskvityanin (journal) 51, 65–8
Mozart, Wolfgang Amadeus (1756–91), composer 16
Muscovite Trickster, The (operetta) 131
Munaretti, dancer and dancing teacher in Moscow 44, 47–9
Murat (Princess) 200
Muravieva, Marfa Nikolaevna [Nikoforovna] (1838–79), ballerina 128, 130–1, 133, 218, 238–9, 251, 254–5, 260, 262–3, 267, 269–72, 274, 279–80, 315
Musset, Louis Charles Alfred de (1810–57), poet, writer and dramatist 320
Muzykal'nyi i teatral'nyi vestnik [Musical and Theatrical Messenger] 187
Myasnikov, [Alexander Konstantinovich], defendant in a forgery trial in St Petersburg in 1872?] 127

Naples 186, 317
Napoleon Bonaparte (1769–1821) 44, 69
Napoleon, Louis [Louis Napoleon Bonaparte] (1808–73), Emperor of France 208
Napravnik, Eduard Frantsevich (1839–1916), conductor 130
Naryshkin, Alexandre Lvovich (1760–1826), Director of Russian Imperial Theatres 1799–1819: 30
Natarova, Anna Petrovna (1835–1917), dancer and artist of the dramatic theatre 114–15, 135–69
Natier, stage machinist for Didelot's *Captive of the Caucasus* 71
Nedremskaya, Alexandra Grigorievna (1863–91), dancer 374
Nemchinov, actor 126
Nemirovich-Danchenko, Vladimir Ivanovich (1858–1943), playwright 312
Nevakhovich, Alexander Lvovich (d. 1850s), supervisor of repertoire in the Petersburg Imperial Theatres 108, 113, 115
New Time, The [Novoe vremya] (newspaper) 317, 350
Niedermeier, Louis (1802–61), composer 209
Nijinsky, Vaslav Fomich (1889?–1950), dancer 423–4
Nikander, tailor of the Petersburg Imperial Theatres 155
Nikiforov, Ivan Grigorievich (1822–82), author, publisher, chief cashier of the Bolshoy Theatre in St Petersburg 259
Nikitina, Natalia Nikitichna, dancer 68
Nikitina, Varvara Alexandrovna (1857–1920), dancer 315, 354, 374
Nikolaev (city) 197
Nikolaeva, Tatyana Vasilievna, dancer 373
Nikolai Alexandrovich (1845–66), Tsarevich [eldest son of Alexandre II] 133
Nikolai Nikolaevich, Grand Duke (1831–91) 131–2
Nikolai Pavlovich [Nikolai I] (1796–1855), Tsar of Russia 118, 121–3, 125, 127, 129, 135, 144, 166–9
Nikulina, Alexandra Nikolaevna (1820–1904), dancer 153, 267
Nilsson, Christine (1843–1921), soprano 315
Northern Bee, The [Severnaya pchela] (newspaper) 86, 173
Nouvel (Nuvel), Valter Fyodorovich (1871–1949), official of the Ministry of the Imperial

Nouvel (*cont.*):
   Court in St Petersburg, friend of Alexandre
   Benois 386
Noverre, Jean-Georges (1727–1810),
   choreographer and theorist 15, 68, 351,
   355
Novitskaya, Nastasia Nikolaevna [Anastasia
   Semenovna Novitskaya] (1790–1822),
   dancer 28, 41, 68
Novoe vremya [The New Time] (newspaper)
   350, 359

Obermiller, Alexandre Leontievich (d. 1892),
   physician 132
Odessa 195–6
Odyssey, The 50
Offenbach, Jacques Levy (1819–80), composer
   284, 318, 320
Ofitserova, Ekaterina Maximovna, theatre
   costumier in the Petersburg Imperial
   Theatres 1863–1902: 360, 393
Ogoleit, Elena Karlovna (1862–1919), dancer
   373
Oldenburgsky, (Prince) Pyotr Georgievich
   (1812–81) 131–2, 134
Olga Nikolaevna (1822–92), Grand Duchess
   [second daughter of Nikolai I] 166
Olimpiev, Alexandre Petrovich, teacher in the
   Petersburg Theatre School 110, 139
Operas:
   Askold's Tomb (Verstovsky) 199
   Barbiere di Siviglia, Il (Rossini) 315
   Caliph of Baghdad, The (Boieldieu) 24
   Demon, The (Rubinstein) 317, 352
   Dieu et la bayadère, Le (Auber) 382
   Enchantress, The [Charodeika]
      (Tchaikovsky) 384
   Eugene Onegin (Tchaikovsky) 383, 386·
   Fenella, ou La muette de Portici (Auber) 34,
      284, 286
   Fidelio (Beethoven) 50
   Fille du Régiment, La (Donizetti) 117, 284
   Fronde, La (Niedermeier) 209
   Gonzago, or The Masquerade (Auber) 129
   Gustave (Auber) 125
   Huguenots, Les (Meyerbeer) 119
   Juif errant, Le (Halévy) 208
   Königin von Saba, Die (Goldmark) 383
   Life for the Tsar, A (Glinka) 124, 273, 293,
      352
   Little Slippers, The [Cherevichki]
      (Tchaikovsky) 382
   Mazepa (Fitinhof-Schell) 128
   Nero (Rubinstein) 311
   Parsifal (Wagner) 374
   Philemon et Baucis (St-Saëns) 311
   Queen of Spades, The (Tchaikovsky) 387
   Queen of Spades, The (von Suppé) 285
   Raoul sire de Créqui (Dalayrac) 50
   Richard Coeur de Lion (Grétry) 50
   Richard III (Salvayre) 311
   Ring of the Nibelung, The (Wagner) 382
   Robert le diable (Meyerbeer) 82
   Roi de Lahore, Le (Delibes) 383
   Rusalka (Dargomyzhsky) 126
   Ruslan and Lyudmila (Glinka) 352, 382
   Traviata, La (Verdi) 315
   Zémire et Azor (Grétry) 15
Orlovsky, Alexandre Osipovich (1777–1832),
   painter of military and folk scenes 51
Ostrovsky, Alexandre Nikolaevich (1823–86),
   dramatist 312
Ovid [Publius Ovidius Naso] (43 BC–AD 18),
   poet 320
Ozerov, transfer student from Moscow to St
   Petersburg 151

Pacifico, assistant to the gymnast Viol 116
Padua 317
Paganini, Nicolò (1782–1840), violinist 84
Panizza, Giacomo, Milanese composer, pianist
   and singing teacher 125
Panov, tutor in the Petersburg Theatre School
   110, 139
Papkov, Alexei Dmitrievich (1829–1903),
   conductor of the Petersburg Imperial Ballet
   1862–90: 274, 308, 321–2
Paris/Parisian 3, 5–7, 16, 22, 30, 41, 69, 82–3,
   126, 128, 134, 152, 185–6, 188, 198–202,
   205–12, 217–18, 236, 258, 269, 271, 280,
   282, 284–5, 311, 316–18, 322, 352, 391,
   423, 430; Opéra 82–3, 126, 200–1, 205–6,
   209–12, 258, 269, 271, 284; Théâtre du
   Châtelet 318; Théâtre Italien 211
Parkacheva, Ekaterina (1834–?), dancer 126–7
Parma 317
Pashkova, Lydia Alexandrovna, scenarist of
   Petipa's *Raymonda* 392
Patrie, La (newspaper) 211
Patriotic Institute (St Petersburg) 123
Patti, Adelina (1843–1919), soprano 251, 284,
   315
Paul, Antoine (1797–1871), French dancer 15
Pavlova, Anna Pavlovna (1881–1931), dancer
   308, 421, 424
Pavlova, artist 121
Pays (newspaper) 207
Pedder, Georgy Ivanovich, wigmaker in the
   Petersburg Imperial Theatres 393
Pension Thibaut (St Petersburg) 123
Permon (Madame) 44
Perrault, Charles (1628–1703), writer of
   fables 360, 373–5, 378, 380, 385, 389
Perrot, Jules (1810–92), dancer and
   choreographer 85, 107, 118–21, 123–5,
   127, 135, 153, 161–4, 166–7, 173–6, 178,
   186, 217, 234, 251, 260–1, 289, 351–2,
   390, 430
Peterburgskaya gazeta [The Petersburg
   Gazette] 235, 250, 420
Peterhof 112, 166–9, 392, 406, 424
Petipa, Jean Antoine (1796–1855), teacher of

Index   441

dance in the Petersburg Theatre School, father of Marius Petipa 151–2, 174
Petipa, Marie Mariusovna (1857–1930), dancer 283, 354, 374–5
Petipa, Marius Ivanovich (1818–1910), dancer and choreographer 85, 87, 125, 128–30, 135, 151, 162, 163, 167, 174, 217–19, 220, 234–9, 250–1, 254, 260–4, 279–89, 291–3, 309, 311–12, 323–4, 350–5, 359–60, 373, 375, 377, 381, 390, 392–3, 405–8, 417–20, 422–3, 429–30
Petipa, Nadezhda Mariusovna (1874–1945), dancer 419
Petipa-Surovshchikova, Maria Sergeyevna (1836–82), dancer 129, 136, 162, 218, 236–7, 251, 254–5, 259–60, 262–3, 280, 283
Petrov, Oleg Alexeyevich, Soviet ballet historian 85, 173
Picheau, Alexandre Nikolaevich (1821–81), dancer 119, 121, 124, 126, 130–1, 274, 288
Picq, Charles le (1749–1806), dancer and choreographer 15, 41, 44
Pimenov, Alexandre (1808–?), dancer and teacher of dance in the Petersburg Theatre School 151
Pinto, [Francisco Antonio Norberto dos Santos (1815–60)?], composer 129
Pisemsky, Alexei Feofilaktovich (1820–81), novelist and playwright 377
Pleshcheyev, Alexandre Alexeyevich (1858–1944), writer and ballet historian 218, 354
Plunkett, Adeline (1824–1910), dancer 206
Pogozhev, Vladimir Petrovich (1851–1935), historian of theatre 135
Poireau, Auguste, *see* Auguste
Popova (Manokhina), transfer student from Moscow to St Petersburg 153
Poppel, dancer from Warsaw performing in St Petersburg 121
Preobrazhenskaya, Olga Iosifovna (1871–1962), dancer 308
Presse, La 207
Prikhunova, Anna Ivanovna (1830–87), dancer 153, 164, 166, 264, 266–7, 274
Priora, Olimpia, dancer 208
Pugni, Cesare (1802–70), composer 120, 124–5, 127, 129, 131, 163, 218, 235, 239, 322, 350, 383
Pushkin, Alexandre Sergeyevich (1799–1837), poet and dramatist 28, 68–70, 72, 82, 112, 379, 406, 408; *Eugene Onegin* 28, Zaretsky 355
Pypin, [Dmitry Alexandrovich?], friend of Alexandre Benois 386

Raab, Wilhelmina Ivanovna, singer in the St Petersburg opera, 1871–85: 317
Racine, Jean (1639–99), poet and dramatist 16, 382
Radina, Lyubov Petrovna (1838–1917), dancer 128, 133, 237, 264, 266, 284, 288
Radina, Sofia Petrovna (1830–70), dancer 166–7
Ramaccini, Giuseppe, Milanese teacher of dance 317
Ramazanov, Pyotr, student of the Petersburg Theatre School 111
Rameau, Jean Philippe (1683–1764), composer 378
Raphael [Raffaello Sanzio] (1483–1520), painter 16, 84
Rappoport, Mavriki Yakovlevich (1824–84), music critic 270
Rech' [Discourse] (newspaper) 424
Regiment of the Nobility 123
régisseurs, 18–19, 120
Reinshausen, Fyodor Andreyevich (1827–c.1900), dancer 114, 116
Revenskaya, flower arranger for the St Petersburg ballet 393
Richard [Lopukhina], Daria Sergeyevna (1802–55), dancer and teacher 152–3, 162, 166, 168
Richard, Zinaida Iosifovna (1832–90), dancer 152
Rimsky-Korsakov, Nikolai Andreyevich (1844–1908), composer 423
Roller, Andrei Adamovich (1805–91), stage designer 88, 90, 117, 120, 127, 155, 166, 169, 235, 238, 252, 288
Rome 117, 317, 353
Roqueplan, Nestor (1804–70), writer, Director of the Paris Opéra 200–1, 205, 207–11
Rosati, Carolina (1826–1905), ballerina 129, 217–18, 234–6, 251
Roussel, French dancer who performed in St Petersburg 174
Rubini, Giovanni Battista (1794–1854), singer 17
Rubinstein, Anton Grigorievich (1829–94), pianist and composer 311, 317, 322, 378
Rumyantseva, Ekaterina Vasilievna, classroom lady in the Petersburg Theatre School 136, 143, 145

Sabat, [Karl Friedrich (1782–1873)?], painter of stage decorations 90
Saburov, Alexandre Matveyevich (1800–31), actor 45
Saburov, Andrei Ivanovich (1797–1866), Director of the Russian Imperial Theatres 1858–62: 218–19
Saint-Aman, dancer and teacher of dance 44
Saint-Georges, Jules-Henri Vernoy de (1801–75), ballet scenarist 127, 129, 218, 220, 234–5, 284
Saint-Léon, Arthur (1821–70), dancer and choreographer 69, 107, 128–31, 133, 186, 206–9, 217–19, 235, 238–9, 251, 260–2, 269–71, 274, 279–82, 390

442 Index

Saint Petersburg/Petersburg 3, 5–7, 16, 20, 22, 24, 26, 30–6, 41, 45, 49–50, 52, 68–70, 81–2, 84, 86–7, 107–10, 116, 119, 123, 126, 129–30, 132, 151–3, 167, 169, 173–4, 185–6, 213, 217–18, 233, 235, 250–2, 263–4, 269, 271, 275, 279–82, 293, 307–8, 312, 314–18, 321–2, 324, 351–2, 354–5, 373, 377, 392–4, 406, 408, 417–18, 423, 425, 430

Saint-Saëns, Charles Camille (1835–1921), composer 311

Salvatore Rosa (drama) 125

Salvayre, Gaston (1847–1916), composer 311, 316

Salvioni, Guglielmina (1842–?), dancer 251

Samoilov, Vasily Vasilievich (1813–87), dramatic actor 125

Sankovskaya, Ekaterina Alexandrovna (1816–78), dancer 116, 154, 191, 378

Sans Souci [Kin-Grust], summer theatre in St Petersburg 317–18

Sapozhnikov, A., balletomane 269

Sarti, Giuseppe (1729–1802), composer 24

Savitskaya, Lyubov Leonidovna (1854–1919), dancer; second wife of Marius Petipa 283

Schiller, Johann Christoph Friedrich von (1759–1805), dramatist, poet and historian 382

Schlegel, August von (1767–1845), scholar and poet 50

Schmidt, [Johann Philippe Samuil (1779–1853)?], composer 127

Schoolteacher, The (play) 113

Schumann, Robert (1810–56), composer 382

Scribe, Augustin Eugène (1791–1861), dramatist and librettist 208

Sebastopol 197

Second Cadet Corps 123

Second Military Konstantinovsky School (St Petersburg) 123

Seifert [Zeifert], [Ivan Ivanovich (1837–?)?], cellist 130

Selezneva, transfer student from Moscow 153

Semenova, Nimfodora Semenovna (1788–1876), singer 45–6

Sergeyev, Nikolai Grigorievich (1876–1951), dancer, régisseur of the Petersburg ballet 1903–18: 422–3

Shakespeare, William (1564–1616) 16, 382

Shakhovskoy, Alexandre Alexandrovich (1777–1846), dramatist and theatre official 5, 30, 113

Shamburgsky, [Alexandre Timofeyevich (1814–58)?], dancer 121–2, 124

Shamburin [Shamburgsky?], dancer 126

Shaposhnikova, Alexandra Vasilievna (1849–1930), dancer 265, 290

Shchetnina, soloist of the Petersburg ballet 266

Shelikhov, Alexei Alexeyevich (1804–58), dancer 68

Shemaeva, Alexandra Antonovna (1799–1855), dancer 124

Sheremetev (Colonel) 135

Sheremetev, Dmitry Nikolaevich (1803–71), Hoffmeister, proprietor of an estate theatre 21

Sheremeteva (Madame) 135

Shishko, [Nikolai Makarovich?], theatre chemist 132, 238

Shishkov, Matvei Andreyevich (1832–97), stage designer 127, 288, 360, 374, 389

Simonova, maker of special accessories for the Petersburg Imperial Theatres 360

Simskaya, Alexandra Ivanovna (1853–1919), dancer 265–6

Skalkovsky, Konstantin Apollonovich (1843–1906), journalist and ballet historian 236, 291–2, 307–9, 353–6, 359, 373

Smirnova, Tatyana Petrovna (1821–71) ballerina 173–4

Smolnyi Institute (St Petersburg) 123, 141

Smorgansky Academy (St Petersburg) 321

Snetkova, Maria Alexandrovna (1831–?), dancer 164

social dances 41–9

Sokolov, rehearsal violinist in the Petersburg Theatre School 163

Sokolov, Pavel Nikolaevich, teacher of singing in the Petersburg Theatre School 143

Sokolova, Evgenia Pavlovna (1850–1925), dancer 251, 267, 315

Sokolova, Maria Petrovna (1832–97), dancer 128, 153, 237, 257, 264, 267, 274

Solich, teacher in the Petersburg Theatre School 110

Solomoni, Giuseppe, dancer and choreographer of the Maddox Theatre in Moscow 44

Some Great Pills; Every Time I Take One, Thanks but No Thanks (operetta-vaudeville) 117

Son of the Fatherland [Syn otechestva] (newspaper) 238

Sorokin, teacher of literature in the Petersburg Theatre School 139–40

Sosnitsky, Ivan Ivanovich (1794–1871) actor 45

Soulié, Frédéric (1800–47), writer 380

Scots, Vasily [Ivanovich (1788–1841)?], theatre censor [and critic?] 52, 71

Sozzo [Zozzo, Zozo], Adelina, Italian dancer at the Zoological Gardens in St Petersburg 355

Stefani, Jan (1746–1829), Polish composer of ballet music 121

Stellovsky, F., typographer and publisher in St Petersburg 239

Stolyarov, Alexei, costumier of the Petersburg Imperial Theatres 155

Stoyunina (pension/gymnasium in St Petersburg) 123–4

Strauss, Johann (1825–99), composer 383

Strekalova, circus artist 115

Stukolkin, Lev Petrovich (1837–95), dancer 129

Stukolkin, Timofei Alexeyevich (1829–94),

Index   443

comic dancer 107–36, 175, 234, 237, 267, 273, 315, 390
Sue, Eugène (1804–57), novelist 380
Sukhovo-Kobylin, Alexandre Vasilievich (1817–1903) dramatist 126
Sunchulei, Circassian prince 72
Suppé, Franz von (1819–95), composer 285
Surovshchikova, Maria Sergeyevna, *see* Petipa-Surovshchikova
Suvorov (Count), governor of Kostroma 192
Svishchev, [Ivan Ivanovich (1837–81)?], dancer 111, 128

Taglioni, Filippo (1777–1871), dancer and choreographer 15, 44, 70, 82–8, 90, 173, 175
Taglioni, Marie (1804–84), ballerina 7, 70, 82–8, 107, 112, 173–5, 201, 207, 209–10, 251, 266, 271, 289–90, 314–15, 430
Tchaikovsky, Pyotr Ilyich (1840–93), composer 107, 322, 354, 359–60, 373, 375, 377–8, 381–8, 390
Tcherepnin, Nikolai Nikolaevich (1873–1945), composer 422–5
Teleshova, Ekaterina Alexandrovna (1804–57), dancer 68
Telyakovskaya, Gurlya, wife of Vladimir Telyakovsky 419
Telyakovsky, Vladimir Arkadievich (1861–1924), Director of the Russian Imperial Theatres 1901–17: 390, 405–6, 417–20, 422–3, 429–30
Termain, maker of theatre headwear 360, 393
Thibaut, teacher of French in the Petersburg Theatre School 142
Thirty Years, or The Life of a Gambler (drama) 125
Thomas, Ambroise (1811–96), composer 386
Titus [Antoine Tetus Doschi] balletmaster and choreographer 44, 81–4, 86, 116, 136–7, 149, 152–3, 173, 430
Tolbecque, Jean Baptiste (1797–1869), composer 208
Tomashevsky, Leonty Ivanovich, physician of the Petersburg Imperial Theatres 131–2
Trefilova, Vera Alexandrovna (1875–1943), dancer 308, 421
Trianon, Henri, ballet scenarist 209
Trois bals, Les (vaudeville) 112
Troitskaya, [Nadezhda Nikolaevna?], dancer 267
Troitsky, Nikolai Petrovich (1838–1903), dancer 129, 131, 267, 273
Tula 193
Turin 317
Tvorogov, theatre-goer 137
Tyazhelov, Vasily Yakovlevich, teacher in the Petersburg Theatre School 110, 139
Tyufyakin, (Prince) Pyotr Ivanovich (1769–1845), Director of the Russian Imperial Theatres 1819–21: 51

Ushakov, Alexandre Pavlovich (1833–74), mineralogist and balletomane 255–7, 259
Uspensky, Pyotr (Father), priest of the Petersburg Theatre School 140–1
Uttermarck (Madame), teacher of piano in the Petersburg Theatre School 143

V teatral' nom mire [In the Theatre World] (Skalkovsky) 307, 309–22, 359
Valberg, A., copyist of Stukolkin's 'Recollections' 107
Valberkh, Ivan Ivanovich (1766–1819), dancer and choreographer 31, 68
Valts, Karl Fyodorovich (1846–1929), theatre machinist 155
Varlamov, Alexandre Egorovich (1801–48), composer 142
Vasiliev, Vasily Mikhailovich [Vasiliev II], singer 317
Vasiliev, [Alexandra Nikolaevna (1843–?)?], dancer 267, 274
Vasilieva, Ekaterina N. (1838–?), circus artist 115
Vasilieva, Moscow artiste, *see* Lavova, Ekaterina Nikolaevna
Vazem, Ekaterina Ottovna (1848–1937), ballerina 129, 176, 219, 236, 251, 266, 270, 279–93, 307, 312, 315, 353, 429
Venzano, Luigi (1815–78) composer 284
Vergina, Alexandra Fyodorovna (1848–1901), dancer 129, 236, 251, 266
Verne, Jules (1828–1905), novelist 318
Versailles 389
Verstovsky, Alexei Nikolaevich (1799–1862), composer and Intendant of the Moscow Theatres 151–2
Vestris, Auguste (1760–1842), dancer and teacher 15, 44, 351
Vidal, Peire (*fl*. 1180–1206), troubadour 319
Viganò, Salvatore (1769–1821), choreographer 50
Vienna 188, 209–12, 317, 352
Viol, L[udovic], gymnast and actor 115–16
Vintulov (General), balletomane 419
Virgil [Publius Vergilius Maro] (70–19 BC), poet 72
Vittoliaro, teacher of singing in the Petersburg Theatre School 142
Voice, The [Golos] (newspaper) 270
Volkonsky, (Prince) Pyotr Mikhailovich (1776–1852), Minister of the Imperial Court 114
Volkonsky, (Prince) Sergei Mikhailovich (1860–1937), Director of the Russian Imperial Theatres 1899–1901: 405–7, 417
Volkov, Fyodor Grigorievich (1729–63), actor, 'father of the Russian theatre' 191
Volkov, [Nikolai Ivanovich (1836–91)?], student at the Petersburg Theatre School 111, 131
Volkova, Varvara Petrovna (1816–98), dancer

444 Index

Volkova (cont.):
   and teacher of dance in the Petersburg Theatre School 153
Vorobieva, Alexandra Alexandrovna (1861–?), dancer 311
Voronova, Evgenia Evgenievna (1859–?), dancer 374
Vsevolozhsky, Ivan Alexandrovich (1835–1909), Director of the Russian Imperial Theatres 1881–99: 307, 350, 353, 359–60, 373, 385, 389, 405–6
Vsevolozhsky, Nikolai Sergeyevich (1772–1857), historian 72

Wagner, Anton Yakovlevich (1810–85), stage designer 127, 288
Wagner, Richard (1813–1883), composer 374, 381–2
Warsaw 121, 186, 212
Wieniawski, Henryk (1835–80), violinist 272
Werther (pension/gymnasium in St Petersburg) 123
Wolf [Volf], A. I., chronicler of the Petersburg theatres 316
Wurm, Vasily Vasilievich (1826–1904), virtuoso trumpeter 130

Yakubovich, (Lukiyan Andreyevich (d. 1839)?], writer and poet 354

Yaroslavl 190–2
Yurkevich, Pyotr Ilyich (d. 1884), journalist and critic 83
Yusupov (Prince) 130

Zadunaisky [(Count) Mikhail Ilarionovich Kutuzov-Smolensky (1745–1803)?, general] 136
Zakhar, valet of Leonty Vasilievich Dubelt 161
Zalivkina (pension/gymnasium in St Petersburg) 123
Zam, menagerie in St Petersburg 115
Zehn Mädchen und kein Mann (vaudeville) 255
Zherebtsov, theatre-goer 137
Zhuchkovsky, Timofei Vasilievich (1785–1838), composer 52, 70
Zhukova, Alexandra Vasilievna (1858–?), dancer 354, 374–5
Zhukovsky, Vasily Andreyevich (1783–1852), poet and translator 88
Zhuleva, Ekaterina Nikolaevna (1830–1905), dramatic actress 140, 153
Zotov, Rafail Mikhailovich (1795–1871), critic and dramatist 50–1, 68, 81, 86, 173–7
Zubova, Vera Andreyevna (1806–53), dancer 68
Zucchi, Virginia (1849–1930), dancer 129, 177, 271, 279, 307, 316–20, 323, 354–5, 381, 385, 391